Sustainable Wastewater Treatment and the Circular Economy

Sustainable Wastewater Treatment and the Circular Economy

Guest Editors

Tao Zhang
Jing Yuan
Huu-Tuan Tran
Muhammad-Jamal Alhnidi

MDPI

Basel • Beijing • Wuhan • Barcelona • Belgrade • Novi Sad • Cluj • Manchester

Guest Editors

Tao Zhang
College of Resources and Environmental Sciences
China Agricultural University
Beijing
China

Jing Yuan
College of Resources and Environmental Sciences
China Agricultural University
Beijing
China

Huu-Tuan Tran
Department of Civil, Environmental, and Architectural Engineering
University of Kansas
Lawrence
United States

Muhammad-Jamal Alhnidi
Institute for Agricultural Engineering
University of Hohenheim
Stuttgart
Germany

Editorial Office
MDPI AG
Grosspeteranlage 5
4052 Basel, Switzerland

This is a reprint of the Special Issue, published open access by the journal *Water* (ISSN 2073-4441), freely accessible at: www.mdpi.com/journal/water/special_issues/YW2D8ZS22J.

For citation purposes, cite each article independently as indicated on the article page online and using the guide below:

Lastname, A.A.; Lastname, B.B. Article Title. *Journal Name* **Year**, *Volume Number*, Page Range.

ISBN 978-3-7258-3654-3 (Hbk)
ISBN 978-3-7258-3653-6 (PDF)
https://doi.org/10.3390/books978-3-7258-3653-6

Contents

About the Editors

Tao Zhang

Dr. Tao Zhang is an Associate Professor at China Agricultural University, China. He received his Ph.D. from Nanjing University in June 2011. His academic background covers agricultural waste utilization, wastewater treatment and nutrient recovery, and biomass utilization. From 2022 to 2024, Dr. Tao Zhang was consecutively selected for the Stanford University team and Elsevier as the World's Top 2% Scientists. He is funded by more than 10 research projects including the National Key Research and Development Program of China, National Natural Science Foundation of China, and so on. He was awarded the Scientific Chinese - Outstanding Young Scientist Award. He has published more than 100 papers in the *Chemical Engineering Journal, Water Research, Journal of Hazardous Materials, Biochar, Green Chemistry*, and so on (total citations = >5,500, H-index = 40), including 14 ESI Highly Cited Papers and 5 ESI Hot Papers. He has published four academic books as the Editor-in-Chief. He has applications for 20 Chinese invention patents.

Jing Yuan

Dr. Jing Yuan works at the College of Resources and Environmental Sciences, China Agricultural University. Dr. Jing Yuan's research interests include waste management, biomass utilization, circular economy, aerobic biotransformation, composting, carbon emission, and risk pollutants.

Huu-Tuan Tran

Dr. Huu-Tuan Tran works in the Department of Civil, Environmental, and Architectural Engineering, University of Kansas. Dr. Huu-Tuan Tran's research interests include bioremediation, soil science, composting, microbiology, waste management, and organic pollutants.

Muhammad-Jamal Alhnidi

Dr. Muhammad-Jamal Alhnidi works at the Institute for Agricultural Engineering, University of Hohenheim. Dr. Muhammad-Jamal Alhnidi's research interests include biomass thermal conversion, nutrient recovery, carbon materials, sustainable waste management, and renewable energy.

Preface

Water is fundamental to life, society, and ecological sustainability, yet its management remains one of the most pressing global challenges. "Sustainable Wastewater Treatment and the Circular Economy" addresses this critical issue, bringing together innovative research and cutting-edge technological developments that aim to revolutionize the way we view and utilize water resources. The subject of this reprint encompasses the significant contemporary issues surrounding water resource management, including wastewater treatment, pollution control, water conservation strategies, and circular economic approaches. The scope spans various methodologies, from the integration of advanced machine learning models for precipitation forecasting to novel biotechnological and chemical methods for wastewater purification. Emphasizing a multidisciplinary perspective, the book explores both theoretical advancements and practical applications.

Our motivation to compile this work stems from the global urgency to mitigate environmental risks and address the growing concerns of water scarcity, pollution, and the demand for sustainable solutions. With increasing pressures from industrialization, urbanization, and agricultural practices, innovative approaches to water resource management have become imperative. This volume synthesizes cutting-edge research to propose effective and sustainable water management practices, aligned with contemporary global sustainability goals.

This reprint targets a wide audience including scientists, engineers, policymakers, environmental managers, and graduate students interested in environmental sciences, water resource management, and sustainable development. The detailed scientific approaches, case studies, and innovative treatment technologies presented herein serve as a valuable reference for professionals aiming to enhance water resource sustainability and efficiency. The chapters of this reprint are contributed to by distinguished researchers and experts from various globally renowned institutions, reflecting a diverse range of insights and expertise. These contributions collectively provide a rich, comprehensive perspective on water management challenges and innovations.

We express our deepest gratitude to all contributing authors for their exceptional work and dedication, and to the peer reviewers whose insightful comments greatly enhanced the manuscript quality. Special acknowledgment is extended to our affiliated institutions and the publication team for their invaluable support during the preparation of this volume. We hope this reprint will inspire further research and practical advancements in sustainable water management, contributing significantly toward global environmental preservation and the achievement of sustainability objectives.

Tao Zhang, Jing Yuan, Huu-Tuan Tran, and Muhammad-Jamal Alhnidi
Guest Editors

water

MDPI

Editorial

Sustainable Wastewater Treatment and the Circular Economy

Tao Zhang

State Key Laboratory of Nutrient Use and Management, Beijing Key Laboratory of Farmland Soil Pollution Prevention and Remediation, Key Laboratory of Plant–Soil Interactions of Ministry of Education, College of Resources and Environmental Sciences, China Agricultural University, Beijing 100193, China; taozhang@cau.edu.cn; Tel.: +86-10-6273-3638

1. Introduction to the Special Issue

At present, the issue of restricted resources and the pressure on the environment are more severe than ever [1,2]. These challenges demand innovative solutions to achieve sustainable development, especially in wastewater treatment [3,4]. A circular economy presents a promising pathway to address these challenges by integrating waste recovery and resource reuse [5–8]. In the context of wastewater treatment, the quantity of biomass waste, such as sewage sludge and organic byproducts, has risen sharply in recent years, resulting in serious environmental pollution and public health concerns [9]. Biomass waste from wastewater treatment poses a dual challenge: it is both a pollutant and a potential resource [10–12]. The improper disposal of this waste can lead to the contamination of soil and water resources, the release of greenhouse gasses, and increased treatment costs. Consequently, the sustainable management and utilization of biomass waste are critical [13,14]. By adopting the principles of the circular economy, this waste can be transformed into valuable resources, providing both environmental and economic benefits [15,16]. Thus, the utilization of biomass waste has practical significance and aligns with the goals of sustainable wastewater management.

Significant progress has been made in recent years in the research on, and the development of technology for, biomass waste recovery. The current approaches primarily focus on thermal conversion, as well as aerobic and anaerobic biotechnologies [8,10,17]. Thermal conversion methods, including pyrolysis and gasification, can produce energy-dense fuels such as biochar, syngas, and bio-oil [18,19]. These products can be used as renewable energy sources, contributing to energy security and reducing dependence on fossil fuels. Similarly, biological processes such as anaerobic digestion and composting can convert organic components of biomass waste into biogas or nutrient-rich fertilizers, fostering resource recovery and reducing environmental pollution [8,17]. Moreover, wastewater treatment biomass waste offers a unique opportunity to recover high-value compounds, such as phosphates, nitrogen, and even rare metals, which are essential for agriculture and industrial applications [10,20,21]. Despite these advancements, challenges remain in scaling up these technologies and integrating them into existing infrastructure [19,22]. Economic feasibility, policy support, and public awareness are critical factors that influence the adoption of circular practices in wastewater management [15,16]. Additionally, the environmental trade-offs of some technologies, such as emissions from thermal conversion, must be carefully managed [18,19].

Therefore, the sustainable utilization of biomass waste from wastewater treatment is crucial for advancing the principles of the circular economy. By leveraging innovative technologies and fostering interdisciplinary collaboration, wastewater treatment can be transformed from a resource-intensive process into a sustainable and resource-generating

check for updates

Received: 13 January 2025
Revised: 18 January 2025
Accepted: 20 January 2025
Published: 24 January 2025

Citation: Zhang, T. Sustainable Wastewater Treatment and the Circular Economy. *Water* **2025**, *17*, 335. https://doi.org/10.3390/w17030335

system. This paradigm shift will not only mitigate environmental pollution but also enhance energy recovery, improve resource efficiency, and ensure long-term sustainability. This Special Issue of *Water* focuses on the sustainable utilization of biomass waste from wastewater treatment to advance the development of a circular economy. Since the call for papers was announced on 14 December 2022, ten original papers were accepted for publication after a rigorous peer-review process (Contributions 1–10). The authors represent diverse countries, including China, Thailand, Spain, Australia, the USA, Mexico, and Egypt. These papers are categorized into four areas: (1) pollutant removal technologies and waterbody restoration; (2) sustainable energy and the advancement of the circular economy; (3) water resource management and the optimization of water-saving efficiency; and (4) ecological conservation within the context of agricultural pollution control. To provide a clearer understanding of this Special Issue, we summarize the key highlights of the published papers below.

2. Overview of the Contributions to This Special Issue

This Special Issue showcases significant advancements in pollutant removal technologies and waterbody restoration, emphasizing the innovative methodologies used to address critical challenges in wastewater treatment. Wang et al. (Contribution 1) investigate an iron-based metal–organic framework (MIL-101(Fe)–Na_2CO_3) adsorbent for the rapid removal of hexavalent chromium (Cr(VI)) from wastewater and elucidate the roles of ion branching and domain-limiting effects in the removal process. The primary objective of this study was to develop an efficient, stable, and cost-effective decontamination strategy for addressing Cr(VI) contamination in water. By modifying the surface of MIL-101(Fe) and optimizing the ratio of metal centers to organic ligands, as well as the synthesis conditions, this study demonstrates that the adsorbent can completely remove Cr(VI) within 20 min, achieving a maximum adsorption capacity of 20 mg/g. the experimental results indicate that the adsorption process follows the Langmuir adsorption isotherm and a pseudo-second-order kinetic model, driven primarily by electro-adsorption and monolayer adsorption mechanisms. Additionally, thermodynamic analyses revealed that the reaction is spontaneous, endothermic, and accompanied by an increase in entropy. The study also found that after alkali (NaOH) elution and acid (HCl) regeneration, MIL-101(Fe)–Na_2CO_3 retained a high adsorption efficiency over multiple cycles, indicating its strong potential for practical applications. Compared with other adsorbents, this material offers substantial advantages in both adsorption capacity and selectivity to Cr(VI), effectively mitigating the interference of coexisting anions. In summary, the MIL-101(Fe)–Na_2CO_3 adsorbent developed in this study shows promising applications in wastewater treatment and provides new insights and technological support for efficient heavy metal remediation. Qin et al. (Contribution 2) investigate a newly isolated halotolerant Bordetella strain and its ability to degrade p-nitrophenol (PNP) in high-salinity wastewater, facilitated by co-metabolites. The primary objective of this study was to overcome the low PNP degradation efficiency in high-salinity wastewater, while exploring novel strategies to enhance biodegradation. The researchers isolated a new Bordetella sp. from seafood-processing wastewater and employed adaptive acclimation, demonstrating a remarkable PNP degradation capacity under high-salinity conditions. Under optimal conditions (30 °C, pH 8.0, 3% NaCl, and an aeration rate of 0.3 $m^3/m^3 \cdot min$), the strain achieved an 85.9% degradation of PNP (initial concentration 350 mg/L) within 72 h. The addition of 30 mg/L pantothenic acid as a co-metabolite further enhanced the PNP degradation rate by 82.5%. GC/MS analyses verified hydroquinone as the key intermediate, indicating that the PNP degradation pathway proceeds via hydroquinone cleavage. Kinetic studies indicated that PNP degradation follows a first-order kinetic model and that elevated PNP concentrations inhibit the strain's degradation efficiency. This study demonstrates that Bordetella sp. exhibits notable halotolerance and PNP degradation capacity, offering an efficient solution for the biological treatment of high-salinity wastewater. Meanwhile, the use of co-metabolites (such as pantothenic acid) offers new possibilities to further optimize the degradation efficiency, thereby laying a solid foundation for future research and applications in industrial wastewater treatment. Liu et al. (Contribution 3) review the status, technological progress, and future challenges of hydrothermal carbonization (HTC) in wastewater treatment, particularly its potential for achieving "carbon peaking" and "carbon neutrality." HTC technology employs a high-temperature, high-pressure hydrothermal environment to convert organic waste into value-added carbon materials, thereby reducing energy consumption and pollutant emissions, and improving carbonization efficiency. The article highlights that HTC technology exhibits remarkable decontamination capabilities in wastewater treatment, including the efficient adsorption of heavy metals, organics, and anions, and plays a particularly prominent role in improving water quality. The hydrochar produced via HTC possesses a large specific surface area, abundant pore structures, and diverse surface functional groups, which can be further modified to improve its pollutant removal performance. Despite the wide-ranging application prospects of HTC technology, challenges persist regarding process optimization, cost management, and environmental impact assessments. Future research should concentrate on developing efficient catalysts, refining reaction mechanisms, and advancing the commercialization of HTC technology with the aid of policy support. In addition, international collaboration remains pivotal for expanding HTC technology applications, addressing global water challenges, and realizing carbon neutrality. Overall, HTC technology offers a novel approach to water environment management, fostering a circular economy and sustainable development while providing tangible support in alleviating water resource crises and enhancing water quality. These studies advance the scientific understanding and practical implementation of cutting-edge wastewater treatment technologies, offering innovative pathways for sustainable waterbody restoration and environmental management.

This Special Issue presents significant advances in sustainable energy and the circular economy, emphasizing innovative wastewater management and energy recovery technologies. Suksaroj et al. (Contribution 4) investigates the potential of biogas codigestion technology to simultaneously treat palm oil mill effluent (POME) and empty fruit bunch (EFB) pressed wastewater, thereby advancing the circular economy. The study aims to optimize wastewater management, improve resource utilization, and foster sustainable clean energy generation in the palm oil production process. The article outlines the experimental approach, in which both batch and semi-continuous fermentation systems are used to examine how varying wastewater mixing ratios and hydraulic retention times (HRT) influence methane production. The results show that under a mixing ratio of 45% POME, 50% inoculum, and 5% EFB wastewater at an HRT of 25 days, total biogas and methane productions reached 18,679 mL/L and 6778 mL/L, respectively, with a methane content at 62%. In addition, a COD removal rate of 67% indicates the effective conversion of organic matter in the wastewater into biogas. The study also included an economic assessment, demonstrating that this codigestion strategy provides a higher internal rate of return and a shorter payback period, rendering it more economically viable compared to traditional processes. The resulting sludge complies with organic compost standards and can be used to enhance soil quality, thereby achieving the high-value utilization of waste. In summary, the proposed biogas codigestion technology delivers substantial environmental and economic benefits while offering practical guidance for sustainable development in the palm oil industry. This research may serve as a reference for other sectors dealing with high-pollution wastewater treatment and energy conversion. Huang (Contribution

5) provide a comprehensive overview of the status of, and emerging trends in, anaerobic digestion (AD) in waste stream treatment, emphasizing its pivotal role in meeting "Dual Carbon" targets. Anaerobic digestion harnesses microbial activity to transform organic waste into biogas and nutrient-rich digestate, thereby reducing greenhouse gas emissions, improving self-sufficiency in energy use, and contributing to the circular economy. The article reviews global advances in anaerobic digestion, encompassing multi-stage anaerobic reactors, membrane bioreactors, and photosynthetically assisted upgrading techniques, all of which have substantially increased biogas yields and enhanced overall system efficiency. Furthermore, the article discusses how co-digestion can optimize the carbon-to-nitrogen ratio and stabilize system operations. In terms of policy, Europe's Green Deal and China's "Dual Carbon" strategy have greatly accelerated the adoption of AD technologies, particularly in wastewater treatment and agricultural waste management. Despite these advancements, the technology continues to encounter significant challenges, such as high initial capital costs, volatile fatty acid (VFA) buildup, and ammonia inhibition. In addition, developing countries face greater barriers to adoption, attributable to insufficient technical support and incomplete policy frameworks. The article proposes that future efforts emphasize technological refinement, cost minimization, and stronger policy incentives, alongside increasing the public acceptance of AD projects. Overall, anaerobic digestion presents substantial potential for waste stream management and resource recovery, acting as a vital instrument for attaining carbon neutrality and sustainable development, and offering scientific support for policy and industrial applications. These studies contribute valuable insights into sustainable waste management and energy recovery, offering scalable solutions for advancing the circular economy and achieving carbon neutrality goals.

This Special Issue contributes significantly to water resource management and water-saving efficiency optimization, offering innovative strategies and practical solutions for sustainable development. Urrea Vivas et al. (Contribution 6) evaluate the feasibility of reusing the water at the El Salitre wastewater treatment plant in Bogotá, Colombia, employing both economic and technical assessments, and highlight its role as a model for circular economy practices. The study initially examines the existing status of the wastewater treatment plant and identifies the need for upgrades, focusing on secondary and tertiary treatments—such as activated sludge and UV disinfection—to improve the quality of the final effluent. The research proposes repurposing treated water for agricultural, industrial, and recreational uses, in alignment with the Colombian regulatory standards, thereby alleviating pressure on conventional water sources. Their economic assessment employs a cost–benefit analysis (ACB) and present net value (NPV) techniques, factoring in externalities like the environmental benefits of diminished river pollution. The findings reveal that reclaimed water can be produced at a cost of EUR 1.23/m^3, offering an economically favorable alternative to traditional water sources and generating approximately EUR 6.52/m^3 in additional profit for industrial users. In addition, the sustained utilization of reclaimed water could curtail reliance on natural water bodies, leading to tangible economic and environmental gains. The study underlines the importance of the circular economy's principles in managing water resources, showcasing how water reuse can facilitate resource efficiency and foster sustainable development. Future directions call for a comprehensive evaluation of technological reliability and strategies to increase the public acceptance of reclaimed water, thereby aiding in the broader deployment of reuse strategies. Cai et al. (Contribution 7) apply a multilayer perception (MLP) neural network and use structural equation modeling (SEM) to investigate the complex factors that shape customers' water-use behavior in hotels, culminating in the proposal of a representative hotel water-use behavior model. The study seeks to uncover the key drivers of water-use behavior and enhance water efficiency within the hotel industry, thereby informing the development of more targeted water conservation strategies. The article examines water-use behavior across diverse dimensions—such as individual attributes, water-saving awareness, and consumer practices. The research demonstrates that different factors distinctly affect specific activities—including washing, handwashing, drinking, showering, and toilet flushing—with gender exerting the largest influence on showering and flushing, whereas length of stay emerges as the key determinant for all water-use behaviors. In addition, the study indicates that education, income, and hotel type each exert varying degrees of influence on customers' water-use habits. A total of 292 valid questionnaires were obtained, and both MLP and SEM analyses were employed to quantify the relative importance of each influencing factor. The findings reveal that individual characteristics significantly shape typical water-use behaviors, whereas consumer behavior exerts a comparatively lesser effect, guiding the future design of more effective water-saving policies and behavior-oriented interventions. In addition to offering theoretical underpinnings for water-saving measures in the hotel industry, this article establishes the groundwork for subsequent inquiries into water-use behavior in other service sectors and serves as a reference point for the development of predictive models aimed at alleviating water scarcity. Wang et al. (Contribution 8) examine the operational performance and water quality of centralized WWTPs in industrial parks along the Yellow River Basin, offering suggestions to enhance their efficiency and sustainability. The study centers on 63 centralized wastewater treatment plants in 54 national- or provincial-level industrial parks, evaluating factors such as treatment capacity, influent properties, energy demand, and operational expenses. The findings indicate that most WWTPs are small to medium in scale (1×10^4–5×10^4 m^3/d), operating at an average hydraulic loading rate of 53.8% and failing to meet their design capacities. Aerobic biological processes prevail, with approximately 55.1% of facilities utilizing AAO (anaerobic–anoxic–oxic), AO (anaerobic–oxic), or oxidation ditch processes, while certain plants have introduced advanced treatments to comply with the ever-more stringent discharge regulations. The influent quality displays uneven distributions of COD, BOD, NH_3-N, TN, and TP, coupled with generally low COD and BOD levels, underscoring the presence of insufficient carbon sources that markedly constrain nitrogen- and phosphorus-removal efficiencies. Furthermore, energy consumption exceeds the national average (approximately 1.1 kWh/m^3), which is mainly attributable to the complex and low-biodegradability profile of industrial wastewater. The article recommends bolstering wastewater segregation and pretreatment within industrial parks, refining treatment processes to optimize carbon utilization, and integrating energy-efficient technologies and renewable sources to reduce overall consumption, thus improving both economic feasibility and long-term sustainability in the Yellow River Basin. Collectively, these studies address critical aspects of water resource management by exploring water reuse, behavioral insights, and treatment plant optimization. Their findings provide a comprehensive approach to enhancing water-saving efficiency, reducing the environmental impacts, and advancing sustainable water management practices in diverse contexts.

The Special Issue makes significant contributions to ecological conservation within the framework of agricultural pollution control, addressing complex challenges through the use of innovative methodologies. Zhang et al. (Contribution 9) introduce a precipitation forecasting method that consolidates CEEMD (complete ensemble empirical mode decomposition), ELM (extreme learning machine), and FFOA (whale optimization algorithm). The proposed model aims to overcome the shortcomings of conventional precipitation forecasting techniques when confronted with nonlinear and intricate time-series data. The central concept involves employing CEEMD to decompose precipitation time series into multiple intrinsic mode functions (IMFs), effectively filtering out noise and extracting informative features. Subsequently, the ELM model—recognized for its rapid training and robust generalization—performs the training and forecasting tasks. To increase its forecasting accuracy, the article integrates the FFOA algorithm to optimize the ELM parameters. The FFOA explores optimal solutions through a whale-foraging paradigm, thus mitigating the risk of converging on local optima. The experimental findings indicate that the CEEMD–ELM–FFOA model surpasses traditional stand-alone approaches in precipitation forecasting, particularly with respect to accuracy and robustness. This method not only adeptly manages the nonlinear characteristics of precipitation data, but also enhances the model's predictive competence, demonstrating substantial practical utility. In conclusion, this integrated forecasting model provides a novel paradigm for precipitation forecasting, making it especially pertinent to the complex nonlinear dynamics of meteorological data. Ge et al. (Contribution 10) address the issue of nonpoint-source phosphorus pollution and its ecological ramifications while exploring the emerging trends in applying amendment materials to mitigate phosphorus runoff. Nonpoint-source phosphorus pollution, predominantly derived from agricultural practices (e.g., fertilization and soil erosion), constitutes a principal driver of water eutrophication. The article first analyzes the pathways by which phosphorus moves from soil to aquatic systems, addressing mechanisms such as the release, migration, and deposition of phosphorus. Phosphorus reaches aquatic environments through runoff and subsurface infiltration, resulting in water quality degradation, algal bloom, and broader ecological disturbances. The study underscores that deciphering the dynamics of phosphorus in soil and curbing its transfer to water systems are paramount for mitigating phosphorus losses. The article then highlights various amendment materials—including organic matter, mineral additives, and chemical remediation agents—that effectively diminish phosphorus loss through bolstering the adsorption of phosphorus by the soil, modifying phosphorus' solubility, or facilitating the immobilization of phosphorus. The article concludes with a comparative assessment of each material's strengths and limitations, advocating for further exploration of innovative amendment options and evaluations of their efficacy across various environmental scenarios. In sum, the article underscores the multifaceted nature of nonpoint-source phosphorus pollution and its implications for aquatic ecosystems, proposing amendment-based strategies to curb phosphorus runoff and offering critical insights and practical foundations for the prevention of water eutrophication. These studies provide valuable insights into predictive modeling and pollution control, advancing ecological conservation efforts. Their findings support sustainable agricultural practices and effective resource management, offering scalable solutions to mitigate the environmental impacts.

3. Conclusions

The advancements highlighted in this Special Issue underscore the transformative potential of integrating circular economy principles into wastewater treatment. By prioritizing the sustainable utilization of biomass waste and leveraging cutting-edge technologies such as anaerobic digestion, thermal conversion, and innovative filtration methods, the field is advancing towards reducing environmental impact and enhancing resource efficiency. These approaches not only mitigate pollution but also generate renewable energy and recover valuable nutrients, aligning with the global sustainability goals. The contributions further emphasize the critical role of interdisciplinary collaboration and policy support in overcoming challenges such as economic feasibility, public acceptance, and technological integration. The research on pollutant removal, sustainable energy recovery, water reuse, and agricultural pollution control demonstrates the multifaceted benefits of these innovations. From achieving carbon neutrality through optimized energy recovery systems to improving water-saving efficiencies in service sectors, these studies provide actionable insights that bridge the gap between scientific discovery and practical implementation.

However, challenges remain in scaling these technologies, ensuring their cost-effectiveness, and addressing environmental trade-offs. Moving forward, greater focus on life-cycle assessments, robust regulatory frameworks, and public awareness campaigns will be pivotal in mainstreaming these solutions.

Ultimately, this Special Issue serves as a call to action for researchers, policymakers, and industry leaders to deepen their commitment to sustainable wastewater management. By transforming wastewater from a burden into a resource, we can build resilient systems that contribute to ecological conservation, energy security, and sustainable development, ensuring a healthier future for both the environment and society.

Funding: The research was sustained by a grant from the National Key Research and Development Program of China "Intergovernmental Cooperation in International Science and Technology Innovation" [Grant number 2023YFE0104700] and the National Natural Science Foundation of China [Grant Number 31401944].

Acknowledgments: As Guest Editor of the Special Issue "Sustainable Wastewater Treatment and the Circular Economy", I would like to express my deep appreciation to all the authors whose valuable work was published in this issue and who have thus contributed to the success of this edition.

Conflicts of Interest: The authors declare no conflicts of interest.

List of Contributions

1. Wang, C.; Chen, J.; Yang, Q. Rapid Removal of Cr(VI) from Wastewater by Surface Ionized Iron-Based MOF: Ion Branching and Domain-Limiting Effects. *Water* **2024**, *16*, 25. https://doi.org/10.3390/w16010025.
2. Qin, L.; Li, H.; Tan, Y.; Yan, X.; Tao, P.; Fan, Z.; Li, T.; Tan, J.; Wang, Y.; Jin, L. Utilizing a Novel Halotolerant Bordetella Bacterium Combined with Co-Metabolites to Boost the Degradation of P-Nitrophenol in High-Salinity Wastewater. *Water* **2024**, *16*, 3360. https://doi.org/10.3390/w16233360.
3. Liu, G.; Xu, Q.; Abou-Elwafa, S.F.; Alshehri, M.A.; Zhang, T. Hydrothermal Carbonization Technology for Wastewater Treatment under the "Dual Carbon" Goals: Current Status, Trends, and Challenges. *Water* **2024**, *16*, 1749. https://doi.org/10.3390/w16121749.
4. Suksaroj, C.; Jearat, K.; Cherypiew, N.; Rattanapan, C.; Suksaroj, T.T. Promoting Circular Economy in the Palm Oil Industry through Biogas Codigestion of Palm Oil Mill Effluent and Empty Fruit Bunch Pressed Wastewater. *Water* **2023**, *15*, 2153. https://doi.org/10.3390/w15122153.
5. Huang, X. The Promotion of Anaerobic Digestion Technology Upgrades in Waste Stream Treatment Plants for Circular Economy in the Context of "Dual Carbon": Global Status, Development Trend, and Future Challenges. *Water* **2024**, *16*, 3718. https://doi.org/10.3390/w16243718.
6. Urrea Vivas, M.A.; Seguí-Amórtegui, L.; Tomás Pérez, C.; Guerrero-García Rojas, H. Technical–Economic Evaluation of Water Reuse at the WWTP El Salitre (Bogotá, Colombia): Example of Circular Economy. *Water* **2023**, *15*, 3374. https://doi.org/10.3390/w15193374.
7. Cai, R.; Bai, X.; Liu, J.; Hu, M. Analysis of Hotel Water-Use Behavior Based on the MLP-SEM Model. *Water* **2023**, *15*, 1534. https://doi.org/10.3390/w15081534.
8. Wang, Y.; Yuan, Y.; Xue, H.; Yu, Y.; Shi, Y.; Wen, H.; Xu, M. Analysis on Operation and Water Quality Characteristics of Centralized Wastewater Treatment Plants of Industrial Parks in Yellow River Basin, China. *Water* **2024**, *16*, 806. https://doi.org/10.3390/w16060806.
9. Zhang, X.; Wu, X. Combined Forecasting Model of Precipitation Based on the CEEMD-ELM-FFOA Coupling Model. *Water* **2023**, *15*, 1485. https://doi.org/10.3390/w15081485.
10. Ge, X.; Chen, X.; Liu, M.; Wang, C.; Zhang, Y.; Wang, Y.; Tran, H.-T.; Joseph, S.; Zhang, T. Toward a Better Understanding of Phosphorus Nonpoint Source Pollution from Soil to Water and the Application of Amendment Materials: Research Trends. *Water* **2023**, *15*, 1531. https://doi.org/10.3390/w15081531.

References

1. Hu, Y.; Su, M.; Wang, Y.; Cui, S.; Meng, F.; Yue, W.; Liu, Y.; Xu, C.; Yang, Z. Food production in China requires intensified measures to be consistent with national and provincial environmental boundaries. *Nat. Food* **2020**, *1*, 572–582. [CrossRef] [PubMed]
2. Liao, S.; Wu, Y.; Wong, S.W.; Shen, L. Provincial perspective analysis on the coordination between urbanization growth and resource environment carrying capacity (RECC) in China. *Sci. Total Environ.* **2020**, *730*, 138964. [CrossRef] [PubMed]
3. Obaideen, K.; Shehata, N.; Sayed, E.T.; Abdelkareem, M.A.; Mahmoud, M.S.; Olabi, A.G. The role of wastewater treatment in achieving sustainable development goals (SDGs) and sustainability guideline. *Energy Nexus* **2022**, *7*, 100112. [CrossRef]
4. Faragò, M.; Damgaard, A.; Madsen, J.A.; Andersen, J.K.; Thornberg, D.; Andersen, M.H.; Rygaard, M. From wastewater treatment to water resource recovery: Environmental and economic impacts of full-scale implementation. *Water Res.* **2021**, *204*, 117554. [CrossRef] [PubMed]
5. Springmann, M.; Clark, M.; Mason-D'Croz, D.; Wiebe, K.; Bodirsky, B.L.; Lassaletta, L.; de Vries, W.; Vermeulen, S.J.; Herrero, M.; Carlson, K.M.; et al. Options for keeping the food system within environmental limits. *Nature* **2018**, *562*, 519–525. [CrossRef] [PubMed]
6. Capodaglio, A.G. Urban wastewater mining for circular resource recovery: Approaches and technology analysis. *Water* **2023**, *15*, 3967. [CrossRef]
7. Cecconet, D.; Capodaglio, A.G. Sewage sludge biorefinery for circular economy. *Sustainability* **2022**, *14*, 14841. [CrossRef]
8. Ghimire, U.; Sarpong, G.; Gude, V.G. Transitioning wastewater treatment plants toward circular economy and energy sustainability. *ACS Omega* **2021**, *6*, 11794–11803. [CrossRef] [PubMed]
9. Rorat, A.; Courtois, P.; Vandenbulcke, F.; Lemiere, S. *Sanitary and Environmental Aspects of Sewage Sludge Management*; Industrial and Municipal Sludge; Butterworth-Heinemann: Oxford, UK, 2019; pp. 155–180.
10. Xie, S.Y.; He, X.Y.; Alshehri, M.A.; Abou-Elwafa, S.F.; Zhang, T. Elevated effect of hydrothermal treatment on phosphorus transition between solid-liquid phase in swine manure. *Results Eng.* **2024**, *24*, 102887. [CrossRef]
11. Capodaglio, A.G.; Callegari, A. Energy and resources recovery from excess sewage sludge: A holistic analysis of opportunities and strategies. *Resour. Conserv. Recycl. Adv.* **2023**, *19*, 200184. [CrossRef]
12. Ye, Y.; Ngo, H.H.; Guo, W.; Chang, S.W.; Nguyen, D.D.; Fu, Q.; Wei, W.; Ni, B.; Cheng, D.; Liu, Y. A critical review on utilization of sewage sludge as environmental functional materials. *Bioresour. Technol.* **2022**, *363*, 127984. [CrossRef] [PubMed]
13. Jin, S.Q.; Zhang, B.; Wu, D.M.; Han, Y.; Hu, C.C.; Ren, C.Z.; Zhang, X.; Wei, Y.; Wu, A.P.J.; Mol, S.; et al. Decoupling livestock and crop production at the household level in China. *Nat. Sustain.* **2021**, *4*, 48–55. [CrossRef]
14. Zheng, W.; Shao, Y.; Qin, S.; Wang, Z. Future directions of sustainable resource utilization of residual sewage sludge: A review. *Sustainability* **2024**, *16*, 6710. [CrossRef]
15. Mannina, G.; Barbara, L.; Cosenza, A.; Wang, Z. *Treatment and Disposal of Sewage Sludge from Wastewater in a Circular Economy Perspective*; Current Developments in Biotechnology and Bioengineering; Elsevier: Amsterdam, The Netherlands, 2023; pp. 11–30.
16. Mainardis, M.; Cecconet, D.; Moretti, A.; Callegari, A.; Goi, D.; Freguia, S.; Capodaglio, A.G. Wastewater fertigation in agriculture: Issues and opportunities for improved water management and circular economy. *Environ. Pollut.* **2022**, *296*, 118755. [CrossRef] [PubMed]
17. Li, H.H.; Zhang, T.; Shaheen, S.M.; Abdelrahman, H.; Ali, E.F.; Bolan, N.S.; Li, G.X.; Rinklebe, J. Microbial inoculants and struvite improved organic matter humification and stabilized phosphorus during swine manure composting: Multivariate and multiscale investigations. *Bioresour. Technol.* **2022**, *351*, 126976. [CrossRef] [PubMed]
18. Zhang, Y.; Zhang, T. Biowaste valorization to produce advance carbon material-hydrochar for potential application of Cr (VI) and Cd (II) adsorption in wastewater: A review. *Water* **2022**, *14*, 3675. [CrossRef]
19. Dong, M.; Jiang, M.; He, L.; Zhang, Z.; Gustave, W.; Vithanage, M.; Niazi, N.K.; Chen, B.; Zhang, X.; Wang, H.; et al. Challenges in safe environmental applications of biochar: Identifying risks and unintended consequence. *Biochar* **2025**, *7*, 12. [CrossRef]
20. Zhang, T.; Wu, X.; Shaheen, S.M.; Zhao, Q.; Liu, X.; Rinklebe, J.; Ren, H. Ammonium nitrogen recovery from digestate by hydrothermal pretreatment followed by activated hydrochar sorption. *Chem. Eng. J.* **2020**, *379*, 122254. [CrossRef]
21. Zheng, M.; Hu, Z.; Liu, T.; Sperandio, M.; Volcke, E.I.P.; Wang, Z.; Hao, X.; Duan, H.; Vlaeminck, S.E.; Xu, K.; et al. Pathways to advanced resource recovery from sewage. *Nat. Sustain.* **2024**, *7*, 1395–1404. [CrossRef]
22. Wu, F.C.; Li, F.B.; Zhao, X.L.; Bolan, N.S.; Fu, P.; Lam, S.S.; Mašek, O.; Ong, H.C.; Pan, B.; Qiu, X.Q.; et al. Meet the challenges in the "carbon age". *Carbon Res.* **2022**, *1*, 1. [CrossRef]

water

MDPI

Review

Hydrothermal Carbonization Technology for Wastewater Treatment under the "Dual Carbon" Goals: Current Status, Trends, and Challenges

Guoqing Liu [1], Qing Xu [1], Salah F. Abou-Elwafa [2], Mohammed Ali Alshehri [3] and Tao Zhang [1,*]

1 Beijing Key Laboratory of Farmland Soil Pollution Prevention and Remediation, Key Laboratory of Plant-Soil Interactions of Ministry of Education, College of Resources and Environmental Sciences, China Agricultural University, Beijing 100193, China
2 Agronomy Department, Faculty of Agriculture, Assiut University, Assiut 71526, Egypt
3 Department of Biology, Faculty of Science, University of Tabuk, Tabuk 71491, Saudi Arabia
* Correspondence: taozhang@cau.edu.cn; Tel.: +86-10-6273-3638

Abstract: Hydrothermal carbonization (HTC) technology transforms organic biomass components, such as cellulose and lignin, into valuable carbon materials, gases and inorganic salts through hydrolysis, degradation and polymerization, with significant advantages over traditional methods by reducing energy consumption, lowering pollutant emissions and enhancing carbonization efficiency. In the context of global climate change, HTC plays a critical role in water environment management by addressing industrial, agricultural, and domestic wastewater challenges. The application of HTC extends to wastewater treatment, where hydrochar effectively adsorbs heavy metals, organic compounds, and anions, thereby improving water quality. However, challenges remain, such as optimizing the process for diverse raw materials, managing economic costs, and addressing environmental and social impacts. Future research and policy support are essential for advancing HTC technology. By enhancing reaction mechanisms, developing catalysts, and promoting international cooperation, HTC can significantly contribute towards achieving carbon neutrality goals and fostering sustainable development.

Keywords: hydrothermal carbonization; hydrochar; wastewater treatment; water environment management; circular economy

Citation: Liu, G.; Xu, Q.; Abou-Elwafa, S.F.; Alshehri, M.A.; Zhang, T. Hydrothermal Carbonization Technology for Wastewater Treatment under the "Dual Carbon" Goals: Current Status, Trends, and Challenges. *Water* **2024**, *16*, 1749. https://doi.org/10.3390/w16121749

Academic Editor: Alejandro Gonzalez-Martinez

Received: 29 May 2024
Revised: 18 June 2024
Accepted: 19 June 2024
Published: 20 June 2024

1. Introduction

As global climate change intensifies, countries around the world, including China, have set carbon reduction targets [1–5]. As the world's largest emitter of carbon, China has formulated a dual-carbon policy, namely achieving carbon peak and carbon neutrality, which not only demonstrates its responsibility as a major nation but also introduces unprecedented challenges and opportunities across various industries [6–11]. To achieve these targets, scientists need to work with industries, policymakers and community stakeholders to reduce net emissions, and concomitantly develop and implement carbon drawdown strategies [12–17]. The emerging field of carbon-based materials boasts new opportunities [15,18–20]. In this context, hydrothermal carbonization, as an efficient and eco-friendly technology for resource utilization, holds broad prospects for application in wastewater treatment [21,22].

In general, (hydro)thermal techniques can be divided into two main categories: thermal treatments (operating under inert atmosphere and dry conditions) and hydrothermal treatments (operating in a closed pressurized system and under wet conditions). Each category can be further divided into carbonization, liquefaction, and gasification treatments, based on the operating temperature and phase partitioning of the products (Figure 1). In this paper, the hydrothermal carbonization treatment will be mainly introduced [23].

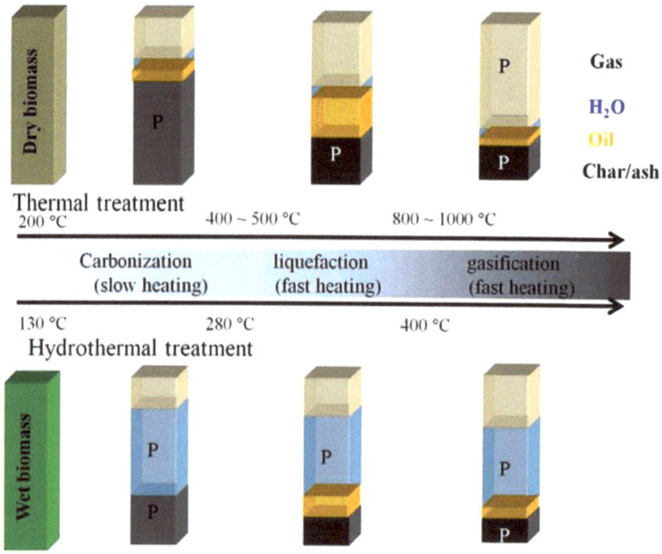

Figure 1. The distribution of products and phosphate in different phases under different thermal treatments [23].

Hydrothermal carbonization (HTC), a recently prominent carbonization technology, is increasingly recognized for its applications in environmental protection [21,24]. This process utilizes water as a thermal medium in a sealed high-temperature, high-pressure environment to transform waste into useful substances such as carbon materials, gases, and inorganic salts [25–27]. Compared to traditional carbonization methods, HTC offers significant advantages, including lower energy consumption, reduced pollutant emissions, higher carbonization efficiency, and greater value-added products [22,27].

As urban modernization progresses rapidly, living standards in cities have significantly improved. Concurrently, China's water environments are increasingly stressed. The discharge of industrial, agricultural, and domestic wastewater has progressively increased, exacerbating water pollution [28–30]. In addition, the development of the pharmaceutical industry, animal husbandry and the unreasonable discharge of wastewater, is causing antibiotic pollution and it is particularly prominent in the water environment [31]. The "China Ecological Environment Statistical Yearbook (2018–2023)" published by the Ministry of Ecology and Environment reports high pollutant levels in discharged wastewater, heavily impacting water environments. Industrial and agricultural wastewater discharges have significantly polluted rivers and lakes, harming the ecosystem [19,20,28,32,33]. Consequently, effectively addressing water environmental issues has become an urgent problem. Figure 2 is a statistical chart of the number of papers on the application of hydrochar in wastewater treatment from 2014 to 2024 in June. In 2014–2019, the number of papers in this area is small, and the number of papers after 2020 increases significantly, indicating that in recent years, people have conducted more and more research on the application of hydrochar for wastewater treatment, which reflects the development of this technology, the increasing attention to water environment problems, and the acceleration of the practical application and transformation of research results. In Figure 3, it can see the main keywords used in articles on applications and use the search term, on the scientific network. The terms "hydrochar yield", "higher heating value", "mechanism", "adsorption" and "functional group" are more controllable terms, which may indicate that the adsorption application of hydrochar is related to the mechanism of hydrochar, the type of functional groups, and the yield and calorific value of hydrochar also have indispensable value for its application.

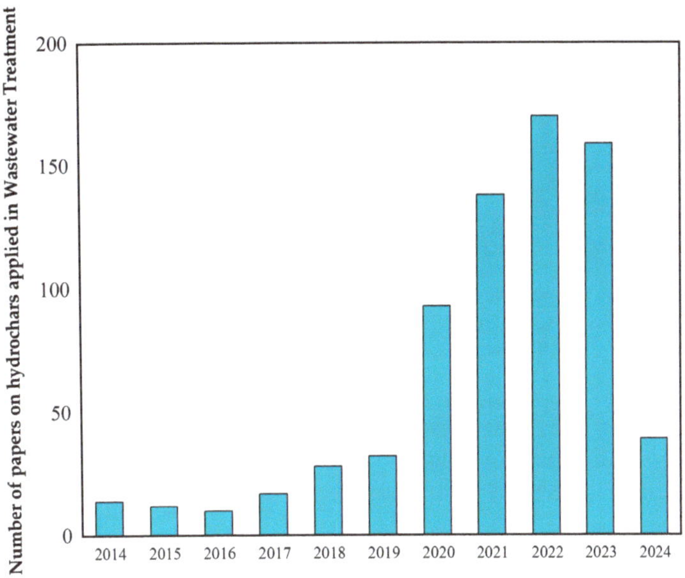

Figure 2. The number of papers on hydrochars applied in wastewater treatment.

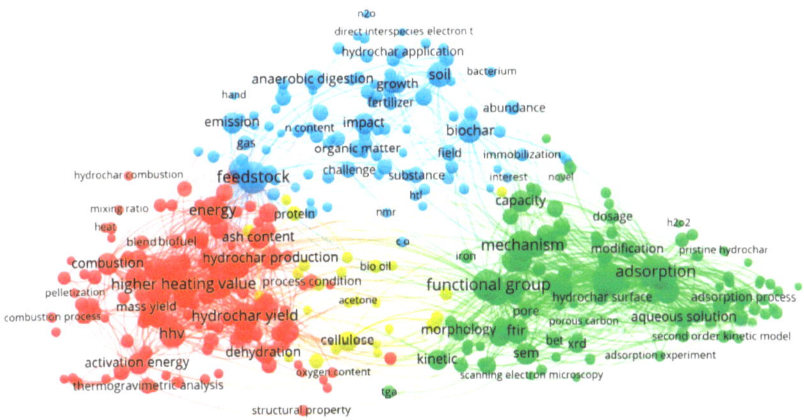

Figure 3. Hop map of keywords for articles on the use of hydrochar using VOSviewer1.6.19 software.

In response to global climate challenges and to achieve the "Dual Carbon" goals of peaking and neutralizing carbon emissions, continuous innovation in wastewater management technologies is essential. HTC, as an emerging treatment technology, offers significant advantages and broad application prospects. This paper explores the current applications of HTC in wastewater treatment, including its effectiveness against various pollutants, technological processes, and engineering examples. We also analyze the principles and historical development of this technology, addressing current challenges, research directions, and policy recommendations. Through this study, we aim to advance the development and application of HTC, thereby significantly contributing to water environment management and protection. We believe that with ongoing technological advancements and policy improvements, HTC will play a crucial role in achieving the "Dual Carbon" goals and promoting sustainable development.

2. HTC Technology

2.1. Basic Principle

HTC is a process that converts biomass into carbon-rich materials using a high-temperature, high-pressure aqueous environment [24,26,27]. This technology operates in the absence of oxygen, preventing combustion and oxidation, thus preserving the carbon content of the raw materials. It not only transforms waste into valuable resources but also reduces environmental pollution. The carbon materials produced take the advantages of low cost, excellent chemical stability, large surface area and abundant pores [34].

HTC is a thermochemical conversion technique that processes biomass with water at specific temperatures (180–250 °C) and pressures (2–10 MPa) into high-value carbon products [32,35], along with significant liquid and minor gaseous byproducts, primarily CO_2 [36,37]. This technology effectively converts fibrous materials found in animal manure and crop residues, including cellulose, hemicellulose, and lignin [27]. HTC involves the decomposition of macromolecules into smaller molecules, which then re-polymerize [37]. The process includes hydrolysis, dehydration, decarboxylation, condensation, and aromatization [38,39]. Cellulose, for example, undergoes hydrolysis where C-O and C-C bonds break, leading to the formation of intermediates like 5-hydroxymethylfurfural, followed by dehydration and aromatization to form hydrophobic microsphere cores and hydrophilic shells through aldol condensation [40]. Similarly, hemicellulose and lignin undergo specific breakdown and reformation processes to produce high-quality, high energy-dense hydrochar [41,42]. HTC's strength lies in the precise control of reaction parameters (temperature, pressure, feedstock particle size, and duration) enabling the production of superior quality carbons with wider application potential [43,44]. In Figure 4, the main mechanism of hydrochar formation is highlighted by describing the HTC reaction pathways from lignocellulosic biomass.

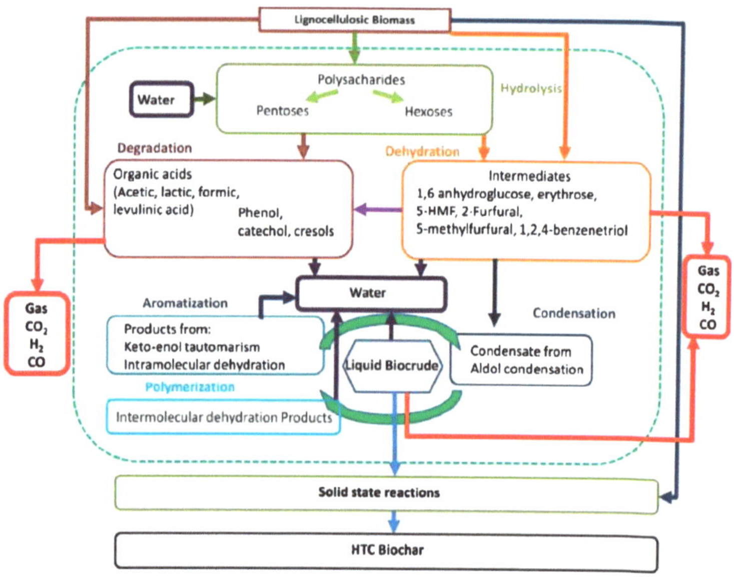

Figure 4. HTC reaction pathways from lignocellulose biomass [45].

HTC traces its origins to the early 19th century when the formation mechanisms of coal began to be studied. In Figure 5, the history of HTC technology is described. In 1913, the German chemist Friedrich Bergius and his colleagues conducted carbonization of cellulose under hydrothermal conditions at temperatures ranging from 250 to 310 °C, resulting in a black carbonaceous material with a significantly reduced oxygen-to-carbon

(O/C) atomic ratio compared to the raw material. Subsequently, researchers expanded the range of feedstocks to other biomass materials and systematically studied HTC processes.

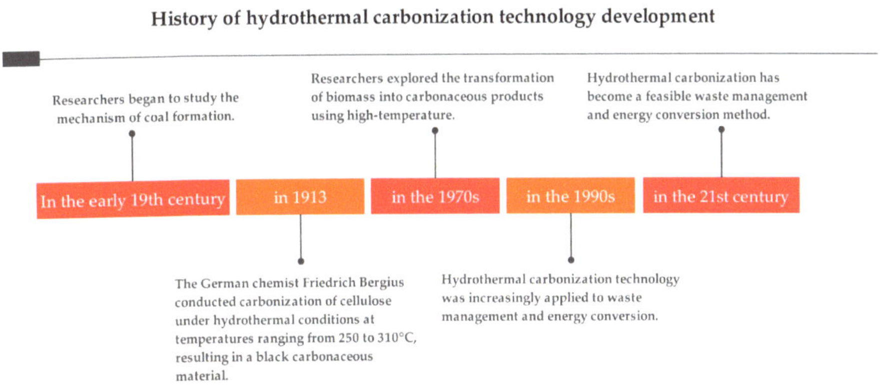

Figure 5. History of HTC technology development.

In the 1970s, HTC technology began to garner attention from scientists and researchers, who explored the transformation of biomass into carbonaceous products using high-temperature, high-pressure water. These studies primarily focused on the thermal decomposition reactions of lignocellulosic biomass [46].

In the 1990s, HTC technology was increasingly applied to waste management and energy conversion. Researchers conducted successful studies on various types of biomasses, including food waste, agricultural residues, and other discarded plant materials, achieving promising results. Since the 21st century, with the increasing prominence of energy and environmental concerns, HTC has emerged as a viable method for waste management and energy conversion. As technology has advanced, there has been a deeper understanding and research into its applications and the value of its products.

2.2. *Main Types and Methods of Hydrochar*

HTC, as a green and sustainable method for carbon material production, has garnered extensive attention and research in China. Based on various process conditions and operational principles, this technology can be categorized into the following main types:

2.2.1. Ordinary HTC

In a closed system, biomass is mixed with water and carbonized under specific temperature and pressure conditions, resulting in a product that is black, solid, rich in oxygenated functional groups, and primarily composed of carbon. The standard HTC process is relatively simple and efficiently utilizes biomass resources.

2.2.2. HTC Using Catalysts

During the resource recovery process from waste, optimizing energy consumption is essential to accommodate variations in biomass feedstocks and enhance the yield and quality of hydrochar [47]. In the reaction, the rate of biomass dehydration and carbon product formation is generally low, particularly when using plant biomass, which results in hydrochar with lower carbon content, surface area, porosity, and thermal stability, limiting its practical applications [48]. Introducing catalysts can improve the efficiency and quality of the carbonization process. Catalysts facilitate the degradation and carbonization of organic materials in biomass, yielding carbon materials with more complex structures and superior properties. Common catalysts include salts, acids, metal oxides, molecular sieves, and hydroxides [47]. In Table 1, some studies are described to better express the effect of using catalysts on HTC technology.

Petrovic et al. [49] utilized grape residue to prepare hydrochar by KOH modification under the condition of 220°C and 1 h of hydrothermal temperature. Obtained results showed that the KOH treatment increased the sorption capacity of hydrochar from 27.8 mg/g up to 137 mg/g at pH 5. Adsorption of lead on either of the materials was achieved through the ion-exchange mechanism, chemisorption and Pb^{2+}-π interaction. The Sips isotherm model gave the best fit with the experimental data obtained for Pb^{2+} sorption using activated hydrochar. Similarly, Jiang et al. [50] prepared hydrochar by using *Sedum Alfredii* Hance as raw material under KOH modification at 180–270 °C for 5 h. Research showed that the adsorption capacity of hydrochar for Cd(II) was greatly enhanced after KOH modification, due to improved specific surface areas and pore structure. The maximum Cd(II) adsorption capacity of modified hydrochar was 25.69 mg/g. Additionally, Qian et al. [51] utilized bamboo sawdust as raw material to prepare hydrochar maintained at 473 K for 24 h and treated with a NaOH modification. The adsorption capacity of hydrochar to methylene blue after the modified treatment is 655.76 mg/g. The high adsorption capacity of hydrochar for methylene blue indicates that alkaline treatment followed by hydrothermal carbonization in acidic media has potential applications for producing highly efficient MB adsorbents for wastewater treatment.

Huang et al. [52] utilized sludge as raw material under hydrothermal conditions at 180 °C for 2 h, with a citric acid dosage ratio (citric acid to dry sludge mass ratio) of 0.1. The produced hydrochar featured a maximum surface area of 59.95 m^2/g, rich in oxygen-containing functional groups, and exhibited the highest equilibrium adsorption capacity for Pb(II). Wang et al. [53] activated hydrochar produced from sawdust at temperatures between 200 and 250 °C and a heating rate of 4 °C/min using acetic acid, which reduced the energy consumption of the HTC reaction. The resultant hydrochar had a higher calorific value and energy density, making it suitable for biofuel applications. Chen et al. [54] used bamboo powder as raw material and activated the product in citric acid solution under hydrothermal conditions at 180 °C for 6 h. Citric acid is an efficient HTC additive to improve the carbonization degree, to optimize the pore structure distribution, and to introduce more oxygen-containing functional groups (OFGs) to hydrochar, which offer an appropriate precursor to be activated by KOH to improve the specific surface area (SSA) level and electrochemical performance of the activated porous carbon.

Huang et al. [55] utilized raw sewage sludge as raw material and modified hydrochar with calcium acetate ($CaAc_2$) and sodium acetate (NaAc) at 160–250 °C for 30 min. This modification increased the surface area and pore volume compared to unmodified samples and achieved a phenol removal rate of 65.7%. The addition of Ca acetate reduces the nitrogen retention in the hydrochar due to enhanced protein hydrolysis and deamination, leading to more nitrogen transformation into the NH_4^+-N in the aqueous product. The addition of Na acetate also slightly enhances protein hydrolysis, thus increasing the polypeptide-N in the aqueous product. Mumme et al. [56] conducted HTC of cow dung and corn digestate at 190–270 °C using natural zeolite as a catalyst. Adding zeolite enhances the carbon, hydrogen, and ash content, improves energy recovery rates, increases the surface area and pore volume of the hydrochar, and significantly boosts the recovery rates of nitrogen and sulfur. Lang et al. [57] activated hydrochar from pig manure with CaO at temperatures of 180, 200, and 220 °C for 10 h, and significantly increased the pH and yield of the hydrochar, although the recovery rates of carbon and nitrogen were slightly decreased. Additionally, the addition of CaO facilitates the transformation of phosphorus from non-apatite inorganic phosphorus (NAIP) to apatite phosphorus (AP), with nearly all the phosphorus in the pig manure recovered as AP in the hydrochar through CaO-assisted HTC.

Table 1. Research on HTC technology using catalysts.

Raw Materials	Hydrothermal Conditions	Catalyst Type	Modification Treatment	Product	References
Grape pomace	220 °C, 1 h	hydroxides	KOH	The KOH treatment increased the sorption capacity for Pb^{2+} from 27.8 mg/g up to 137 mg/g at pH 5.	[49]
Sedum Alfredii Hance	180–270 °C, 5 h	hydroxides	KOH	The adsorption capacity of hydrochar for Cd(II) was greatly enhanced after KOH modification, and the maximum Cd(II) adsorption capacity was 25.69 mg/g.	[50]
Bamboo sawdust	473 K, 24 h	hydroxides	NaOH	The adsorption capacity of hydrochar to methylene blue after modified treatment is 655.76 mg/g.	[51]
Sludge	180 °C, 2 h	acids	citric acid	Featured a maximum surface area of 59.95 m^2/g, and exhibited the highest equilibrium adsorption capacity for Pb(II).	[52]
Sawdust	200–250 °C	acids	CH_3COOH	Had a higher calorific value and energy density.	[53]
Bamboo powder	180 °C, 6 h	acids	citric acid	Citric acid could improve the carbonization degree of hydrochar, and optimize the pore structure distribution.	[54]
Raw sewage sludge	160–250 °C, 30 min	salts	$CaAc_2$, NaAc	The addition of calcium acetate can reduce the retention of nitrogen in the hydrate, and the addition of sodium acetate can slightly enhance the hydrolysis of the protein.	[55]
Cow dung and corn digestate	190–270 °C	molecular sieves	natural zeolite	Increases the surface area and pore volume of the hydrochar, and significantly boosts the recovery rates of nitrogen, sulfur and energy.	[56]
Pig manure	180, 200, and 220 °C, 10 h	metal oxides	CaO	Facilitates the transformation of phosphorus from non-apatite inorganic phosphorus (NAIP) to apatite phosphorus (AP).	[57]

2.2.3. Multi-Step HTC

During the HTC process, raw materials are mixed with a suitable solvent such as water in a sealed container. The reaction conditions are manipulated through multiple stages of varying temperatures and durations to produce hydrochar with unique porous structures and surface characteristics, enhancing its utility across various fields. Specifically, the two-step method involves an initial stage under low temperatures and high pressure to degrade organic matter in biomass into smaller organic molecules. The second stage operates at high temperatures and lower pressure to further carbonize these molecules into carbon materials. This multi-step approach enhances carbonization efficiency, optimizes the structure of the carbon materials, and reduces energy loss.

Zhang et al. [58] synthesized a novel arsenic adsorption material, iron-modified hydrochar, by using different iron species, i.e., $FeCl_3 \cdot 6H_2O$, $FeSO_4 \cdot 7H_2O$, and $Fe(NO_3)_3 \cdot 9H_2O$. These hydrochars were prepared through a one-step hydrothermal carbonization process at 220 °C. The results indicated that compared with $FeCl_3 \cdot 6H_2O$ and $FeSO_4 \cdot 7H_2O$, hydrochar modified with $Fe(NO_3)3 \cdot 9H_2O$ exhibited a maximum iron retention rate of 84.2% and a maximum arsenic adsorption capacity of 11.19 mg/g. Jiao et al. [59] proposed an effective strategy for synergistic production of high value-added xylooligosaccharides (XOS) and humic-like acid (HLA) from vinegar residue based on a two-step hydrothermal pretreatment. During the first-step hydrothermal pretreatment (170 & DEG; C, 50 min), 29.1% of XOS (X-2-X-6) was obtained. The XOS yield was further improved to 36.2% with endoxylanase hydrolysis, thereby increasing the value of (X-2-X-4)/XOS from 0.8 to 1.0. Subsequently, the second-step hydrothermal pretreatment was investigated to produce

HLA from the solid residue of the first-step hydrothermal pretreatment. The highest HLA yield was 15.3% in the presence of 0.6 mol/L of KOH at 210 °C for 13 h. In addition, 31.7% of hydrochar by-product was obtained. The mass balance results showed that 1000 g of vinegar residue produced 67.9 g of XOS, 91.6 g of HLA, and 189.5 g of hydrochar. Tan et al. [60] presented a novel approach to synthesize nitrogen-doped porous carbon materials via a three-step fabrication process using citric acid as the carbon source and urea as the nitrogen source. Firstly, hydrochar was synthesized by a microwave-assisted hydrothermal method using citric acid and urea and as the reactants. The hydrochar was then subjected to high-temperature carbonization in an Ar atmosphere followed by KOH activation, resulting in nitrogen-doped porous carbon materials. The as-prepared porous carbon possesses a high BET surface area of 2397 m^2/g and an average pore size of 1.8 nm. Such N-rich porous carbon shows outstanding capacitive performance (365 F/g at 0.5 A/g), good rate capacitive behavior, and excellent cycling stability, indicating a great potential for supercapacitors.

2.2.4. HTC Used in the Hydrothermal Process of Liquid Circulation

During the HTC process, the liquid by-products are recycled to enhance the biomass conversion rate and the production efficiency of carbon materials. The recycling of the process liquid significantly improves the resource recovery efficiency of the hydrothermal fluid, maximizing the value of the products and enhancing water resource utilization. The use of hydrothermal fluid as a recyclable medium not only reduces water consumption, but also offers benefits such as the reuse of thermal energy, increasing the efficiency of energy recovery, and reducing the costs of HTC processing [61]. Furthermore, recycling hydrothermal fluid significantly reduces the volume of wastewater produced, decreases the costs of wastewater treatment and contributes to environmental protection [62].

Stemann et al. [62] conducted 19 cycles of process water under hydrothermal conditions at 220 °C for 4 h. During the first five cycles, the total organic carbon (TOC) concentration was increased, and then stabilized. The accumulation of organic acids in the liquid phase enhances the calorific value and carbon concentration of the solids, thereby increases energy density. Uddin et al. [63] used *Pinus taeda* at hydrothermal temperatures of 200–260 °C with a residence time of 5 min and a water-to-biomass dry basis mass ratio of 5. They recycled the process at 200 and 230 °C for 9 cycles and 260 °C for 5 cycles. The biomass carbon yield was 5–10% higher than the initial cycle at each investigated temperature, with the higher heating value (HHV) remaining essentially unchanged. The TOC concentration in the aqueous phase was gradually concentrated with each cycle, reaching equilibrium.

2.2.5. Co-HTC

Co-HTC involves mixing two or more types of biomasses and processing them together. This method leverages the differences in properties and densities of the materials, resulting in a synergistic effect that significantly enhances combustion. This synergy allows the mixture to burn more completely, releasing greater amounts of heat and making more efficient use of waste biomass materials [47,64,65]. Additionally, co-carbonization can lower the carbonization temperature required for individual materials, thus improving resource utilization. The co-treatment extends the temperature interval between hydrolysis and pyrolysis, reducing the activation energy required for pyrolysis. Furthermore, it significantly improves ignition and combustion characteristics, shifting the peak combustion rate and its temperature range to higher temperatures, thereby lowering the activation energy of the combustion reaction [47]. In Table 2, several research are carried out to better understand co-HTC technology.

Zhu et al. [64] mixed food waste digestate (FWD) with wood waste (WW) as feedstock for co-HTC at 250 °C for 2 h. The results showed that the combination strategy enhanced the fuel properties of the hydrochar. In particular, the ash content of hydrochar was reduced to a minimum of 6.3%, thereby increasing the heating value by nearly two folds. The comprehensive combustion and combustion stability indices were improved with the maximum values of 3.98×10^{-9} $\%^{2}{}^{\circ}C^{-5}$ and 4.22×10^{2} $\%{}^{\circ}C^{-3}$, respectively. Wang et al. [65] used food waste (FW) and woody sawdust (WS) as feedstock for conducting co-HTC at 180–260 °C for 1 h with a stirrer speed of 100 r/min. Results suggested that hydrochar yield consistently decreased with an increase in both the FW ratio and HTC temperature. The C retention from 260 °C hydrochar was low (approximately 65%), but more microsphere structures were formed due to the enhanced carbonization degree of the hydrochar. Li et al. [66] mixed livestock manure (SM) with corn cob (CC) as feedstock for co-HTC at 180–260 °C for 2 h. Compared with HTC of SM, the addition of CC could effectively reduce the ash content, enhance the N recovery to 38.95–47.61% (SM:CC = 1:1), and increase the surface porous structure, making the Co-hydrochars suitable as fertilizers. Under the optimal hydrothermal conditions of 240 °C, 2 h, and mixing ratio of 1:1, the hydrochar yield was as high as 36.72%, and the TNC (6.341%), N recovery rate (47.61%) and P recovery rate (86.41%) were all suitable for fertilizer. Sharma and Dubey [67] mixed food waste (FW) and yard waste (YW) for co-HTC at 220–260 °C for 1–4 h. They found that the calorific value of blended raw feedstock was 13.5 MJ/kg which increased to 27.6 MJ/kg after Co-HTC at 220 °C for 1 h. The energy yield and fuel ratio calculated was 45% and 0.65, respectively. The hydrochar produced demonstrated a stable combustion profile as compared to reactive combustion profile for raw samples. Shen et al. [68] used corn straw and chlorella as raw materials for co-HTC at 240 °C for 1 h. They observed that moderate reaction conditions favored nitrogen enrichment and increased porosity in the hydrochar. Under these conditions, the optimal nitrogen content and surface area of the product were 3.50% and 5.91 m^2/g, respectively.

Table 2. Research on co-HTC.

Raw Materials	Hydrothermal Conditions	Product	References
Food waste digestate (FWD), wood waste (WW)	220 °C, 4 h	The comprehensive combustion and combustion stability indices were improved with the maximum values of 3.98×10^{-9} %^{2}°C^{-5} and 4.22×10^2 %°C^{-3}, respectively.	[64]
Food waste (FW), woody sawdust (WS)	180–260 °C, 1 h, and a stirrer speed of 100 r/min	Hydrochar yield consistently decreased with the increase in both the FW ratio and HTC temperature, and the C retention from 260 °C hydrochar was low (approximately 65%).	[65]
Livestock manure (SM), corn cob (CC)	180–260 °C, 1–4 h	The addition of CC could effectively reduce the ash content, and enhance the N recovery to 38.95–47.61% (SM:CC = 1:1).	[66]
Food waste (FW), yard waste (YW)	220–260 °C, 1–4 h	The sulfur in SS and CS gradually converted into thiophenic sulfur and sulfates.	[67]
Corn straw, chlorella	240 °C, 1 h	Moderate reaction conditions favored nitrogen enrichment and increased porosity in the hydrochar.	[68]

2.2.6. HTC at Different Temperatures

In the HTC process, the temperature significantly impacts the properties and structure of the carbonized products. Variations in hydrothermal temperatures yield diverse products, and studying these differences is crucial for developing high-performance carbon materials for practical applications. Higher temperatures (above 300 °C) facilitate the effective degradation of organic materials in biomass, resulting in high-purity carbon materials with greater potential application value. Conversely, lower temperatures (below 300 °C) are more conducive to preserving carbon. In Table 3, some studies were performed to better understand the effects of different temperatures on HTC technology.

Oxidized hydrothermal biochar was prepared by hydrothermal carbonization of Spartina alterniflora biomass (240 °C for 4 h) and subsequent oxidization (240 °C for 10 min) under air. Oxidized hydrochar achieved a Fe(III) reducing capacity of 2.15 mmol/g at pH 2.0 with 120 h, which is 1.2 times higher than un-oxidized hydrochar. Low temperature oxidization increases the contents of carboxyl and carbonyl groups on the hydrochar surface.

This study reveals that low temperature oxidization is an effective way to improve and restore the abiotic reducing ability of hydrochar [69]. The hydrochar yield and carbon retention in processing the livestock manure were decreased when increasing the reaction temperatures between 180 and 240 °C for 1 h [70]. Nguyen et al. [71] performed the reaction with grape marc at 180–260 °C for 30 min to prepare the hydrochar slurries for ignition experiments. It has shown that the 260 °C solid hydrochar exhibited the shortest ignition delay time (0.2 s) and the lowest ignition temperature (179 °C).

Gao et al. [72] used eucalyptus bark as the raw material in the reaction for 2–10 h in the range of 220–300 °C to obtain the hydrochar. With the increase in temperature, the yield of hydrochar decreased slightly from 46.4% at 220 °C to 40.0% at 300 °C. The atomic ratio of O/C and H/C decreased from 1.69 and 0.80 to 0.83 and 0.23, respectively, and the oxygen-containing functional groups decreased with increasing temperature. Peng et al. [73] used glucose as the raw material and kept it at 160–220 °C for 1–12 h to make the hydrochar. The experimental study showed that the adhesion of hydrochar was different: the lower the temperature, the greater the adhesion. Hydrochar is mainly composed of furan domains and aromatic clusters, and its surface is rich in oxygen-containing functional groups. The increase in the reaction temperature (160–220 °C) enhanced the aromatization degree of the hydrochar. Li et al. [74] prepared hydrochar with chicken, dairy, and swine manures at 200–350 °C for 2 h. The results showed that high temperature resulted in low yield and decreased H/C, O/C and volatiles of hydrochar. The high temperature results in a higher fixed carbon, fuel ratio and calorific value of the hydrochar, which indicates that hydrochar from animal manures can be used as a substitute for fuel. The carbon retention rate decreases with the increase in temperature. High temperature improves the aromatics of hydrochar and enhances its thermal oxidation resistance.

Table 3. Research on HTC at different temperatures.

Raw Materials	Hydrothermal Conditions	Product	References
Spartina alterniflora biomass	240 °C, 4 h	The contents of carboxyl group and carbonyl group on the surface of hydrochar were increased by low temperature oxidation.	[69]
Livestock manure	180–240 °C, 1 h	Both hydrochar yield and carbon retention decreased with increasing reaction temperatures.	[70]
Grape marc	180–260 °C, 30 min	The 260 °C solid hydrochar exhibited the shortest ignition delay time (0.2 s) and the lowest ignition temperature (179 °C).	[71]
Eucalyptus bark	220–300 °C, 2–10 h	The oxygen-containing functional group decreased with the increase in temperature.	[72]
Glucose	160–220 °C, 1–12 h	The increase in the reaction temperature (160–220 °C) enhanced the aromatization degree of hydrochar.	[73]
Chicken, dairy, swine manures	200–350 °C, 2 h	With the increase in temperature, the carbon retention rate decreased, and the aromatics and thermal oxidation resistance is increased.	[74]

In practical applications, the choice of carbonization methods and operational conditions varies based on specific needs and circumstances. For example, in waste management, HTC can transform waste into hydrochar and steam, extracting valuable substances and energy. In the field of biomass energy conversion, this technology can process biomass into high-purity, high energy-dense hydrochar, which can be further utilized in the production of fuels, chemicals, and electrode materials. In Table 4, the advantages of the methods for the different types of HTC technology are described.

In HTC, the conversion of biomass into carbon materials involves several critical steps. In Figure 6, the general steps of HTC technology are described. Initially, biomass constituents such as cellulose, sugars, and lignin are mixed with water to create a slurry. This mixture is then transferred to a sealed reactor or autoclave, where it is heated to a predetermined temperature for a specific duration. Under these high temperature and pressure conditions, the biomass undergoes a series of chemical reactions including hydrolysis, degradation, and polymerization, leading to the formation of carbon materials. Once the reaction is completed, the reactor is cooled to room temperature, and the solid carbonaceous product is removed, washed, and dried. Finally, to enhance the properties of the carbon material, further post-treatment steps such as activation or surface modification may be employed.

Table 4. Methods for different types of HTC.

Type	Methods	Advantage
HTC	Biomass is mixed with water and directly carbonized under specific temperature and pressure conditions.	Makes full use of biomass resources, less process, and low difficulty to operate.
HTC with catalyst	A certain amount of catalyst was added to modify the hydrochar.	Complex structure and excellent performance, which improves the HTC yield and energy recovery efficiency.
Multi-step HTC	The reaction conditions are manipulated through multiple stages of varying temperatures and durations.	Has unique pore structure and surface characteristics, which optimizes the structure, improves the carbonization efficiency and reduces the energy loss.
Liquid phase circulating HTC	Repeatedly recycle by-products of hydrothermal fluid.	Improves the conversion rate of biomass and utilization efficiency of water resources and reduces the cost of carbonization process and wastewater treatment.
Co-HTC	Mix a variety of raw materials together and carry out HTC at the same time.	Reduces the carbonization temperature, and improves the material performance, combustion performance and resource utilization rate.
High temperature HTC	The hydrothermal temperature is higher in the carbonization process.	High calorific value, and excellent product stability, with long-term carbon sequestration potential.
Low temperature HTC	The hydrothermal temperature is low in the carbonization process.	Has better hydrochar yield, carbon recovery and energy recovery, and has an obvious short-term carbon sequestration effect.

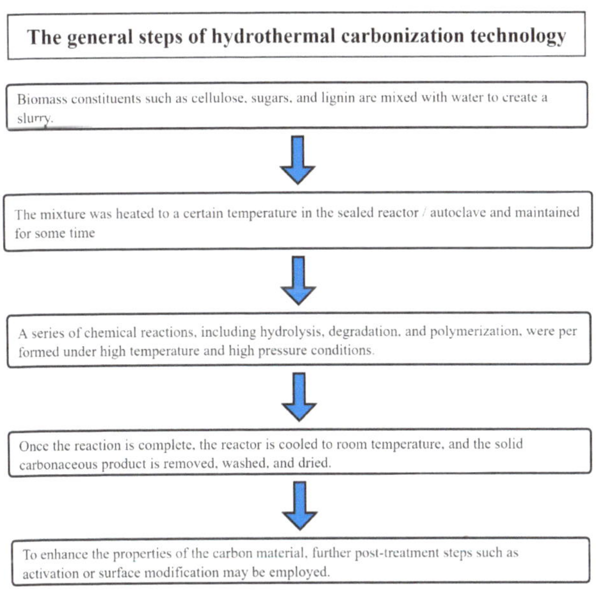

Figure 6. The general steps of HTC technology.

3. Application of HTC in Water Environment Management

Water environment management is essential for global sustainable development [29,30,33]. HTC technology represents an emerging environmentally friendly approach for revolutionizing global water environmental management. Primarily, this technology leverages the heat and pressure conditions in aqueous environments to transform organic materials into carbon materials with outstanding properties. As a novel green technology, HTC holds broad prospects for application in water environmental governance. This paper will explore in detail the applications of HTC in managing water environments, including wastewater treatment, surface water restoration, groundwater purification, and other areas of use.

3.1. Wastewater Treatment

HTC products are widely used in wastewater treatment, such as heavy metals removal, organic pollutants, and anionic wastewater. Hydrochar offers several advantages in wastewater treatment, including high efficiency, versatility, low cost, and simple operation, thus realizing significant environmental and economic benefits through the "waste-to-wealth" approach. Due to its high functional group content, porosity, and surface charge, hydrochar can effectively adsorb heavy metals, organic compounds, and anions in wastewater [75]. By modifying hydrothermal conditions and adding modifiers, the quantity and types of oxygen-containing functional groups on the surface of hydrochar can be increased, thereby further optimizing the availability of active sites. In Figure 7, HTC's liquid and solid processes for treating wastewater are described.

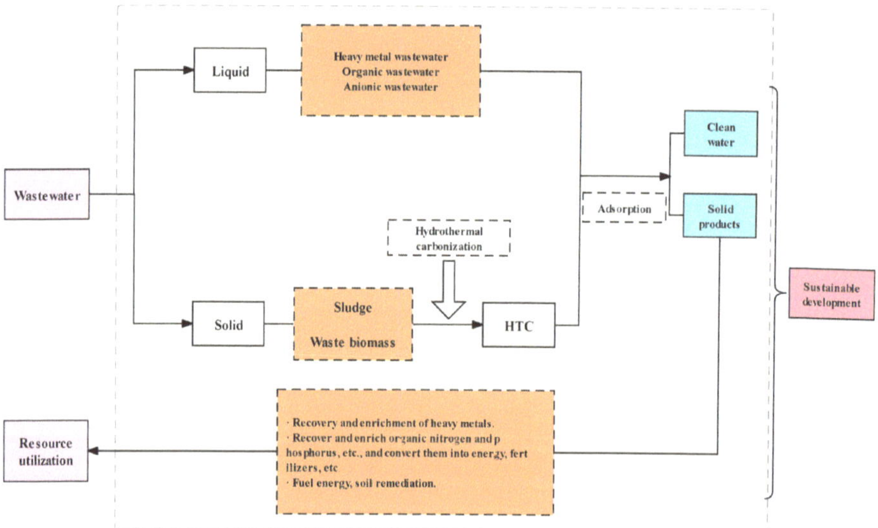

Figure 7. Schematic diagram of the HTC process for treating wastewater—liquid and solid flows.

3.1.1. Use of Hydrochar for Heavy Metal Ions

For heavy metal ions, the adsorption mechanisms of hydrochar primarily include electrostatic attraction, ion exchange, complexation, redox reactions, and coprecipitation as chemical adsorption processes, along with physical adsorption [76–78]. Redox adsorption can facilitate chemical reactions between various anions and mineral components on the surface of hydrochar and heavy metal ions, resulting in precipitation [77]. In addition to chemical adsorption, physical adsorption is also an important mechanism for hydrochar adsorption of heavy metal ions. Physical adsorption is mainly due to the van der Waals force or hydrophobic force of heavy metal ions with the hydrochar surface. When dealing with heavy metal pollution, appropriate hydrochar materials and optimized preparation conditions can be selected according to the specific application scenarios and the characteristics of the target heavy metal ions to improve the adsorption efficiency. At the same time, the adsorption performance of hydrochar can be further improved by means of surface modification and doping with other functional materials. In Figure 8, the main Ce(III) adsorption mechanism onto hydrochar surface is described. In Table 5, some studies show the effect of hydrochar on the adsorption of heavy metals.

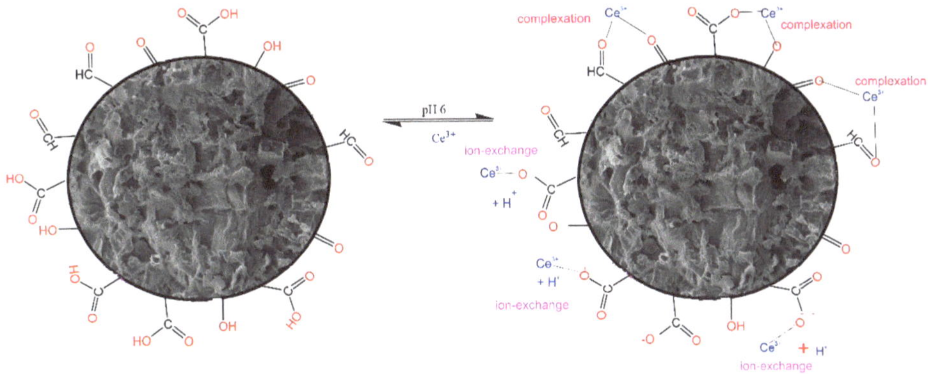

Figure 8. Main Ce(III) adsorption mechanism onto hydrochar surface [79].

Amine-modified hydrochar synthesized from black liquor using hexamethylenediamine (HMDA) at hydrothermal temperatures of 150, 180, 200, and 220 °C, and a duration of 0.5 h exhibited a maximum adsorption capacity of Cr(VI) removal of 741.74 mg/g at 45 °C and pH 2 [80]. Hydrochar microspheres produced from glucose at 180 °C for 48 h enables highly selective separation of U(VI). At 333.15 K and pH 4.5, the maximum adsorption capacity for U(VI) was 408.36 mg [81]. Similarly, hydrochar microspheres synthesized from sucrose at 190 °C for 11 h, activated with KOH demonstrated a maximum adsorption capacity of 704.2 mg/g for methylene blue. The adsorption process was endothermic [82]. Phosphoric acid-modified hydrochar produced from banana peels to at 230 °C over 2 h was evaluated for adsorption capabilities of Pb^{2+} [83]. The evaluation revealed lead adsorption capacities of 359 mg/g and 193 mg/g for dehydrated and fresh banana peel hydrochar, respectively. Moreover, hydrochar produced from willow twigs via HTC at 300 °C for 30 min used as an adsorbent to remove Cu^{2+} and Cd^{2+} from aqueous solutions exhibited adsorption capacities of 34 mg/g (0.313 mmol/g) for Cd^{2+} and 31 mg/g (0.503 mmol/g) for Cu^{2+} [84]. Furthermore, hydrochar from pine sawdust (SD) at 260 °C for 2 h, was then activated with H_2O_2 to enhance the Pb^{2+} removal capacity [85]. The maximum adsorption capacity of Pb^{2+} reached 92.80 mg/g at pH 5.0 and 298 K, which is more than 42 times higher than that of the original hydrochar (2.20 mg/g). Hydrochar from synthesized Pseudomonas aeruginosa shells at 200 °C for 20 min was found to be more effective in adsorbing Pb^{2+} and Cd^{2+} compared to biochar. The adsorption kinetics followed a pseudo-second-order model, describable by the Langmuir isotherm [86]. After reaction at 180–220 °C for 9 h, citric acid catalyzed arecanut husk to produce hydrochar. The results showed that compared with the parent biomass, the mass yield of hydrochar was 58.7%, and the fixed carbon increased from 17% to 39.7%, compared with Zn^{2+}, Cr^{6+} and Ni^{2+}, in the aqueous solution with the initial concentration of 100 mg/L. Hydrochar at a dose of 0.1% shows a maximum adsorption rate of Pb^{2+} (79.86 mg/g), and hydrochar can be used to remove Pb^{2+} from wastewater as it shows a maximum removal efficiency of 95.08% at 25 mg/L [87]. Additionally, sulfide-modified magnetic hydrochar (MHC-S_4) was prepared by simultaneously supporting the synthesis of nanoparticles with Fe_3O_4 and grafting a sulfide-containing group onto a carboniferous water prepared by reacting with pinecone at 200 °C for 5 h. MHC-S_4 can effectively adsorb Cd (II)/Pb (II) over a wide pH range and achieve rapid absorption equilibrium within 25 min. In addition, MHC-S_4 has excellent adsorption properties in a single system, and the maximum single-layer adsorption capacity of Cd (II) and Pb (II) is 62.49 and 149.33 mg/g, respectively [88].

Table 5. Research on adsorption of heavy metal ions by hydrochar.

Raw Materials	Hydrothermal Conditions	Modification Treatment	Adsorbates	Product	References
Black liquor	150, 180, 200, 220 °C, and 0.5 h	hexamethylenediamine (HMDA)	50 mL of Cr(VI) solutions	The maximum adsorption capacity of Cr(VI) was 741.74 mg/g at 45 °C and pH 2.	[80]
Glucose	180 °C, 48 h	—	20.00 mL of pure uranyl ion (UO_2^{2+}) solutions	The highly selective separation of U(VI) was achieved, and the maximum adsorption capacity reached 408.36 mg at 333.15 K and pH 4.5.	[81]
Sucrose	190 °C, 11 h	KOH activated	25.0 mL of methylene blue solutions (500 mg/L)	The maximum adsorption capacity of methylene blue was 704.2 mg/g.	[82]
Banana peels	230 °C, 2 h	phosphoric acid	40 mL of Pb^{2+} stock solutions (2000 mg/L)	The adsorption capacities of Pb^{2+} for dehydrated and fresh banana peel hydrochar were 359 mg/g and 193 mg/g, respectively.	[83]
Willow twigs	300 °C, 30 min	—	50 mL aqueous copper or cadmium solutions (40 mg/L)	The hydrochar exhibited adsorption capacities of 34 mg/g (0.313 mmol/g) for Cd^{2+} and 31 mg/g (0.503 mmol/g) for Cu^{2+}.	[84]
Pine sawdust	260 °C, 2 h	H_2O_2	25 mL of Pb^{2+} solutions (1–200 mg/L)	The maximum adsorption capacity of Pb^{2+} reached 92.80 mg/g at pH 5.0 and 298 K.	[85]
Pseudomonas aeruginosa shells	200 °C, 20 min	—	50 mL of Pb^{2+} or Cd^{2+} aqueous solution (10–100 mg/L)	It was found to be more effective at adsorbing Pb^{2+} and Cd^{2+} compared to biochar.	[86]
Arecanut husk	180–220 °C, 9 h	citric acid	Zn^{2+}, Cr^{6+}, Ni^{2+}, Pb^{2+} solutions (0–100 mg/L)	The maximum removal efficiency of Pb^{2+} in wastewater was 95.08% at 25 mg/L.	[87]
Pinecone	200 °C, 5 h	$FeCl_3 \cdot 6H_2O$, CH_3COONa, $Na_3C_6H_5O_7$	100 mL of Cd(II)/Pb(II) solutions (10–300 mg/L)	The maximum single-layer adsorption capacity of MHC-S_4 for Cd (II) and Pb (II) was 62.49 and 149.33 mg/g, respectively.	[88]

3.1.2. Use of Hydrochar for Organic Pollutants

Hydrochar serves as an effective adsorbent for organic pollutants such as fuels and antibiotics in wastewater. The adsorption mechanisms of hydrochar involve both physical and chemical interactions, including electrostatic interactions, hydrophobic effects, hydrogen bonding, pore filling, partitioning, and interactions between aromatics and cations [76,77,89]. Physical adsorption utilizes the surface characteristics of hydrochar, such as porosity and surface area, to adsorb pollutants on its surface or within its micropores [90]. Chemical adsorption involves the formation of chemical bonds between the adsorbent's functional groups, such as hydroxyl, carboxyl, and amino groups, and the adsorbates, facilitating effective removal of heavy metal ions through stable chelation or ionic and covalent bonds. The dominant mechanism depends on the structural and physicochemical properties of the organic pollutants and the biomass charcoal. Modifications and preparations of hydrochar alter its physicochemical properties, enhancing its adsorption performance for organic compounds [91]. In Figure 9, the adsorption mechanism of hydrochar for organic pollutants was emphasized by describing the mechanism of TC adsorption of hydrochar, modified-hydrochar, and hydrochar-derived ACs. In Table 6, some studies show the effect of hydrochar on the adsorption of organic pollutants.

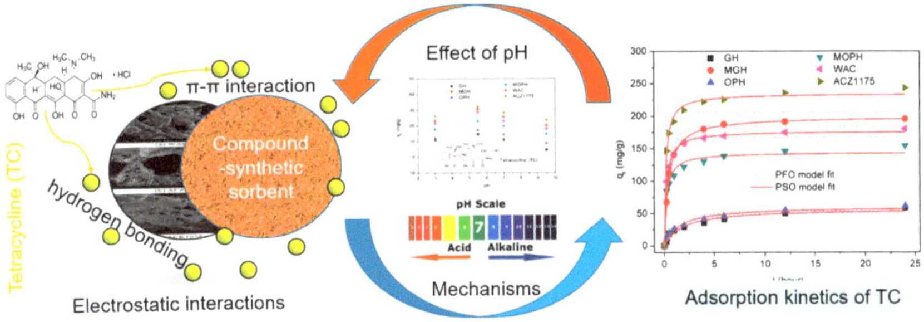

Figure 9. Proposed mechanism for TC adsorption by hydrochar, modified-hydrochar, and hydrochar-derived ACs [92].

Table 6. Research on adsorption of organic pollutants by hydrochar.

Raw Materials	Hydrothermal Conditions	Modification Treatment	Adsorbates	Product	References
Municipal sludge	600 °C, 2 h	Zn/Fe	0.10–0.30 g/L TC, 0.20–0.50 g/L CIP	The product demonstrated maximum adsorption capacities of 145.0 mg/g for TC and 74.2 mg/g for CIP.	[93]
Bamboo shavings	200 °C, 3 h	acid assisted	30 mL of Congo red or 2-naphthol solutions (0.5 mg/mL)	The resultant material exhibited maximum adsorption capacities of 90.51 mg/g for Congo Red and 72.93 mg/g for 2-naphthol.	[94]
Montmorillonite and rice husks	180 °C, 16 h	KOH	100 mL of estrogen stock solutions (2.5 mg/mL)	KOH improved the adsorption capacity of estrogen.	[91]
Rice straw	160–200 °C, 40–70 min	——	30 mL of Congo red, berberine hydrochloride, 2-naphthol, Zn^{2+} and Cu^{2+} solutions (0.5 mg/mL)	The maximum adsorption capacities for Congo Red, berberine hydrochloride, 2-naphthol, Zn^{2+}, and Cu^{2+} were 222.1 mg/g, 174.0 mg/g, 48.7 mg/g, 112.8 mg/g, and 144.9 mg/g, respectively.	[95]
Coffee husks	210 °C, 243 h	——	25.0 mL of the MB solutions (300 mg/L)	The adsorption capacity of methylene blue was 34.85 mg/g, mainly by physical adsorption.	[96]
Moroccan nut shells (ANS)	180, 200 °C, and 6 h.	——	200 mL of BPA (60 mg/L) and diuron (40 mg/L) solutions	The hydrochar generated at 200 °C removed about 92% and 95% of BPA and diuron, respectively, with high adsorption efficiency.	[97]
Fructose	180 °C, 2 h	phloroglucinol	25 mL of tetracycline solutions	The adsorption capacity of the product to tetracycline in water was 274.7 mg/g.	[98]

Hydrochar from synthesized municipal sludge at 600 °C for 2 h with Zn/Fe modification demonstrated maximum adsorption capacities of 145.0 mg/g for tetracycline (TC) and 74.2 mg/g for ciprofloxacin (CIP) [93]. Moreover, hydrochar from produced bamboo shavings at 200 °C for 3 h, assisted by acid treatment, exhibited maximum adsorption capacities of 90.51 mg/g for Congo Red and 72.93 mg/g for 2-naphthol [94]. Tian et al. [91] evaluated the adsorption effects of KOH-modified and unmodified hydrochar produced from montmorillonite and rice husks at 180 °C for 16 h on estrogens in water. They found that a 1% mass ratio of KOH to hydrochar maximized the adsorption efficiency. Li et al. [95] prepared hydrochar from rice straw at temperatures ranging from 160 °C to 200 °C and durations from 40 to 70 min. At 298 K and an initial concentration of 0.5 mg/mL, the maximum adsorption capacities for Congo Red, berberine hydrochloride, 2-naphthol, Zn^{2+}, and Cu^{2+} were 222.1 mg/g, 174.0 mg/g, 48.7 mg/g, 112.8 mg/g, and 144.9 mg/g, respectively. Furthermore, hydrochar produced from coffee husks at 210 °C for 243 h achieves an adsorption capacity of 34.85 mg/g for methylene blue, predominantly through physical adsorption [96]. Zbair et al. [97] synthesized hydrochar from Moroccan nut shells (ANS) at temperatures of 180 °C and 200 °C for 6 h. Additionally, hydrochar produced from fructose at 180 °C for 2 h modified with phloroglucinol showed an adsorption capacity of 274.7 mg/g for tetracycline in water [98].

3.1.3. Use of Hydrochar for Inorganic Anions

For inorganic anions, hydrochar typically undergoes chemical adsorption through chemically formed bonds on the surface [77], exhibiting good adsorption efficiency for anions such as phosphate, arsenate, and fluoride in wastewater. For example, the hydroxyl group on the surface of the hydrochar can form hydrogen bonds with the phosphate ions or undergo coordination interactions to remove the phosphate from the wastewater. Similarly, the carboxyl group on the surface can coordinate with the metal ions in arsenic acid to form a stable complex to effectively remove arsenic acid. For fluoride ions, metal oxides such as aluminum and iron on the surface of hydrochar can exchange ions with fluoride ions to reduce the concentration of fluoride ions in water. In addition, good hydrochar chemical stability and corrosion resistance can be in a wider range of pH to maintain good adsorption performance, which makes the hydrochar become the ideal material for dealing with the wastewater containing inorganic anions. In Table 7, some studies show the effect of hydrochar on the adsorption of inorganic anions.

Anoxygenically digested cattail hydrochar was produced by reacting with pure water, acetic acid assisted, and sodium hydroxide assisted HTC at 200–300 °C for 1 h. The study showed that when the HTC temperature increased from 200 °C to 300 °C, the yield using water, acetic acid and sodium hydroxide decreased from 40.2% to 31.6%, 37.5% to 28.3% and 35.7% to 22.7%, respectively. The adsorption capacities of NH_4^+-N and PO_4^{3-}-P of these hydrochar are 92.6–122.4 mg/g and 1.6–15.8 mg/g, respectively, which fit the Freundlich model well [99]. Hydrothermal modification reduces the number of acidic functional groups, which is beneficial for anion adsorption. Hydrochar synthesized from chicken feathers at 150–170 °C after 1–3 h has a strong affinity for phosphate, and has a good adsorption efficiency for phosphorus removal under acidic conditions [100]. The adsorption isotherms for hydrochar suited the Langmuir model better, with the maximum adsorption capacity (q_m) of hydrochar being 21.70 mg/g at 30 °C. Hydrochar synthesized from microalgae at 348 K for 40 min modified with magnesium exhibited a strong affinity for phosphate, with a maximum adsorption capacity of 89.61 mg/g [25]. Additionally, the magnesium-containing hydrochar adsorbed phosphate from water mainly through ion exchange. Magnetic hydrochar modified with Fe_3O_4 was synthesized from sodium alginate at 210 °C for 5 h [101]. The results showed that at a dosage of 2 g/L, the removal rates of arsenic and fluoride were around 85%, with maximum adsorption capacities of 26.06 mg/g and 15.64 mg/g, respectively, in individual systems.

Table 7. Research on adsorption of inorganic anions by hydrochar.

Raw Materials	Hydrothermal Conditions	Modification Treatment	Adsorbates	Product	References
Anaerobically digested cattails	200, 250 and 300 °C, 1 h	CH_3COOH and NaOH	1470 mg/L COD, 50 mg/L NH_4^+-N and 151 mg/L TP	The adsorption capacities of NH_4^+-N and PO_4^{3-}-P were 92.6–122.4 mg/g and 1.6–15.8 mg/g, respectively.	[99]
Chicken feathers	150–170 °C, 1–3 h	—	0–150 mg/L KH_2PO_4	The maximum adsorption capacity (q_m) of hydrochar was 21.70 mg/g at 30 °C.	[100]
Microalgae	348 K, 40 min	Mg	20 mL of P solutions (20–15,000 mg/L)	The modified hydrochar exhibited strong affinity for phosphate, with a maximum adsorption capacity of 89.61 mg/g.	[25]
Sodium alginate	210 °C, 5 h	Fe_3O_4	As (V) solutions (1–100 mg/L), F solutions (1–50 mg/L)	At a dosage of 2 g/L, the maximum adsorption capacity of arsenic and fluoride was 26.06 mg/g and 15.64 mg/g, respectively.	[101]

3.2. Surface Water Restoration

Surface water restoration is a primary focus in current water environment management [102], with HTC technology playing a significant role, particularly in sediment improvement and water quality enhancement in rivers and lakes. Among them, surface water restoration refers to the treatment and restoration of polluted or damaged surface water bodies through a series of hydrothermal carbonization technical measures and management measures to achieve the purpose of improving water quality and restoring water ecosystem functions. By thermally decomposing and carbonizing sediments and deposits, this technology effectively reduces nutrient release from sediments, thereby controlling eutrophication trends in lakes. Additionally, the hydrochar produced exhibits high adsorptive properties, enabling the selective absorption of harmful substances in the water, significantly enhancing water quality and purifying the water body. Following treatment with HTC technology, rivers and lakes gradually restore their natural ecosystems, providing favorable habitats for diverse aquatic life forms and yielding substantial economic and environmental benefits [103]. In Table 8, studies through surface water restoration highlight the role of hydrochar.

Using Prunus serrulata bark as raw material, hydrochar was prepared by a reaction at 200 °C for 6 h, and the product has high efficiency and low precursor material cost, which can remove atrazine from river water. When the acidic pH = 3 is favorable for adsorption, the ideal adsorbent dose is 0.8 g/L, and kinetic studies show that the concentration does not affect the system equilibrium time, which is reached at 240 min. The Langmuir model presented the greatest compliance to the isotherm data and indicated a higher affinity between atrazine and hydrochar, reaching a maximum adsorption capacity of 63.35 mg/g [104]. Additionally, extending the reaction time proved to be beneficial for improving the quality of the hydrothermal biochar. Water hyacinth was employed as a feedstock to produce hydrochar at 240 °C and various durations from 0.5 to 24.0 h [105]. The results revealed that the specific surface area of the hydrochar was initially increased and then decreased over time. This process not only utilizes the invasive water hyacinth but also achieves ecological restoration. Hydrochar prepared from glucose at 180 °C for 10 h and composited with $FeCl_3{\cdot}6H_2O$ and $MnCl_2{\cdot}4H_2O$ to successfully load $MnFe_2O_4$ on its surface could rupture algal cells and disrupt their photosynthetic systems, demonstrating its capability to treat organic matter [106]. Water hyacinth was modified with $MgCl_2{\cdot}6H_2O$ and $AlCl_3{\cdot}6H_2O$, and reacted at 150°C for 24 h to obtain Mg/Al-layered double hydroxides modified water hyacinth hydrochar (MgAl@WH) with layered surface and many adsorption active sites.

The maximum adsorption capacity is 311.0 mg/g, and the adsorption efficiency is about 98.0% after 10 cycles [107]. Moreover, research on algal raw materials offers a potential solution for managing eutrophication in lakes and rivers.

Simultaneously, an appropriate amount of hydrochar can enhance the water and soil environment, thereby promoting plant growth and facilitating ecological restoration. Numerous studies have demonstrated that the carbon and magnesium content in hydrochar can influence the uptake of critical nutrients, such as ammonium salts and phosphates [108]. Additionally, hydrochar can enhance sulfur levels in the soil and decrease the available nitrogen, and thereby enhance soybean growth [109]. However, excessive application of hydrochar may inhibit crop growth, particularly during the seed germination and seedling stages [109–111].

3.3. Groundwater Purification

Groundwater, as a crucial water resource, is the focus of intense protection and management efforts. Using HTC technology in groundwater purification aims to remove heavy metals and organic pollutants, reduce their bioavailability, and thus inhibit microbial growth and reproduction, ensuring groundwater purity and safety. HTC technology is utilized to prepare hydrochar with diverse adsorption capabilities for various pollutants. Employed as an adsorbent with specific particle sizes, it facilitates efficient underground transport. The impact of carbon-based materials on pollutant removal depends on their physicochemical properties, which are determined by the type of raw material and the pyrolysis conditions used for production. The increase in additives can change the physicochemical properties of carbon-based materials, including biochar and hydrochar, to remove different types of pollutants, which is helpful to develop remediation technologies for contaminated groundwater [112]. This enables the removal of organic substances and heavy metal ions from groundwater, safeguarding water source safety and promoting sustainable groundwater use. In Table 8, studies through groundwater purification highlight the role of hydrochar. Hydrochar prepared from switchgrass at 300 °C for 30 min was utilized to remove U (VI) contaminants from groundwater [113]. The removal of pathogenic rotavirus (RV) and human adenovirus (HAdV) was investigated by hydrothermal treatment of swine feces (2 h at 180 °C and 7 h at 230 °C) under two conditions, using a 10 cm bed height sand column and with or without an aqueous charcoal supplement (1.5%, *w*/*w*). The results showed that the removal efficiency of the two viruses in the improved hydration column was >3 logarithm (complete removal). Although different HTC conditions lead to different characteristics of 180 HTC and 230 HTC, there is no significant difference in virus removal performance between the two hydrates. Hydrochar derived from fecal waste can be used as a capacity grade virus adsorbent [114]. Prepared by anaerobic digestion of swine manure at 180 °C for 45 min, hydrochar has been shown to stimulate the growth of denitrifies without toluene degradability, including Candidatus Competibacter and Ferrovibrio, and increase NO_3^--N removal, which was mainly attributed to the partial denitrification [112]. Following ball milling treatment, it exhibited excellent underground transport properties, making it effective for removing TCE from groundwater.

Table 8. Research on the ecological protection of hydrochar.

Raw Materials	Hydrothermal Conditions	Modification Treatment	Product	Type of Protection	References
Prunus serrulata bark	200 °C, and 6 h	—	The maximum adsorption capacity for atrazine in the river is 63.35 mg/g.	surface water restoration	[104]
Water hyacinth	240 °C, and 0.5–24.0 h	—	With the increase in residence time, the higher heating value in all hydrochar products was 16.83 MJ/kg to 20.63 MJ/kg.	surface water restoration	[105]

Table 8. *Cont.*

Raw Materials	Hydrothermal Conditions	Modification Treatment	Product	Type of Protection	References
Glucose	180 °C, 10 h	$FeCl_3 \cdot 6H_2O$, $MnCl_2 \cdot 4H_2O$	The product can break algal cells and destroy the photosynthetic system of algal cells, and deal with organic substances.	surface water restoration	[106]
Water hyacinth	150 °C, 24 h	Mg/Al-layered double hydroxide modification	The maximum adsorption capacity of the product for mordant brown (anionic dye) was 311.0 mg/g.	surface water restoration	[107]
Switchgrass	300 °C, 30 min	——	The hydrochar-formed permeable reaction wall demonstrated rapid removal of U (VI).	groundwater purification	[113]
Swine feces	180 °C, 2 h; 230 °C, 7 h	H_2SO_4	The removal efficiency of pathogenic RV and HAdV in the hydrated modified column was >3 logarithms (complete removal).	groundwater purification	[114]
Anaerobic digestion of swine manure	180 °C, 45 min	——	Hydrochar enhance toluene removal from groundwater, stimulate the growth of denitrifiers without toluene degradability, and increase NO_3^--N removal.	groundwater purification	[112]

4. Research Advances in the Context of the "Dual Carbon" Goals

4.1. Influence of HTC Technology on Carbon Emission

Currently, HTC technology is a research hotspot under the "dual carbon" goals. This technology utilizes high-temperature and high-pressure hydrothermal conditions to transform organic matter into carbonaceous materials [21–24]. Compared to traditional carbonization methods, HTC offers advantages such as lower energy consumption, minimal pollution, and enhanced performance of carbon materials. More importantly, HTC can convert carbon in organic wastes into stable carbonaceous materials and steam, significantly reduce carbon emissions from the transformation of organic matter into carbon materials and avoid greenhouse gas emissions. This process effectively sequesters carbon from biomass, which is crucial for achieving the "dual carbon" goals. Additionally, HTC technology can recycle and utilize the energy from organic wastes, further enhancing energy efficiency.

4.2. Comparison with Other Environmental Governance Technologies

HTC technology exhibits unique advantages in environmental management. Compared to traditional environmental management technologies, it is more eco-friendly and can significantly reduce secondary pollutants during the treatment of organic wastes, which is fully aligned with current green and low-carbon development concepts [25–27]. Additionally, this technology facilitates resource utilization by converting organic wastes into high-value products such as activated carbon and carbon black [32]. This not only enhances the economic value of wastes but also yields significant social benefits. Notably, HTC processes organic wastes under high temperatures and pressures, achieving high processing efficiency and energy recovery rates, which underscores its effectiveness in environmental management. These advantages underscore the broad potential of HTC technology in environmental management.

Traditional sewage treatment technologies, such as the activated sludge method and biofilm method, can effectively remove organic matter and nutrients in wastewater, but the treatment process may produce secondary pollution, such as odor, noise and chemical residue. The HTC technology can significantly reduce these secondary pollutants, which is fully in line with the current concept of green and low-carbon development. Anaerobic fermentation technology is mainly used to treat high concentration of organic wastewater, and the biogas produced is a kind of clean energy. However, for some wastewater with low concentrations and a complex composition, the treatment effect of anaerobic fermentation technology may not be ideal. HTC technology can deal with organic waste of various components to achieve high processing efficiency and energy recovery. In Table 9, different water environmental treatment technologies are compared in terms of advantages, Shortcoming and energy consumption.

However, the HTC technique also has some limitations. High-temperature and high-pressure conditions of HTC technology may increase equipment costs compared to conventional wastewater treatment technologies, and the procurement and processing of biomass or carbon-containing waste may pose environmental challenges. In order to fully exploit the advantages of HTC technology and mitigate its weaknesses, further research and practice are needed. This includes research on technology optimization, reduced equipment costs, and a sustainable supply of biomass or carbon-containing waste.

Table 9. Comparison of different water environment treatment technologies.

Type of Technology	Environmental Management Effectiveness	Advantage	Shortcoming	Energy Consumption
Adsorption	Effective removal of organic matter and certain inorganic substances	Easy to operate, high removal efficiency can be achieved. Adaptable, adsorbents can be reused.	The cost of adsorbents can be high, and energy is required to regenerate or replace adsorbents.	Medium to high
Membrane technology	Separation and purification of particulates and dissolved substances in water bodies	Modular design, easy to expand, high separation efficiency, and high degree of automation.	Membrane materials can be expensive, with membrane fouling and membrane replacement costs, and may require chemical cleaning.	Medium to high
Precipitation	Removal of suspended solids and certain dissolved solids from water	The processing capacity is large, the operating cost is relatively low, and the application range is wide.	A large amount of sedimentation tank space is required and may produce a large amount of sludge that is difficult to control.	Low to moderate
Enzyme technology	Decomposes and transforms organic matter	Efficient for specific pollutants, pollutants can be converted into harmless substances, reducing sludge production.	Enzyme stability may be low, requiring appropriate Ph and temperature conditions, and initial enzyme cost.	Medium to low
HTC	Efficient treatment of organic waste into high-value products	Environmental protection, reduce secondary pollution, good resource effect, high treatment efficiency under high-temperature and high-pressure conditions.	Equipment costs are higher, dependence on biomass or carbon-containing waste, and energy consumption can be higher.	high

4.3. Experiments with HTC Technology

Using the Nanjing Qiaobei Sewage Treatment Plant as an example, Wang et al. [115] applied HTC technology to sludge treatment, combined with physicochemical pretreatment, namely Fenton oxidation (FO) and HTC technology to utilize SSL, and found that moderate oxidation could improve the output and performance of the produced hydrochar. The results showed that the pretreatment affected the surface structure and organic composition of SSL and promoted the carbonization of intermediate products. Moreover, compared with the yield of hydrochar (50.7%) obtained via the direct HTC treatment, the yield of the hydrochar obtained using the combined process increased to 55.0–65.2% (depending on the pretreatment condition). Hydrochar properties were enhanced using the combined process, and the energy density of hydrochar slightly decreased after pretreatment (1.2–13.1%);

however, the energy yields increased by 0.6–30.1% due to the enhanced hydrochar yield. The carbonization degree of hydrochar was improved; the carbon in the feedstock distributed to the hydrochar increased from 33.40% to 46.09%. The treated sludge not only was effectively reduced but also enabled the recovery and reuse of energy and materials, thereby enhancing the waste's resource utilization. Furthermore, it adsorbs heavy metal ions in wastewater, contributing to water environmental protection. The application of this technology not only reduces carbon emissions but also effectively addresses challenges in sludge treatment, achieving resource recycling. This provides a valuable reference for urban sludge disposal and resource utilization in China.

During the research process, several successful experiments have provided valuable insights. For example, Shi et al. [116] utilized HTC technology to process large particles (>2 cm) of kitchen waste, successfully converting them into carbon materials. This not only addressed wastes disposal issues but also created a new supply source for the carbon materials market. This experiment demonstrates that technological innovation and application can facilitate a win-win situation for economic development and environmental protection.

Experimental studies serve as an effective means to verify the application effects of HTC technology. Scholars have explored the application of HTC technology in various domains, including agricultural wastes, urban trash, and industrial wastewater [117]. These experimental studies demonstrate that while HTC technology is effective in practical applications, it also faces several challenges that need to be addressed. Challenges include the pretreatment and post-treatment of biomass or carbon-containing wastes and scaling up and industrializing the equipment.

5. Challenges and Countermeasures

5.1. Technical Challenges

Although the principle of HTC technology is relatively simple, its practical application requires precise control of parameters such as temperature, pressure, and time to achieve optimal carbonization effects [26,118,119]. Additionally, the adaptability of this technology to various raw materials poses a technical challenge, as organic wastes of different origins and properties may require tailored processing conditions. Therefore, further optimizing the technological process and enhancing carbonization efficiency are critical challenges that need to be addressed [118,119]. Currently, this technology remains in the research and development stage and requires further optimization and enhancements to achieve more efficient and stable production. Furthermore, the widespread adoption and application of this technology must overcome several obstacles, including challenges related to equipment design, manufacturing, operation, and maintenance.

5.2. Economic Costs

Economic cost is a significant challenge facing the application of innovative technology [120–122]. While the long-term economic benefits of HTC technology are substantial, the initial investment is considerable, encompassing equipment purchases, installation, commissioning, and personnel training. Additionally, operational energy consumption and maintenance costs are non-negligible. These factors make it prohibitive for economically disadvantaged areas, potentially limiting its adoption in certain regions. Therefore, it is essential to explore strategies to reduce production costs and enhance economic efficiency, enabling broader adoption of this technology by more regions and enterprises.

5.3. Environmental and Social Impact

Environmental and social impact will be closely related to innovative technology application [123,124]. When used to process organic wastes, HTC technology can reduce greenhouse gas emissions and enhance resource utilization efficiency. However, it may also generate by-products and pollutants, such as carbonization residues and wastewater. Improper management of these can lead to secondary environmental pollution. Therefore, in promoting this technology, it is crucial to fully assess its environmental and social impacts and establish comprehensive mechanisms for managing by-products and pollutants to ensure real societal benefits and welfare.

5.4. Policy and Management

The policy and management play a crucial role in the application of innovative technology [125,126]. Governments should establish policies and regulations to encourage and support the research, development, and application of this technology. Additionally, it is essential to establish regulatory mechanisms, define technical standards, and set regulatory requirements to ensure compliance with national and international standards and prevent potential environmental and social risks, thereby providing a robust legal framework for the technology's application.

6. Conclusions and Perspective

6.1. Summary of the Main Research Results

In the research and application of HTC technology, we have achieved remarkable results. Compared with other wastewater treatment techniques, this technique has been successfully applied to various wastewater treatments, showing effective effects and stability. Moreover, HTC technology has obvious advantages in improving the efficiency of resource recovery extracted from wastewater. It not only transforms organic substances into valuable materials but also can produce high-value carbon materials suitable for different fields, realizing the recycling and reuse of resources.

The application of HTC technology in water environment management provides a new idea and method for water environment management in the world. HTC technology can not only effectively deal with water environment pollution and improve the quality of water environment, but also has obvious advantages in improving the efficiency of resource recovery, reducing the pressure of the resource shortage in China. In addition, the application of HTC technology can also promote the development of the environmental protection industry and promote the realization of a green economy.

6.2. Suggestions for Future Research Directions

Although HTC technology has achieved remarkable results in water environment management, there are still some aspects that need further research and improvement. In my perspective, the study of the HTC reaction mechanism is crucial, which helps to improve the reaction efficiency and optimize the process parameters, so as to ensure the product quality. Therefore, it is necessary to organize more in-depth research for a comprehensive understanding and grasp of the HTC reaction mechanism. In addition, the application scope of HTC technology also needs to be further expanded to adapt to various water quality and flow conditions. To this end, the HTC technology is needed to be optimized and improved to make it suitable for more kinds of wastewater treatment, including industrial wastewater, urban sewage, etc. Finally, enhanced integration with other wastewater treatment technologies is the key to developing more efficient and environmentally friendly wastewater treatment systems. This can not only improve processing efficiency, reduce operating costs, but also better cope with complex water quality problems.

To effectively promote and apply HTC technology, decision makers should provide necessary support and guidance. On the one hand, decision makers need to increase their investment in HTC technology research and development, and encourage the cooperation between enterprises, universities and research institutions, to jointly promote the development of HTC technology. This can not only accelerate the process of technology research and development, but also promote technological innovation and upgrading. On the other hand, decision makers need to develop environmental policies to encourage enterprises to adopt environmentally friendly wastewater treatment technologies and improve wastewater treatment standards. This can not only reduce environmental pollution, but also promote sustainable economic development. Finally, the establishment of a set of comprehensive standards for the sewage treatment industry is crucial to standardize its development and promote the wide application of HTC technologies. Through these measures, it can effectively promote the development and application of HTC technology and provide strong technical support for solving the water environment problems in China.

In general, the research and application of HTC technology in organic wastewater treatment and resource recovery provides a new solution for water environment management in China and has broad application prospects and important research value. In the future, we should continue to study HTC technology, optimize its treatment effect, reduce its operating cost, so as to better cope with the challenges of water environment management, promote the development of China's water environment protection cause, and make contributions to the construction of a beautiful China.

Author Contributions: G.L.: investigation, data curation, writing—original draft; Q.X.: writing—review and editing, visualization; S.F.A.-E.: writing—review and editing; M.A.A.: writing—review and editing; T.Z.: writing—review and editing, supervision, project administration, funding acquisition. All authors have read and agreed to the published version of the manuscript.

Funding: The research was sustained by the grant from the Undergraduate Research Program of China Agricultural University, the National Key Research and Development Program of China "Intergovernmental Cooperation in International Science and Technology Innovation" [Grant number 2023YFE0104700], the National Natural Science Foundation of China [Grant Number 31401944].

Conflicts of Interest: The authors declare that they have no known competing financial interests or personal relationships that could have appeared to influence the work reported in this paper.

References

1. Zhu, Z.; Zhang, X.; Dong, H.; Wang, S.; Reis, S.; Li, Y.; Gu, B. Integrated livestock sector nitrogen pollution abatement measures could generate net benefits for human and ecosystem health in China. *Nat. Food* **2022**, *3*, 161–168. [CrossRef] [PubMed]
2. Hu, Y.; Su, M.; Wang, Y.; Cui, S.; Meng, F.; Yue, W.; Liu, Y.; Xu, C.; Yang, Z. Food production in China requires intensified measures to be consistent with national and provincial environmental boundaries. *Nat. Food* **2020**, *1*, 572–582. [CrossRef]
3. Springmann, M.; Clark, M.; Croz, D.M.D.; Wiebe, K.; Bodirsky, B.L.; Lassaletta, L.; de Vries, W.; Vermeulen, S.J.; Herrero, M.; Carlson, K.M.; et al. Options for keeping the food system within environmental limits. *Nature* **2018**, *562*, 519–525. [CrossRef] [PubMed]
4. Raihan, A.; Voumik, L.C.; Mohajan, B.; Rahman, M.S.; Zaman, M.R. Economy-energy-environment nexus: The potential of agricultural value-added toward achieving China's dream of carbon neutrality. *Carbon Res.* **2023**, *2*, 43. [CrossRef]
5. Qiu, L.; Yu, R.; Hu, F.; Zhou, H.; Hu, H. How can China's medical manufacturing listed firms improve their technological innovation efficiency? An analysis based on a three-stage DEA model and corporate governance configurations. *Technol. Forecast. Soc. Chang.* **2023**, *194*, 122684. [CrossRef]
6. Jin, S.; Zhang, B.; Wu, B.; Han, D.; Hu, Y.; Ren, C.; Zhang, C.; Wei, X.; Wu, Y.; Mol, A.P.J.; et al. Decoupling livestock and crop production at the household level in China. *Nat. Sustain* **2021**, *4*, 48–55. [CrossRef]
7. Li, F.; Cheng, S.; Yu, H.; Yang, D. Waste from livestock and poultry breeding and its potential assessment of biogas energy in rural China. *J. Clean. Prod.* **2016**, *126*, 451–460. [CrossRef]
8. Jing, Q.; Luo, W.; Bai, H.; Xu, H. A method for city-level energy-related CO_2 emission estimation. *J. Environ. Sci.* **2018**, *38*, 4879–4886.
9. Wang, Y.; Quan, S.; Tang, X.; Hosono, T.; Hao, Y.; Tian, J.; Pang, Z. Organic and inorganic carbon sinks reduce long-term deep carbon emissions in the continental collision margin of the southern Tibetan plateau: Implications for Cenozoic climate cooling. *J. Geophys. Res.-Solid Earth* **2024**, *129*, e2024JB028802. [CrossRef]
10. Zhao, S.; Zhang, L.; An, H.; Peng, L.; Zhou, H.; Hu, F. Has China's low-carbon strategy pushed forward the digital transformation of manufacturing enterprises? Evidence from the low-carbon city pilot policy. *Environ. Impact Assess. Rev.* **2023**, *102*, 107184. [CrossRef]
11. Zhao, S.; Zhang, L.; Peng, L.; Zhou, H.; Hu, F. Enterprise pollution reduction through digital transformation? Evidence from Chinese manufacturing enterprises. *Technol. Soc.* **2024**, *77*, 102520. [CrossRef]
12. Li, H.; Zhang, T.; Shaheen, S.M.; Abdelrahman, H.; Ali, E.F.; Bolan, N.S.; Li, G.; Rinklebe, J. Microbial inoculants and struvite improved organic matter humification and stabilized phosphorus during swine manure composting: Multivariate and multiscale investigations. *Bioresour. Technol.* **2022**, *351*, 126976. [CrossRef] [PubMed]
13. Shang, M.; Luo, J. The Tapio decoupling principle and key strategies for changing factors of Chinese urban carbon footprint based on cloud computing. *Int. J. Environ. Res. Public Health* **2021**, *18*, 2101. [CrossRef] [PubMed]
14. Sun, C.; Chen, J.; He, B.; Liu, J. Digitalization and carbon emission reduction technology R&D in a Stackelberg model. *Appl. Econ. Lett.* **2024**, 1–6. [CrossRef]

15. Wu, F.; Li, F.; Zhao, X.; Bolan, N.S.; Fu, P.; Lam, S.S.; Mašek, O.; Ong, H.C.; Pan, B.; Qiu, X.; et al. Meet the challenges in the "carbon age". *Carbon Res.* **2022**, *1*, 1. [CrossRef]
16. Zhang, T.; Wu, X.; Shaheen, S.M.; Abdelrahman, H.; Ali, E.F.; Bolan, N.S.; Ok, Y.S.; Li, G.; Tsang, D.C.W.; Rinklebe, J. Improving the humification and phosphorus flow during swine manure composting: A trial for enhancing the beneficial applications of hazardous biowastes. *J. Hazard. Mater.* **2022**, *425*, 127906. [CrossRef] [PubMed]
17. Zhang, T.; Li, H.; Yan, T.; Shaheen, S.M.; Niu, Y.; Zhang, Y.; Abdelrahman, H.; Ali, E.F.; Bolan, N.S.; Rinklebe, J. Organic matter stabilization and phosphorus activation during vegetable waste composting: Multivariate and multiscale investigation. *Sci. Total Environ.* **2023**, *891*, 164608. [CrossRef] [PubMed]
18. Fang, C.; Zhang, T.; Li, P.; Jiang, R.; Wu, S.; Nie, H.; Wang, Y. Phosphorus recovery from biogas fermentation liquid by Ca-Mg loaded biochar. *J. Environ. Sci.* **2015**, *29*, 106–114. [CrossRef]
19. He, X.; Zhang, T.; Ren, H.; Li, G.; Ding, L.; Pawlowski, L. Phosphorus recovery from biogas slurry by ultrasound/H_2O_2 digestion coupled with HFO/biochar adsorption process. *Waste Manag.* **2017**, *60*, 219–229. [CrossRef]
20. Zhang, T.; Xu, H.; Li, H.; He, X.; Shi, Y.; Kruse, A. Microwave digestion-assisted HFO/biochar adsorption to recover phosphorus from swine manure. *Sci. Total Environ.* **2018**, *621*, 1512–1526. [CrossRef]
21. He, X.; Zhang, T.; Xue, Q.; Zhou, Y.; Wang, H.; Bolan, N.S.; Jiang, R.; Tsang, D.C.W. Enhanced adsorption of Cu(II) and Zn(II) from aqueous solution by polyethyleneimine modified straw hydrochar. *Sci. Total Environ.* **2021**, *778*, 146116. [CrossRef]
22. Zhang, T.; Wu, X.; Shaheen, S.M.; Zhao, Q.; Liu, X.; Rinklebe, J.; Ren, H. Ammonium nitrogen recovery from digestate by hydrothermal pretreatment followed by activated hydrochar sorption. *Chem. Eng. J.* **2020**, *379*, 122254. [CrossRef]
23. Huang, R.; Fang, C.; Lu, X.; Jiang, R.; Tang, Y. Transformation of phosphorus during (hydro)thermal treatments of solid biowastes: Reaction mechanisms and implications for P reclamation and recycling. *Environ. Sci. Technol.* **2017**, *51*, 10284–10298. [CrossRef] [PubMed]
24. Shi, Y.; Zhang, T.; Ren, H.; Kruse, A.; Cui, R. Polyethylene imine modified hydrochar adsorption for chromium (VI) and nickel (II) removal from aqueous solution. *Bioresour. Technol.* **2018**, *247*, 370–379. [CrossRef]
25. Deng, Y.; Zhang, T.; Sharma, B.K.; Nie, H. Optimization and mechanism studies on cell disruption and phosphorus recovery from microalgae with magnesium modified hydrochar in assisted hydrothermal system. *Sci. Total Environ.* **2019**, *646*, 1140–1154. [CrossRef] [PubMed]
26. Deng, Y.; Zhang, T.; Clark, J.; Aminabhavi, T.; Kruse, A.; Tsang, D.C.W.; Sharma, B.K.; Zhang, F.; Ren, H. Mechanisms and modelling of phosphorus solid–liquid transformation during the hydrothermal processing of swine manure. *Green Chem.* **2020**, *22*, 5628–5638. [CrossRef]
27. Wang, T.; Zhai, Y.; Zhu, Y.; Li, C.; Zeng, G. A review of the hydrothermal carbonization of biomass waste for hydrochar formation: Process conditions, fundamentals, and physicochemical properties. *Renew. Sust. Energy Rev.* **2018**, *90*, 223–247. [CrossRef]
28. Zhang, T.; Ding, L.; Ren, H.; Guo, Z.; Tan, J. Thermodynamic modeling of ferric phosphate precipitation for phosphorus removal and recovery from wastewater. *J. Hazard. Mater.* **2010**, *176*, 444–450. [CrossRef]
29. Zhang, T.; Li, Q.; Ding, L.; Ren, H.; Xu, K.; Wu, Y.; Sheng, D. Modeling assessment for ammonium nitrogen recovery from wastewater by chemical precipitation. *J. Environ. Sci.* **2011**, *23*, 881–890. [CrossRef]
30. Fang, C.; Zhang, T.; Jiang, R.; Ohtake, H. Phosphate enhance recovery from wastewater by mechanism analysis and optimization of struvite settleability in fluidized bed reactor. *Sci. Rep.* **2016**, *6*, 32215. [CrossRef]
31. Xu, X.; Feng, F.; Wan, Z.; Wang, Y.; Yu, M.; Han, X.; Wu, G.; Xing, W. Rapid electron transfer reinforced by interfacial Co-O bonding in MOF/COF hybrids for highly efficient degrade tetracycline by activating peroxymonosulfate. *Colloid Surf. A-Physicochem. Eng. Asp.* **2024**, *689*, 133686. [CrossRef]
32. Zhang, Y.; Zhang, T. Biowaste valorization to produce advance carbon materials-hydrochar for potential application of Cr (VI) and Cd (II) adsorption in wastewater: A review. *Water* **2022**, *14*, 3675. [CrossRef]
33. Zhang, T.; Li, P.; Fang, C.; Jiang, R. Phosphate recovery from animal manure wastewater by struvite crystallization and CO_2 degasification reactor. *Ecol. Chem. Eng. S.* **2014**, *21*, 89–99. [CrossRef]
34. Wang, Y.; Liu, Z.; Huang, W.; Lu, J.; Luo, S.; Czech, B.; Li, T.; Wang, H. Capture-reduction mechanism for promoting Cr(VI) removal by sulfidated microscale zerovalent iron/sulfur-doped graphene-like biochar composite. *Carbon Res.* **2023**, *2*, 11. [CrossRef]
35. Mumme, J.; Eckervogt, L.; Pielert, J.; Diakité, M.; Rupp, F.; Kern, J. Hydrothermal carbonization of anaerobically digested maize silage. *Bioresour. Technol.* **2011**, *102*, 9255–9260. [CrossRef] [PubMed]
36. Wang, L.; Chang, Y.; Li, A. Hydrothermal carbonization for energy-efficient processing of sewage sludge: A review. *Renew. Sust. Energ. Rev.* **2019**, *108*, 423–440. [CrossRef]
37. Wu, Y.; Li, W.; Wu, Q.; Liu, S. Preparation, properties and applications of hydrochar. *Prog. Chem.* **2016**, *28*, 121–130.
38. Funke, A.; Ziegler, F. Hydrothermal carbonization of biomass: A summary and discussion of chemical mechanisms for process engineering. *Biofuels Bioprod. Biorefining* **2010**, *4*, 160–177. [CrossRef]
39. Kruse, A.; Funke, A.; Titirici, M.M. Hydrothermal conversion of biomass to fuels and energetic materials. *Curr. Opin. Chem. Biol.* **2013**, *17*, 515–521. [CrossRef]
40. Sevilla, M.; Fuertes, A.B. The production of carbon materials by hydrothermal carbonization of cellulose. *Carbon* **2009**, *47*, 2281–2289. [CrossRef]

41. Kang, S.; Li, X.; Fan, J. Characterization of hydrochars produced by hydrothermal carbonization of lignin, cellulose, d-xylose, and wood meal. *Ind. Eng. Chem. Res.* **2012**, *51*, 9023–9031. [CrossRef]
42. Liu, W.; Jiang, H.; Yu, H. Thermochemical conversion of lignin to functional materials: A review and future directions. *Green Chem.* **2015**, *17*, 4888–4907. [CrossRef]
43. Liu, Q.; Zhang, G.; Kong, G.; Liu, M.; Cao, T.; Guo, Z.; Zhang, X.; Han, L. Valorizing manure waste into green coal-like hydrochar: Parameters study, physicochemical characteristics, combustion behaviors and kinetics. *Renew. Energy* **2023**, *216*, 119103. [CrossRef]
44. Yan, W.; Hoekman, S.K.; Broch, A.; Coronella, C.J. Effect of hydrothermal carbonization reaction parameters on the properties of hydrochar and pellets. *Environ. Prog. Sustain. Energy* **2014**, *33*, 676–680. [CrossRef]
45. Reza, M.T.; Andert, J.; Wirth, B.; Busch, D.; Pielert, J.; Lynam, J.G.; Mumme, J. Hydrothermal carbonization of biomass for energy and crop production. *Appl. Bioenergy* **2014**, *1*, 11–29. [CrossRef]
46. Lynam, J.G.; Reza, M.T.; Yan, W.; Vásquez, V.R.; Coronella, C.J. Hydrothermal carbonization of various lignocellulosic biomass. *Biomass Convers. Biorefinery* **2015**, *5*, 173–181. [CrossRef]
47. Djandja, O.S.; Liew, R.K.; Liu, C.; Liang, J.; Yuan, H.; He, W.; Feng, Y.; Lougou, B.G.; Duan, P.-G.; Lu, X.; et al. Catalytic hydrothermal carbonization of wet organic solid waste: A review. *Sci. Total Environ.* **2023**, *873*, 162119. [CrossRef]
48. Zhao, H.; Lu, X.; Wang, Y.; Sun, B.; Wu, X.; Lu, H. Effects of additives on sucrose-derived activated carbon microspheres synthesized by hydrothermal carbonization. *J. Mater. Sci.* **2017**, *52*, 10787–10799. [CrossRef]
49. Petrovic, J.T.; Stojanovic, M.D.; Milojkovic, J.V.; Petrovic, M.S.; Sostaric, T.D.; Lausevic, M.D.; Mihajlovic, M.L. Alkali modified hydrochar of grape pomace as a perspective adsorbent of Pb^{2+} from aqueous solution. *J. Environ. Manag.* **2016**, *182*, 292–300. [CrossRef]
50. Jiang, L.; Li, K.; Xia, L.; Gao, J.; Tang, L.; Jia, Y. KOH-modified hydrochar produced from Cd/Zn hyperaccumulator *Sedum alfredii* Hance for aqueous Cd(II) removal: Behavior and mechanism. *J. Environ. Chem. Eng.* **2023**, *11*, 110925. [CrossRef]
51. Qian, W.; Luo, X.; Wang, X.; Guo, M.; Li, B. Removal of methylene blue from aqueous solution by modified bamboo hydrochar. *Ecotoxicol. Environ. Saf.* **2018**, *157*, 300–306. [CrossRef] [PubMed]
52. Huang, Y.; Shen, D.; Wang, Z. Preparation of citric acid-sewage sludge hydrochar and its adsorption performance for Pb(II) in aqueous solution. *Polymers* **2022**, *14*, 968. [CrossRef] [PubMed]
53. Wang, T.; Zhai, Y.; Zhu, Y.; Peng, C.; Xu, B.; Wang, T.; Li, C.; Zeng, G. Acetic acid and sodium hydroxide-aided hydrothermal carbonization of woody biomass for enhanced pelletization and fuel properties. *Energy Fuels* **2017**, *31*, 12200–12208. [CrossRef]
54. Chen, F.; Zhang, Y.; Zheng, M.; Xiao, Y.; Hu, H.; Liang, Y.; Liu, Y.; Dong, H. Preparation of high-performance porous carbon materials by citric acid-assisted hydrothermal carbonization of bamboo and their application in electrode materials. *Energy Fuels* **2022**, *36*, 9303–9312. [CrossRef]
55. Huang, J.; Wang, Z.; Qiao, Y.; Wang, B.; Yu, Y.; Xu, M. Transformation of nitrogen during hydrothermal carbonization of sewage sludge: Effects of temperature and Na/Ca acetates addition. *Proc. Combust. Inst.* **2021**, *38*, 4335–4344. [CrossRef]
56. Mumme, J.; Titirici, M.M.; Pfeiffer, A.; Lueder, U.; Reza, M.T.; Masek, O. Hydrothermal carbonization of digestate in the presence of zeolite: Process efficiency and composite properties. *ACS Sustain. Chem. Eng.* **2015**, *3*, 2967–2974. [CrossRef]
57. Lang, Q.; Zhang, B.; Liu, Z.; Jiao, W.; Xia, Y.; Chen, Z.; Li, D.; Ma, J.; Gai, C. Properties of hydrochars derived from swine manure by CaO assisted hydrothermal carbonization. *J. Environ. Manag.* **2019**, *233*, 440–446. [CrossRef] [PubMed]
58. Zhang, Y.; Chen, D.; Xing, Y.; Liu, B.; Zhou, Y.; Lu, P. Role of nitrate in the production of iron-modified hydrochar for arsenic removal. *Water Air Soil Pollut.* **2024**, *235*, 233. [CrossRef]
59. Jiao, N.; Zhu, Y.; Li, H.; Yu, Y.; Xu, Y.; Zhu, J. Two-step hydrothermal pretreatments for co-producing xylooligosaccharides and humic-like acid from vinegar residue. *Fermentation* **2023**, *9*, 589. [CrossRef]
60. Tan, J.; Chen, H.; Gao, Y.; Li, H. Nitrogen-doped porous carbon derived from citric acid and urea with outstanding supercapacitance performance. *Electrochim. Acta* **2015**, *178*, 144–152. [CrossRef]
61. Leng, S.; Leng, L.; Chen, L.; Chen, J.; Chen, J.; Zhou, W. The effect of aqueous phase recirculation on hydrothermal liquefaction/carbonization of biomass: A review. *Bioresour. Technol.* **2020**, *318*, 124081. [CrossRef] [PubMed]
62. Stemann, J.; Putschew, A.; Ziegler, F. Hydrothermal carbonization: Process water characterization and effects of water recirculation. *Bioresour. Technol.* **2013**, *143*, 139–146. [CrossRef] [PubMed]
63. Uddin, M.H.; Reza, M.T.; Lynam, J.G.; Coronella, C.J. Effects of water recycling in hydrothermal carbonization of loblolly pine. *Environ. Prog. Sustain. Energy* **2014**, *33*, 1309–1315. [CrossRef]
64. Zhu, X.; He, M.; Xu, Z.; Luo, Z.; Gao, B.; Ruan, R.; Wang, C.-H.; Wong, K.-H.; Tsang, D. Combined acid pretreatment and co-hydrothermal carbonization to enhance energy recovery from food waste digestate. *Energy Conv. Manag.* **2022**, *266*, 115855. [CrossRef]
65. Wang, T.; Si, B.; Gong, Z.; Zhai, Y.; Cao, M.; Peng, C. Co-hydrothermal carbonization of food waste-woody sawdust blend: Interaction effects on the hydrochar properties and nutrients characteristics. *Bioresour. Technol.* **2020**, *316*, 123900. [CrossRef]
66. Li, C.; Cai, R.; Hasan, A.; Lu, X.; Yang, X.; Zhang, Y. Fertility assessment and nutrient conversion of hydrochars derived from co-hydrothermal carbonization between livestock manure and corn cob. *J. Environ. Chem. Eng.* **2023**, *11*, 109166. [CrossRef]
67. Sharma, H.B.; Dubey, B.K. Co-hydrothermal carbonization of food waste with yard waste for solid biofuel production: Hydrochar characterization and its pelletization. *Waste Manag.* **2020**, *118*, 521–533. [CrossRef] [PubMed]

68. Shen, Q.; Zhu, X.Q.; Peng, Y.; Xu, M.; Huang, Y.; Xia, A.; Zhu, X.; Liao, Q. Structure evolution characteristic of hydrochar and nitrogen transformation mechanism during co-hydrothermal carbonization process of microalgae and biomass. *Energy* **2024**, *295*, 131028. [CrossRef]
69. Xu, Y.; Lou, Z.; Yi, P.; Chen, J.; Ma, X.; Wang, Y.; Li, M.; Chen, W.; Liu, Q.; Zhou, J.; et al. Improving abiotic reducing ability of hydrothermal biochar by low temperature oxidation under air. *Bioresour. Technol.* **2014**, *172*, 212–218. [CrossRef]
70. Zhou, S.; Han, L.; Yang, Z.; Ma, Q. Influence of hydrothermal carbonization temperature on combustion characteristics of livestock and poultry manures. *Trans. Chin. Soc. Agric. Eng.* **2017**, *33*, 233–240.
71. Nguyen, D.; Zhao, W.X.; Mäkelä, M.; Alwahabi, Z.T.; Kwong, C.W. Effect of hydrothermal carbonisation temperature on the ignition properties of grape marc hydrochar fuels. *Fuel* **2022**, *313*, 122668. [CrossRef]
72. Gao, P.; Zhou, Y.; Meng, F.; Zhang, Y.; Liu, Z.; Zhang, W.; Xue, G. Preparation and characterization of hydrochar from waste eucalyptus bark by hydrothermal carbonization. *Energy* **2016**, *97*, 238–245. [CrossRef]
73. Peng, J.; Kang, X.; Zhao, S.; Zhao, P.; Ragauskas, A.J.; Si, C.; Xu, T.; Song, X. Growth mechanism of glucose-based hydrochar under the effects of acid and temperature regulation. *J. Colloid Interface Sci.* **2023**, *630*, 654–665. [CrossRef] [PubMed]
74. Li, F.; Jiang, Z.; Ji, W.; Chen, Y.; Ma, J.; Gui, X.; Zhao, J.; Zhou, C. Effects of hydrothermal carbonization temperature on carbon retention, stability, and properties of animal manure-derived hydrochar. *Int. J. Agric. Biol. Eng.* **2022**, *15*, 124–131. [CrossRef]
75. Maniscalco, M.P.; Volpe, M.; Messineo, A. Hydrothermal carbonization as a valuable tool for energy and environmental applications: A review. *Energies* **2020**, *13*, 4098. [CrossRef]
76. Chen, H.; Gao, Y.; Li, J.; Fang, Z.; Bolan, N.; Bhatnagar, A.; Gao, B.; Hou, D.; Wang, S.; Song, H.; et al. Engineered biochar for environmental decontamination in aquatic and soil systems: A review. *Carbon Res.* **2022**, *1*, 4. [CrossRef]
77. Liu, Z.; Wang, Z.; Chen, H.; Cai, T.; Liu, Z. Hydrochar and pyrochar for sorption of pollutants in wastewater and exhaust gas: A critical review. *Environ. Pollut.* **2021**, *268*, 115910. [CrossRef] [PubMed]
78. Li, D.; Cui, H.; Cheng, Y.; Xue, L.; Wang, B.; He, H.; Hua, Y.; Chu, Q.; Feng, Y.; Yang, L. Chemical aging of hydrochar improves the Cd^{2+} adsorption capacity from aqueous solution. *Environ. Pollut.* **2021**, *287*, 117562. [CrossRef] [PubMed]
79. Dos Reis, G.S.; Schnorr, C.E.; Dotto, G.L.; Vieillard, J.; Netto, M.S.; Silva, L.F.O.; De Brum, I.A.S.; Thyrel, M.; Lima, É.C.; Lassi, U. Wood waste-based functionalized natural hydrochar for the effective removal of Ce(III) ions from aqueous solution. *Environ. Sci. Pollut. Res.* **2023**, *30*, 64067–64077. [CrossRef]
80. Tian, Y.; Yin, Y.; Liu, H.; Zhou, H. One-step hydrothermal carbonization of amine modified black liquor and lignin for efficient Cr(VI) adsorption. *J. Water Process. Eng.* **2022**, *46*, 102583. [CrossRef]
81. Song, Q.; Ma, L.; Liu, J.; Bai, C.; Geng, J.; Wang, H.; Li, B.; Wang, L.; Li, S. Preparation and adsorption performance of 5-azacytosine-functionalized hydrothermal carbon for selective solid-phase extraction of uranium. *J. Colloid Interface Sci.* **2012**, *386*, 291–299. [CrossRef]
82. Bedin, K.C.; Martins, A.C.; Cazetta, A.L.; Pezoti, O.; Almeida, V.C. KOH-activated carbon prepared from sucrose spherical carbon: Adsorption equilibrium, kinetic and thermodynamic studies for methylene blue removal. *Chem. Eng. J.* **2016**, *286*, 476–484. [CrossRef]
83. Zhou, N.; Chen, H.; Xi, J.; Yao, D.; Zhou, Z.; Tian, Y.; Lu, X. Biochars with excellent Pb(II) adsorption property produced from fresh and dehydrated banana peels via hydrothermal carbonization. *Bioresour. Technol.* **2017**, *232*, 204–210. [CrossRef]
84. Regmi, P.; Moscoso, J.L.G.; Kumar, S.; Cao, X.; Mao, J.; Schafran, G. Removal of copper and cadmium from aqueous solution using switchgrass biochar produced via hydrothermal carbonization process. *J. Environ. Manag.* **2012**, *109*, 61–69. [CrossRef] [PubMed]
85. Xia, Y.; Yang, T.; Zhu, N.; Li, D.; Chen, Z.; Lang, Q.; Liu, Z.; Jiao, W. Enhanced adsorption of Pb(II) onto modified hydrochar: Modeling and mechanism analysis. *Bioresour. Technol.* **2019**, *288*, 121593. [CrossRef]
86. Elaigwu, S.E.; Rocher, V.; Kyriakou, G.; Greenway, G.M. Removal of Pb^{2+} and Cd^{2+} from aqueous solution using chars from pyrolysis and microwave-assisted hydrothermal carbonization of prosopis africana shell. *J. Ind. Eng. Chem.* **2014**, *20*, 3467–3473. [CrossRef]
87. Ramesh, S.; Sundararaju, P.; Banu, K.S.P.; Karthikeyan, S.; Doraiswamy, U.; Soundarapandian, K. Hydrothermal carbonization of arecanut husk biomass: Fuel properties and sorption of metals. *Environ. Sci. Pollut. Res.* **2019**, *26*, 3751–3761. [CrossRef] [PubMed]
88. Zhang, Y.; Qu, J.; Yuan, Y.; Song, H.; Liu, Y.; Wang, S.; Tao, Y.; Zhao, Y.; Li, Z. Simultaneous scavenging of Cd(II) and Pb(II) from water by sulfide-modified magnetic pinecone-derived hydrochar. *J. Clean Prod.* **2022**, *341*, 130758. [CrossRef]
89. Zhang, Z.; Zhu, Z.; Shen, B.; Liu, L. Insights into biochar and hydrochar production and applications: A review. *Energy* **2019**, *171*, 581–598. [CrossRef]
90. Xia, M.; Chen, Z.; Li, Y.; Li, C.; Ahmad, N.M.; Cheema, W.A.; Zhu, S. Removal of Hg(II) in aqueous solutions through physical and chemical adsorption principles. *RSC Adv.* **2019**, *9*, 20941–20953. [CrossRef]
91. Tian, S.-R.; Liu, Y.-G.; Liu, S.-B.; Zeng, G.-M.; Jiang, L.-H.; Tan, X.-F.; Huang, X.-X.; Yin, Z.-H.; Liu, N.; Li, J. Hydrothermal synthesis of montmorillonite/hydrochar nanocomposites and application for 17β-estradiol and 17α-ethynylestradiol removal. *RSC Adv.* **2018**, *8*, 4273–4283. [CrossRef]
92. Ngoc, D.M.; Hieu, N.C.; Trung, N.H.; Chien, H.H.; Thi, N.Q.; Hai, N.D.; Chao, H. Tetracycline removal from water by adsorption on hydrochar and hydrochar-derived activated carbon: Performance, mechanism, and cost calculation. *Sustainability* **2023**, *15*, 4412. [CrossRef]
93. Ma, Y.; Li, M.; Li, P.; Yang, L.; Wu, L.; Gao, F.; Qi, X.; Zhang, Z. Hydrothermal synthesis of magnetic sludge biochar for tetracycline and ciprofloxacin adsorptive removal. *Bioresour. Technol.* **2021**, *319*, 124199. [CrossRef] [PubMed]

94. Li, Y.; Meas, A.; Shan, S.; Yang, R.; Gai, X.; Wang, H.; Tsend, N. Hydrochars from bamboo sawdust through acid assisted and two-stage hydrothermal carbonization for removal of two organics from aqueous solution. *Bioresour. Technol.* **2018**, *261*, 257–264. [CrossRef] [PubMed]
95. Li, Y.; Tsend, N.; Li, T.; Liu, H.; Yang, R.; Gai, X.; Wang, H.; Shan, S. Microwave assisted hydrothermal preparation of rice straw hydrochars for adsorption of organics and heavy metals. *Bioresour. Technol.* **2019**, *273*, 136–143. [CrossRef] [PubMed]
96. Ronix, A.; Pezoti, O.; Souza, L.S.; Souza, I.P.A.F.; Bedin, K.C.; Souza, P.S.C.; Silva, T.L.; Melo, S.A.R.; Cazetta, A.L.; Almeida, V.C. Hydrothermal carbonization of coffee husk: Optimization of experimental parameters and adsorption of methylene blue dye. *J. Environ. Chem. Eng.* **2017**, *5*, 4841–4849. [CrossRef]
97. Zbair, M.; Bottlinger, M.; Ainassaari, K.; Ojala, S.; Stein, O.; Keiski, R.L.; Bensitel, M.; Brahmi, R. Hydrothermal carbonization of argan nut shell: Functional mesoporous carbon with excellent performance in the adsorption of bisphenol a and diuron. *Waste Biomass Valorization* **2020**, *11*, 1565–1584. [CrossRef]
98. Bai, C.-X.; Shen, F.; Qi, X.-H. Preparation of porous carbon directly from hydrothermal carbonization of fructose and phloroglucinol for adsorption of tetracycline. *Chin. Chem. Lett.* **2017**, *28*, 960–962. [CrossRef]
99. Ghimire, S.; Wang, L.; Zhang, B.; Li, X.; Shahbazi, A. Production and modification of hydrochar from anaerobically digested cattail for adsorbing ammonium and phosphorous in wastewater. *Water Sci. Technol.* **2021**, *84*, 1678–1692. [CrossRef]
100. Akter, J.; Islam, M.A.; Kibria, K.Q.; Islam, M.A. Adsorption of phosphate ions on chicken feather hydrochar and hydrochar-soil mixtures. *Water Air Soil Pollut.* **2021**, *232*, 413. [CrossRef]
101. Fu, J.; Jin, J.; Ji, H.; Liu, Y.; Fu, X. Adsorption characteristics of arsenic and fluoride in water by magnetic hydrothermal carbon. *J. Jiangsu Univ. Sci. Technol. (Nat. Sci.)* **2019**, *40*, 423–430.
102. Ge, X.; Chen, X.; Liu, M.; Wang, C.; Zhang, Y.; Wang, Y.; Tran, H.T.; Joseph, S.; Zhang, T. Toward a better understanding of phosphorus nonpoint source pollution from soil to water and the application of amendment materials: Research trends. *Water* **2023**, *15*, 1531. [CrossRef]
103. Wang, G.; Zeng, W.; Li, S. Adsorption characteristics of phosphate on cerium modified water hyacinth biochar. *Environ. Sci.* **2021**, *42*, 4815–4825.
104. Netto, M.S.; Georgin, J.; Franco, D.; Mallmann, E.S.; Foletto, E.L.; Godinho, M.; Pinto, D.; Dotto, G.L. Effective adsorptive removal of atrazine herbicide in river waters by a novel hydrochar derived from *Prunus serrulata* bark. *Environ. Sci. Pollut. Res.* **2022**, *29*, 3672–3685. [CrossRef] [PubMed]
105. Gao, Y.; Wang, X.; Wang, J.; Li, X.; Cheng, J.; Yang, H.; Chen, H. Effect of residence time on chemical and structural properties of hydrochar obtained by hydrothermal carbonization of water hyacinth. *Energy* **2013**, *58*, 376–383. [CrossRef]
106. Li, M.; Xiang, P.; Jiang, W.; Zhang, Z.; Mo, J. Performance and mechanism of $MnFe_2O_4$@HTC activated persulfate system for algae removal. *J. Environ. Sci.* **2021**, *41*, 3535–3544.
107. Waly, S.M.; El-Wakil, A.M.; Abou El-Maaty, W.M.; Awad, F.S. Hydrothermal synthesis of Mg/Al-layered double hydroxide modified water hyacinth hydrochar for remediation of wastewater containing mordant brown dye. *RSC Adv.* **2024**, *14*, 15281–15292. [CrossRef]
108. Takaya, C.A.; Fletcher, L.A.; Singh, S.; Anyikude, K.U.; Ross, A.B. Phosphate and ammonium sorption capacity of biochar and hydrochar from different wastes. *Chemosphere* **2016**, *145*, 518–527. [CrossRef]
109. Yu, Y.; Han, L.; Jiang, X. Production, properties and environmental application of hydrochar. *Environ. Chem.* **2018**, *37*, 1232–1244.
110. Marco, K.; Peter, S.N.; Mark, G.J.; Markus, K. Dynamic molecular structure of plant biomass-derived black carbon (biochar). *Environ. Sci. Technol.* **2010**, *44*, 1247–1253.
111. Vozhdayev, G.V.; Spokas, K.A.; Molde, J.S.; Heilmann, S.M.; Wood, B.M.; Valentas, K.J. Response of maize germination and growth to hydrothermal carbonization filtrate type and amount. *Plant Soil* **2015**, *396*, 127–136. [CrossRef]
112. Ni, P.-Y.; Zhang, X.; Ye, M.; He, R. Biochar enhanced the stability of toluene removal in extracted groundwater amended with nitrate under microaerobic conditions. *Chemosphere* **2024**, *353*, 141551. [CrossRef]
113. Kumar, S.; Loganathan, V.A.; Gupta, R.B.; Barnett, M.O. An assessment of U(VI) removal from groundwater using biochar produced from hydrothermal carbonization. *J. Environ. Manag.* **2011**, *92*, 2504–2512. [CrossRef]
114. Chung, J.W.; Breulmann, M.; Clemens, A.; Fühner, C.; Foppen, J.W.; Lens, P. Simultaneous removal of rotavirus and adenovirus from artificial ground water using hydrochar derived from swine feces. *J. Water Health* **2016**, *14*, 754–767. [CrossRef] [PubMed]
115. Wang, C.; Gui, B.; Wu, C.; He, C.; Li, L.; Ling, X.; Zuo, X. Hydrothermal carbonization of sewage sludge coupled with fenton oxidation pretreatment: Moderate oxidation to enhance hydrochar yield and properties. *J. Environ. Chem. Eng.* **2023**, *11*, 110788. [CrossRef]
116. Shi, Y.; Li, C.; Chai, R.; Wu, J.; Wang, Y. Effect of different hydrothermal parameters on calorific value and pyrolysis characteristics of hydrochar of kitchen waste. *Energies* **2023**, *16*, 3561. [CrossRef]
117. Marx, S.; van der Merwe, K. Utilization of hydrochar derived from waste paper sludge through hydrothermal liquefaction for the remediation of phenol contaminated industrial wastewater. *Water Pract. Technol.* **2021**, *16*, 756–771. [CrossRef]
118. He, X.; Zhang, T.; Niu, Y.; Xue, Q.; Ali, E.F.; Shaheen, S.M.; Tsang, D.C.W.; Rinklebe, J. Impact of catalytic hydrothermal treatment and Ca/Al-modified hydrochar on lability, sorption, and speciation of phosphorus in swine manure: Microscopic and spectroscopic investigations. *Environ. Pollut.* **2022**, *299*, 118877. [CrossRef] [PubMed]

119. Zhang, T.; Wu, X.; Fan, X.; Tsang, D.C.; Li, G.; Shen, Y. Corn waste valorization to generate activated hydrochar to recover ammonium nitrogen from compost leachate by hydrothermal assisted pretreatment. *J. Environ. Manag.* **2019**, *236*, 108–117. [CrossRef] [PubMed]
120. Luo, J.; Zhou, W.; Xu, B. A deep neural network-based assistive decision method for financial risk prediction in carbon trading market. *J. Circuits Syst. Comput.* **2023**, *33*, 2450153. [CrossRef]
121. Luo, J.; Zhou, W.; Liu, S.; Xu, B. The Optimization of carbon emission prediction in low carbon energy economy under big data. *IEEE Access* **2024**, *12*, 14690–14702. [CrossRef]
122. Zheng, C.; Chen, H. Revisiting the linkage between financial inclusion and energy productivity: Technology implications for climate change. *Sustain. Energy Technol. Assess.* **2023**, *57*, 103275. [CrossRef]
123. Jia, B.; Zhou, G. Estimation of global karst carbon sink from 1950s to 2050s using response surface methodology. *Geo-Spat. Inf. Sci.* **2023**, 2165974. [CrossRef]
124. Zheng, S.; Hai, Q.; Zhou, X.; Stanford, R.J. A novel multi-generation system for sustainable power, heating, cooling, freshwater, and methane production: Thermodynamic, economic, and environmental analysis. *Energy* **2024**, *290*, 130084. [CrossRef]
125. Wang, Q.; Zhang, T.; He, X.; Jiang, R. Assessment of phosphorus recovery from swine wastewater in Beijing, China. *Sustainability* **2017**, *9*, 1845. [CrossRef]
126. Yang, C.; Nutakki, T.U.K.; Alghassab, M.A.; Alkhalaf, S.; Alturise, F.; Alharbi, F.S.; Elmasry, Y.; Abdullaev, S. Optimized integration of solar energy and liquefied natural gas regasification for sustainable urban development: Dynamic modeling, data-driven optimization, and case study. *J. Clean Prod.* **2024**, *447*, 141405. [CrossRef]

water

MDPI

Review

Toward a Better Understanding of Phosphorus Nonpoint Source Pollution from Soil to Water and the Application of Amendment Materials: Research Trends

Xiaofei Ge [1], Xingyu Chen [1], Mingxin Liu [1], Chensi Wang [1], Yingyu Zhang [1], Yukai Wang [1], Huu-Tuan Tran [2], Stephen Joseph [3] and Tao Zhang [1,*]

[1] Beijing Key Laboratory of Farmland Soil Pollution Prevention and Remediation, Key Laboratory of Plant-Soil Interactions of Ministry of Education, College of Resources and Environmental Sciences, China Agricultural University, Beijing 100193, China
[2] Department of Civil, Environmental and Architectural Engineering, University of Kansas, Lawrence, KS 66045, USA
[3] School of Materials Science and Engineering, University of New South Wales, Sydney, NSW 2052, Australia
* Correspondence: taozhang@cau.edu.cn; Tel.: +86-10-6273-3638

Abstract: Phosphorus (P) nonpoint source pollution from soil to water is increasing dramatically, leading to the eutrophication of water bodies. Using amendment materials for P retention in soil is a promising strategy for environmental restoration and nonpoint source pollution management. This strategy has attracted significant attention because of its highly effective P retention. This study reviews management strategies of P nonpoint pollution from soil to water, including the basic P forms and accumulation situation in soil and P loss from soil to water. Recent advances in the use of amendment materials, such as inorganic, organic, and composite amendment materials, to mitigate P pollution from soil to water have also been summarized. Environmental risks of reloss of P retention in soil with different soil properties and water conditions have also been investigated. This review improves the understanding of P nonpoint source pollution from soil to water, providing an innovative perspective for the large-scale application of amendment materials to control water eutrophication.

Keywords: phosphorus; nonpoint source pollution; amendment materials; water

Citation: Ge, X.; Chen, X.; Liu, M.; Wang, C.; Zhang, Y.; Wang, Y.; Tran, H.-T.; Joseph, S.; Zhang, T. Toward a Better Understanding of Phosphorus Nonpoint Source Pollution from Soil to Water and the Application of Amendment Materials: Research Trends. *Water* **2023**, *15*, 1531. https://doi.org/10.3390/w15081531

Academic Editor: Yung-Tse Hung

Received: 27 February 2023
Revised: 5 April 2023
Accepted: 10 April 2023
Published: 13 April 2023

1. Introduction

In an era of rapid economic and technological development, resource scarcity has become an inevitable problem for global sustainable development [1–4]. Phosphorus (P) is an indispensable element for organisms to conduct their life activities [5–9]. The use of P fertilizer can supplement the effective P in soil, increase crop yield and quality, and maintain food security and sustainable development [10–12]. The dependence of modern agriculture on inorganic phosphate fertilizer will inevitably increase the demand for nonrenewable phosphate rock [13,14]. According to the data from the United States Geological Survey in 2022 (Figure 1), the global base reserves of phosphate rock are 69 billion tons. China's current phosphate rock reserves are 3.2 billion tons, accounting for less than 5% of the global total reservation amounts, and approximately 70% are recognized as refractory phosphate rock. As the country with the largest consumption of inorganic fertilizers in the world, China uses 80–90% of the mined phosphate ore as mineral phosphate fertilizers in food production, and crops absorb only approximately 20% of this P amount, which eventually enters the food consumed by people [15]. Meanwhile, the high demand for inorganic fertilizers, along with low nutrient utilization efficiency [16], has put significant pressure on the limited P rock reserves and the inorganic fertilizer market that depends on these reserves.

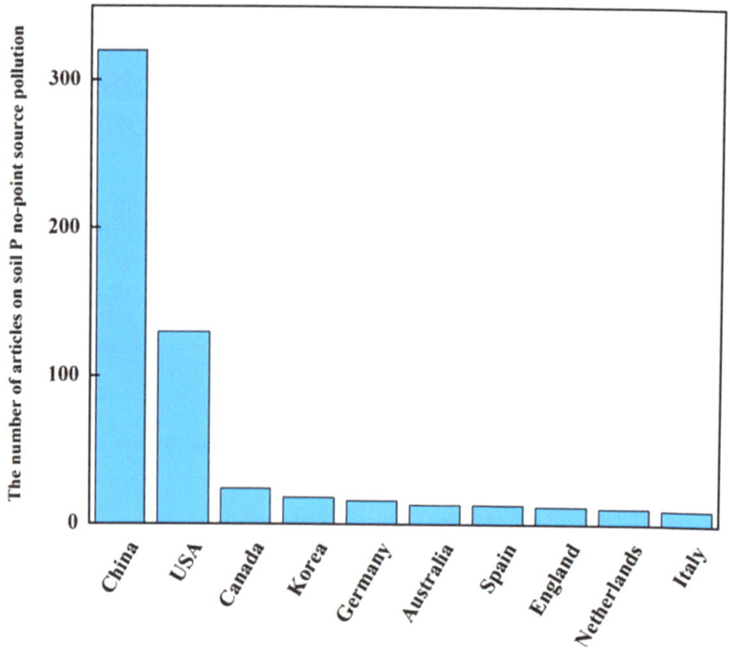

Figure 1. The number of articles on P nonpoint source pollution from soil to water.

In addition, due to the wide range and prevalence of agricultural activities, a large amount of P accumulation results in farmland nonpoint source pollution, which is considered one of the main sources of water environmental pollution risks [17,18]. Phosphate fertilizer applied is easily absorbed by soil particles or forms precipitates with calcium, magnesium, iron, and aluminum plasma in soil. However, the P absorption capacity of soil with excessive P is close to saturation, which increases unstable P content in the soil and, thus, increases the risk of P loss to water (Figure 2) [1]. The transfer of P from the farmland to the water environment results in severe agricultural nonpoint source pollution [19]; it can dominate the eutrophication of aquatic ecosystems [20].

For a long time, the P load of freshwater ecosystems has been increasing with the process of urbanization and industrialization. A large amount of nitrogen, P, and other nutrients enter water bodies, causing serious environmental pollution problems and leading to water degradation and reduction of aquatic biodiversity. The eutrophication degree of water bodies is improved, which provides nutrients for algae and other plankton, especially cyanobacteria. A large number of algae will reduce the oxygen content of water bodies and release algal toxins into water bodies, threatening other aquatic organisms. Eutrophication and harmful algal blooms (HAB) are widely considered two of the biggest water quality problems at present. Eutrophication can be divided into natural eutrophication and artificial eutrophication. Under natural conditions, lake sediments are constantly increasing, leading to nutrient accumulation and eutrophication. This process usually takes hundreds or even thousands of years. Therefore, an important measure to reduce water eutrophication is to reduce anthropogenic nitrogen and P emissions.

A large amount of applied P fertilizer is directly lost into river water bodies without plant absorption, leading to a series of water ecological pollution problems, such as the destruction of water ecosystem biodiversity and serious eutrophication of water bodies. The main reason is the management omission of P fertilizer application in agricultural production. The efficiency of P fertilizer utilization in many rural areas is generally low. River nonpoint source TP pollution area is widely distributed and is mainly concentrated in the agricultural production and living areas where soil erosion is more serious. After soil erosion, the massive input of chemical fertilizers and pesticides and poor management, as well as rural livestock and poultry-breeding discharges, have made the problem of nonpoint source TP pollution in river water bodies increasingly serious, and nonpoint source pollution has become one of the most important causes of river water quality deterioration.

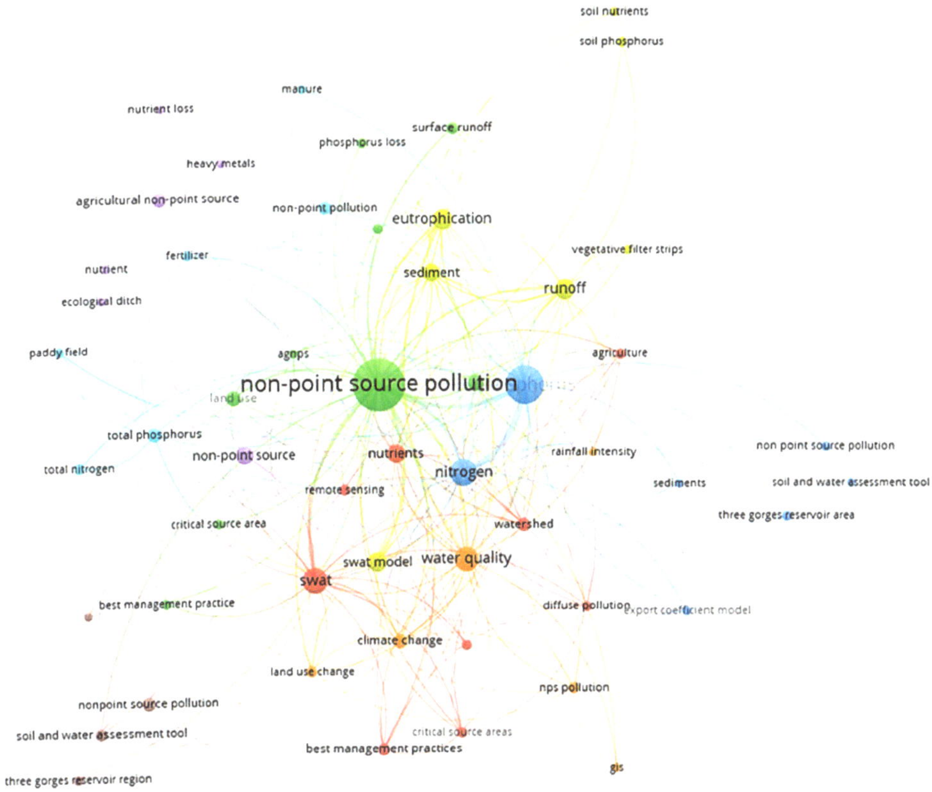

Figure 2. Hop map of keywords for articles on P nonpoint source pollution from soil to water using VOSviewer1.6.18 software.

China is a largely agricultural country with a fertilization amount of 100 million tons. The average fertilization amount is 2.6 times the world average, but the fertilization utilization rate is only 30–50% of the world average. A great amount of phosphorus enters water bodies through various channels. Usually, the contribution of P in water bodies from various external sources in descending order is agricultural drainage, domestic sewage, and industrial sources of pollution [21].

Numerous studies have shown that nutrient losses from agricultural production are the main source of nutrient surpluses in agricultural ecosystems, water bodies and wetland ecosystems, and, therefore, the largest cause of phosphorus pollution in water. TP emissions from agricultural sources amounted to 284,700 t, accounting for 67.27% of the total national emissions. According to the latest data from the Food and Agriculture Organization of the United Nations (FAO), China was the country that applied the most nitrogen and P fertilizers in the world during 2002–2017, exceeding the sum of the fertilizer applied by the second- and third-ranked countries. China's arable land area only accounts for about 7% of global arable land, but the amount of P fertilizers applied to agriculture accounted for 30.09% of the world's fertilizer utilization (2002). It is noteworthy that the in-season utilization rate of fertilizers in Chinese agricultural production is low, with the in-season utilization rate of P fertilizer (P_2O_5) being around 20%. This means that some of the nitrogen and P nutrients that are not absorbed and used by crops and leave the cultivated soil will be further lost to surface water, groundwater, and the atmosphere through surface runoff, leaching, ammonification, nitrification, and denitrification. This will cause environmental impacts such as nitrate pollution of groundwater and eutrophication of water bodies.

In agricultural ecosystems, fertilizer and animal manure are the main sources of P input, so the two main sources of P emissions from agricultural sources can be judged to be excessive fertilizer application and animal manure. In terms of source composition, based on the national multiyear average, the proportion of P emissions from agricultural sources is 45.73% from overfertilization and 54.27% from livestock and poultry manure, indicating that the primary cause of agricultural surface pollution in China is livestock and poultry manure emissions.

When overfertilization or manure overload occurs in farmland (or grassland) ecosystems, nitrogen and P in different compositional forms enter surface and groundwater mainly through surface runoff and subsurface leaching, which, in turn, leads to surface pollution such as eutrophication of water bodies [22].

Farmland nonpoint source pollution affects 30–50% of the global land area [1]. In the United States, nonpoint pollution accounts for two-thirds of the total pollution [23]. Fertilizer and manure inputs have exceeded crop removal rates by up to 50% in many agricultural regions [24]. The input of TP from agricultural nonpoint sources to the North Sea in Europe accounts for 25% of the flux to the sea. Among various sources, TP from agricultural nonpoint sources in the Netherlands accounts for 40–50% of water pollution. P loads introduced by nonpoint sources in 270 rivers in Denmark accounted for 52% of total pollution [25]. Therefore, an important way to control P nonpoint source pollution from soil to water is to control P loss in soil.

Recently, many scholars have researched P nonpoint source pollution from soil to water and the application of amendment materials. We selected "soil P nonpoint source pollution" as the keyword for screening on the Web of Science website, and selected the literature data from 2013 to 2023 to analyze with the country region; the results are shown in Figure 1. China has published approximately 56.4% of the publications on soil P nonpoint source pollution. As shown in Figure 2, the keyword "P nonpoint source pollution" in articles on the Web of Science database is strongly related to the keywords "nitrogen" and "water quality." This explains the consistency of the global agricultural nonpoint pollution situation with the research hotspots described in the Web of Science database. This shows that the management of P nonpoint source pollution from soil to water is receiving increasing attention in China.

Research has been conducted on applying environment-friendly amendment materials that can absorb large amounts of P to convert unstable P into medium-stable P in soil [26–28]. Adding iron, aluminum, and calcium to P improves soil P retention ability and reduces soil P loss to water [29]. This effective and environmentally sound treatment method that can avoid the ecological risk of P loss. In this study, P pollution regulation from soil to water using specific amendments has been discussed and summarized. This review provides critical information about P pollution regulation and amendment material application, which is useful for agricultural nonpoint source pollution management.

2. Phosphorus Nonpoint Source Pollution Management Strategies

2.1. *Phosphorus Forms and Accumulation in Soil*

P exists in two forms in soil: inorganic P and organic P. Most P in arable soils is mainly inorganic P, which plants can absorb and use [30]. Organic P can be converted to inorganic P after mineralization. Inorganic P exists in soil in three forms: orthophosphate, pyrophosphate, and polyphosphate. Among these three forms, orthophosphate is the main component. Soil P can also be divided into mineral P, adsorbed P, and water-soluble P, according to its binding characteristics with the soil matrix. Mineral P is the most difficult

P form to be absorbed and utilized by plants, mainly occurring in apatite. Adsorbed P is mainly adsorbed by soil clay, Fe-Al oxide, hydrated oxide, carbonate, or organic matter by physical actions in the form of $H_2PO_4^-$ and HPO_4^{2-}. Water-soluble P can be directly absorbed and utilized by plants, and its content is generally low, which is affected by the dissolution of mineral phosphorus and the release of adsorbed phosphorus. The difference in soil types frequently leads to different organic P proportions, which generally account for 20–80% of soil TP [31]. Soil organic P can be divided into combined organophosphorus, adsorbed organophosphorus, and microbial P, including phosphoric acid, phosphorus ester, stable nucleic acid, phosphoprotein, and microbial P. Among them, inositol phosphate is the main organic P form, accounting for approximately 50% of the total organic P.

2.2. An Important Source of Phosphorus to Water: Soil Phosphorus Loss

P loss from soil to water occurs in three ways: surface runoff, soil erosion, and leaching, all of which end up in the water and cause eutrophication. The amount of P loss is related to rainfall, rain intensity, soil texture, and P background value in soil. The results of the first survey of pollution sources in China show that from 2012 to 2014, the average total P loss in China was 1.08×10^5 t [32].

Algae are very sensitive to P in water, and a small amount of soil P in water (0.05–0.10 $mg \cdot L^{-1}$) can lead to water eutrophication. P migration from soil to water includes water-soluble P and soil particle-combined P. Generally, the low solubility of soil P, the high adsorption capacity of clay for P, and the strong combination of P and soil organic matter make the water-soluble P content in the soil low. The P migration is low, and most soil P is in the form of particle-combined P. However, in soil containing excessive P, the absorption capacity of soil for P is close to saturation, and water-soluble P content increases [33]. Therefore, the risk of P loss from soil to water due to leaching increases. McDowell et al. [34] also concluded that leaching causes as much, and sometimes more, P loss from farmland soil to water as surface runoff and soil erosion.

In recent years, P loss from soil has seriously affected P levels in water bodies, and many scholars have paid increasing attention to P loss in arable land. Different indicators have been used to evaluate P loss from soil to water, mainly including direct monitoring and soil P index prediction [35,36]. Direct monitoring is based on simulated rainfall, laboratory soil column leaching, and long-term monitoring tests, which evaluate P loss from soil to water by measuring TP in runoff or groundwater and leaching solutions. The risk of P loss from soil to water is assessed by predicting and assessing the soil P index. There are two methods: single-index prediction and multi-index combination prediction.

For single indicator prediction, different soil test P (STP), such as $CaCl_2$-P, Olsen-P, Bray-P, and M3-P, is used to determine the P content in a specific part of the soil, and then various P contents are evaluated according to the established threshold value. Beyond the threshold value, the risk of P loss threatening water quality increases significantly. Based on the relationship between soil P content and water environment P, the piecewise linear model is generally adopted, and the discontinuity point is the threshold of P loss from soil to water. With an increase in the STP concentration, the P concentration in leachates increases slowly at first and then increases sharply when it exceeds the threshold value [37]. For example, the long-term Broad balk test showed that when soil Olsen-P content was less than 60 $mg \cdot kg^{-1}$, the corresponding total P concentration in the soil drainage decreased (<0.15 $mg \cdot L^{-1}$), but when soil Olsen-P content was greater than 60 $mg \cdot kg^{-1}$, the total P concentration increased linearly. The threshold value of soil environmental P varies regionally, which is related to soil texture, planting methods, soil management methods, and other factors. In addition, soil $CaCl_2$-P is a capacity index that can be easily released into the water and can directly reflect the risk of P loss from soil to water. Therefore, the threshold of P loss from soil to water can also be determined on the basis of the linear relationship between other STP and $CaCl_2$-P.

Multi-index combination prediction relates the P loss risk to the intensity factor (the amount of soil P extracted using a particular method) and the capacity factor (the maximum amount of P that soil can absorb), i.e., the soil adsorption degree of phosphorus saturation (DPS), in which the intensity and capacity factors are measured in different ways. The capacity factor can be obtained from the P adsorption characteristic curve and the combination of weak crystal hydroxyl ferric oxide and hydroxyl alumina (Fe_{ox}, Al_{ox}), measured using the ammonium oxalate-oxalate solution extraction method. The intensity factors include M3-P, Olsen-P, Pox, etc. [38]. The threshold value of DPS is typically 25%, and the risk of P loss from soil to water increases rapidly when more significant than this value. In acidic soils, 12.5% is commonly used as the threshold of DPS [39]. In addition, some studies have used Fe^{2+}:P to predict the risk of P loss from soil to water, but it was mainly based on soil pore water [40].

2.3. The Management Strategies for Soil Phosphorus Loss to Water Body

In the 1980s, China identified the problem of soil P loss to water. There are several methods for reducing the risk of P loss from agricultural fields: reducing P inputs, reducing P loss to water, conservation tillage, and buffer zone construction. Increased and stabilized crop output is achieved through the yield response method, fertilization management, and nutrient balance method, reducing environmental problems caused by excessive fertilization [20]. Regarding P management, European Union researchers have proposed several strategies for achieving the goal of global sustainable P use, including adjusting P input, reducing P loss to water, increasing the cycle of P biological resources, recycling P products, and P transfer in the food chain.

Reducing P inputs includes reducing fertilization, precise fertilization, and limiting the use of organophosphorus pesticides. Soil organophosphorus loss is an important factor in water eutrophication and is caused by organophosphorus pesticides.

Under specific soil and runoff conditions, P is mainly lost through surface runoff. To a certain extent, preventing surface runoff can increase soil resistance to erosion and reduce the risk of soil P loss to water, including covering the soil surface with crop residues and setting buffer zones and grassland waterways [33]. However, these strategies effectively reduce particulate P transport but are ineffective for reducing dissolved P in high-P residual soil. In addition, these strategies take a long time for soil P concentrations to decrease, during which large amounts of dissolved P are lost [41].

The loss of P to water can be reduced by changing the form of P in soil. A quick and effective method is to add amendment materials to the soil to change the unstable P component in the soil, which is prone to P loss from soil to water, into a stable P component, thereby reducing the risk of P loss from soil to water.

3. Application of Amendment Materials for Phosphorus Nonpoint Source Pollution from Soil to Water

The transformation of P is in a dynamic equilibrium, including adsorption–desorption, precipitation–dissolution, and the action of soil microorganisms. Generally, the role of soil microorganisms is relatively weak, whereas chemical precipitation and physicochemical adsorption play a key role. For example, in acidic soils, iron, aluminum oxides, and hydroxides are the main substances for P retention. Clay minerals can immobilize P by the same mechanism. P retention in alkaline and calcareous soils is typically associated with the formation of calcium-bound P and can be fixed by Al^{3+} or $A1(OH)_3$ in clay minerals. Inorganic matter, organic matter, and composite matter can be extracted from the waste and applied to the soil as modified materials to control P loss from soil to water. For example, the alumina could be extracted from alum sludge using different chemical treatment methods after being subjected to different thermal treatments [42]. Red mud is a byproduct of the Bayer process to extract alumina from bauxite [43]. Wheat straw biochar and corn straw biochar can be produced from wheat rice and corn straw, respectively. Fly ash is solid waste after the combustion of pulverized coal in coal-fired power plants [44]. These are modified materials converted from solid waste.

P amendment materials can accelerate the transformation of soil P from an unstable state to a medium or highly stable state, thereby reducing the risk of P loss from soil to water. According to their properties, amendment materials can be divided into three categories: (1) inorganic materials, including natural minerals, chemical materials, and industrial waste; (2) organic materials; and (3) composite materials. P retention is primarily affected by soil type, soil P background value, material properties, and material addition amount. Although amendment materials are effective in soil P retention based on laboratory-scale research, long-term P retention information about amendment materials at the farmland scale is lacking.

3.1. Inorganic Amendment Materials

P retention in the soil strongly correlates with extractable iron, aluminum, and $CaCO_3$ content in acidic soils [45]. The P retention mechanism of inorganic materials is primarily by increasing the concentration of Fe, Al, Ca, and Mg ions in soil and, thus, transforming the unstable P in the soil into medium and high stable P under the action of adsorption, precipitation, ligand exchange, and electrostatic attraction. In addition, some materials with good pore structure and significant specific electrostatic force and ion exchange properties, such as bentonite and zeolite, can effectively adsorb phosphate in soil, thereby reducing P loss. According to their composition, inorganic materials are classified into iron and aluminum materials, calcium and magnesium materials, clay minerals, and other materials.

3.1.1. Calcium and Magnesium Inorganic Amendment Materials

Calcium and magnesium materials, including dolomite, desulfurized gypsum, gypsum, lime, calcite, and so on, can be applied for P retention from soil to water (Table 1). There are two mechanisms for P retention from soil to water when using calcium materials: (1) an increase in the Ca^{2+} concentration in soil that can promote the formation of calcium phosphate precipitation, thereby reducing the solubility of P; and (2) increasing soil pH. Adding dolomite can promote the adsorption or precipitation of unstable P in soil on the surface of calcium ions and increase stable calcium and P compounds by increasing soil pH [46,47]. Adding desulphurized gypsum can increase the content of insoluble calcium phosphate (Ca_8-P and Ca_{10}-P) in soil, and then effectively control the dissolved P in the soil [47]. The amount of material added is an important factor affecting soil P retention. When the amount of lime is small, the release of calcium and magnesium ions is insufficient to promote the formation of phosphate precipitation, thereby increasing P content during leaching. For example, 2% addition (*w*/*w*) of slaked lime dust can reduce P loss by 77.2%, whereas 1.5% can increase P loss by 236% [46]. Calcium and magnesium can also be applied to soil to improve soil porosity and aggregate strength, thereby improving gas exchange and increasing crop yields.

Table 1. Different Ca/Mg inorganic amendment materials for P retention from soil to water.

Ca/Mg Inorganic Materials	Soil Type	pH	Addition Amount (*w*/*w*)	Retention Situation	References
calcium carbonate	red soil	5.42	2.0%	the soil Olsen-P contents increased by 33.9%	[48]
dolomite	red soil	5.42	2.0%	the soil Olsen-P contents increased by 66.3%	[48]
dolomite	Calcareous soil	7.56	2.0%	the soil $CaCl_2$-P content was reduced by 57.8%	[48]
dolomite	Calcareous soil	7.6	5.0%	soil available P decreased by 3.10%	[49]
dolomite	Calcareous soil	7.9	1.25%	the total dissolved P of leachate decreased by 68.4%	[50]
magnesia	Sandy	7.1	2.0%	soluble P was reduced by 78.6%	[51]

3.1.2. Iron and Aluminum Inorganic Amendment Materials

Iron and aluminum amendment materials for P retention from soil to water, include alum, red mud, ferrous sulfate, aluminum sulfate, fly ash, wastewater treatment residues, and so on (Table 2). Iron and P can form inner- and outer-sphere complexes through ligand exchange and electrostatic attraction to effectively regulate the transport of P in soil [51]. Red mud can fix phosphate on the existing surface by complexation, and then, through the adsorption of metal cations, such as Al^{3+}, Fe^{3+}, and Ca^{2+}, it generates metal oxide hydrate sites for further phosphate adsorption, eventually leading to surface precipitation or multilayer adsorption [52]. The establishment of multilayer adsorption depends on the availability of polyvalent metal ions during red mud leaching, mineral dissolution rate, surface species, solution pH, and total surface area [53]. Brennan et al. [52] found that in sandy, clay, and loam soils, adding 5-t·hm^{-2} red mud could effectively reduce water-soluble P in soil, and the effect was the most significant in high-P soil. Aluminum is an element with a strong affinity for P, and the addition of 20-g·kg^{-1} $KAl(SO_4)_2 \cdot 12H_2O$ can significantly reduce the content of unstable P in soil while increasing the content of moderately unstable P 30. At different soil pH values, the mechanism of P retention of aluminum-based materials differs. In calcareous soil, alum mainly forms poorly crystallized hydroxyl aluminum to adsorb unstable P, whereas, in acidic soil, it is fixed by the precipitation of Al^{3+} and unstable P [54]. The main components of fly ash are silica, alumina, and iron oxide, among which rich iron and aluminum can increase the adsorption and precipitation of P in acidic soil. After the application of fly ash, inorganic P in soil is generally converted into Ca-P (H_2SO_4-P), NaOH-Pi, and residual P [26]. It can also reduce soil swelling and clay dispersion, as well as P loss to water, by reducing soil erosion.

Table 2. Different Fe/Al inorganic amendment materials for P retention from soil to water.

Fe/Al Inorganic Materials	Soil Type	pH	Addition Amount (*w/w*)	Retention Situation	References
alum	Calcareous soil	7.56	2.0%	the soil CaCl2-P content was reduced by 77.0%	[48]
alum	red soil	6.04	2.0%	the soil CaCl2-P content was reduced by 93.8%	[48]
alum		7.6	5.0%	the soil CaCl2-P content was reduced by 91.9%	[49]
aluminum sulfate	red soil	7.25	0.2%	the total dissolved P of leachate decreased by 80.6%	[55]
ferrous sulfate	red soil	7.25	0.2%	the total dissolved P of leachate decreased by 80.6%	[55]
Al-WTR	peat soil	3.8	10%	P adsorption maxima was increased by 11%	[56]

3.1.3. Clay Mineral Inorganic Amendment Materials

P retention clay minerals include bentonite, zeolite, hydrotalcite, and so on (Table 3). The P retention mechanism by clay minerals is mainly by chemical adsorption. The higher the calcium, aluminum, and iron content in the elemental composition, the stronger the P adsorption capacity, but the phosphate integration performance is poor. As a result, these clays need to be modified physically and chemically to make them easier to absorb phosphate. Layered double hydroxide (LDH), also known as hydrotalcite compounds or anionic clay, is a promising P adsorption material [28]. It is composed of two-dimensional layered mixed hydroxides with the advantages of a permanent positive charge between layers, high anion exchange capacity, large specific surface area, and water resistance of the structure [57]. The mechanisms of phosphate adsorption include electrostatic attraction, ligand exchange, hydrogen bonding interaction, and ion exchange [58]. The direct application of LDHs through tillage on soils with a high-P application rate could fix P in soil through its strong adsorption capacity and reduce P transport to nearby water systems.

Table 3. Different clay mineral inorganic amendment materials for P retention from soil to water.

Clay Mineral Inorganic Materials	Soil Type	pH	Addition Amount (w/w)	Retention Situation	References
Mg-Al LDHs		7.08	2%	the P effluent mass balance decreased 82.7%	[28]
natural zeolite	inceptisol	6.4	5%	P was removed from the solution by 3.6%	[59]
lanthanum modified zeolite	aquatic soil	7.9		the soluble reactive P in decreased by 86.9%	[60]
CFL-Z	aquatic soil			the P content in overlying water was reduced by 97.3%	[61]
montmorillonite	sandy clay loam	7.8	1%	the calcium chloride-extractable P content was reduced by 62.8%	[62]
zeolite	sandy clay loam	7.8	3%	the water-extractable P content was reduced by 9.9%	[62]
vermiculite	sandy clay loam	7.8	3%	the Olsen-extractable P content was reduced by 79.8%	[62]
bentonite	Sandy loam soil	8.28	10%	P maximum sorption capacity increased by 42.7%	[63]
kaolinite	Sandy loam soil	8.28	10%	P maximum sorption capacity increased by 77.5%	[63]
zeolite	Sandy loam soil	8.28	5%	P maximum sorption capacity increased by 70.0%	[63]

3.1.4. Waste Inorganic Amendment Materials

Some inorganic waste materials can be used as P retention inorganic amendment materials (Table 4). Irshad et al. [64] evaluated the effects of waste inorganic amendment materials, coal ash, and wood ash on P retention from soil to water. It found that coal ash and wood ash can reduce water-soluble P in soil. Faridullah et al. [65] investigated the effects of waste inorganic amendment materials, wood ash, and sawdust, on P retention from soil to water.

Table 4. Different waste inorganic amendment materials for P retention from soil to water.

Waste Inorganic Materials	Soil Type	pH	Addition Amount (w/w)	Retention Situation	References
fly ash	inceptisol	6.4	5%	P was removed from the solution by 97.0%	[59]
coal ash	loamy sand	7.6	10%	water-soluble P was reduced by 22.3%	[64]
wood ash	sandy loam	7.4	10%	water-soluble P was reduced by 16.5%	[64]
wood ash	silt loam	7.6		P concentration was reduced by 55.6%	[65]
sawdust	sandy soil	7.8		P concentration was reduced by 58.1%	[65]
bauxite residues			4%	the water-extractable P was reduced 95%	[66]

3.2. *Organic Amendment Materials*

Compared with inorganic materials, organic materials have a positive effect on reducing the risk of P loss from soil to water [67]. The main organic amendment materials used are biochar and polyacrylamide (PAM) (Table 5).

Biochar is produced by biomass pyrolysis, and the raw materials used for biochar production include corn stover scale, rice straw, peanut husk, bamboo waste, bagasse, soybean straw, animal feces, etc. [68,69]. More recently, sewage sludge/biosolids biochar from nonindustrial treatment plants are being viewed as a way of recycling P and providing great P uptake efficiency. Biochar has a large specific surface area and abundant functional groups [70], which can change the cycle and availability of P by changing the adsorption and desorption performance of soil for P and can promote the adsorption and fixation of free PO_4^{3-} in soil [71,72]. On the other hand, biochar can indirectly promote P retention by changing the structure of soil aggregates. Peng et al. [73] reported that the use of alkaline straw biochar in P-rich calcareous soil could effectively promote the conversion of unstable P to medium–high steady-state P, thereby reducing the availability of P. Xu et al. [74] reported that tree-derived biochar is conducive to the accumulation of thermally stable carbon, primarily aromatic carbon, which can provide additional adsorption sites for P and lower alkalinity, promoting the adsorption of P and inhibiting the effectiveness of P.

PAM is a type of polymer material with a crosslinked structure, which is polymerized from acrylamide and can promote the mutual condensation of fine soil particles to form stable aggregates [75]. There are three main mechanisms for P retention from soil to water by PAM [76]: (1) through the interaction between soil particles and P to reduce the mobility of P in soil, which reduces the risk of P loss; (2) formed hydrogen bonds in soil solutions and high hydrophilicity that can reduce P loss from soil to water caused by soil erosion; and (3) through flocculation to convert soil colloidal P into soil particulate P, thereby reducing the migration of colloidal P from soil to water.

Table 5. Different organic amendment materials for P retention from soil to water.

Organic Amendment Materials	Applied Soil Type	P Release Reduction	References
polyacrylamide		the total P concentration of the leachate was decreased by 32.4%	[51]
anionic polyacrylamide	tea soil	total P reduced by 54%	[75]
maize stover biochar	corn-growing soil	inorganic P reduced by 3.3–59%	[77]
polyacrylamide modified biochar	paddy soil	total P reduced by 41.1%	[78]
Sugarcane-Derived Biochar	calcareous soil	/	[79]
wheat straw biochar	paddy-wheat rotation soil	the P utilization rate is increased by 38–230%	[80]
Rice-residue waste biochar	paddy soil	/	[81]
reed-biochar	paddy soil	total P reduced by 5.3–13.3%	[82]
maple and hickory sawmill waste biochar	forest soil	increases the absorption of a small amount of soluble P	[83]

3.3. Composite Amendment Materials

Negatively charged materials and inherent low-adsorption capacity materials have limited P adsorption due to electrostatic repulsion and limited adsorption ability [57,84]. When using these materials, the addition of minerals or metallic or cationic surfactants to improve P retention efficiency is typically required [85,86]. Iron, aluminum, and magnesium oxides are common modified metal oxides, among which iron and aluminum are common metal oxides in soil with a large specific surface area and abundant functional groups. They show superior adsorption capacity for P by forming stable bonds with P(Fe-P, Al-P and P-O-Al/Fe). Metal oxides have been added to the surface of materials to fabricate composite amendment materials for reducing P loss from soil to water. Chen et al. [87] loaded magnesium onto cow manure biochar and found that magnesium-modified biochar could reduce the leaching amount of P by 89.25%, and the adsorption of P on the biochar surface also improved the oxidation resistance of biochar. Similarly, Zhao et al. [88] loaded the rare earth element lanthanum onto biochar and found that adding lanthanum-modified biochar could improve the adsorption capacity of soil for P and enhance the binding force between soil and P and that the adsorption capacity was stable and less affected by soil pH, which helps control the leaching of P from soil to water. Feng et al. [89] loaded cerium oxide onto corn stalk biochar (Ce-MSB) and found that adding Ce-MSB could reduce the TP concentration in surface water by 27.33%, which was 52.05% lower than that of MSB treatment.

4. Soil Retention Phosphorus Reloss to Water Environmental Risk

Soil properties, such as soil pH, soil inorganic and organic matter, and water condition, can cause the reloss process of soil P loss to the water environment.

4.1. Influence of Soil pH

P adsorbed in aluminum (hydrogen) oxides (Al-P) or calcium phosphate precipitates (Ca-P) is highly sensitive to pH changes. When the soil pH < 7, the protonated hydroxide radical generates positive charges on the surface of soil minerals, inducing the adsorption of negatively charged substances on the mineral surface through the formation of surface complexes, such as phosphate groups or phosphorus-containing organic molecules/colloids. When pH becomes neutral or slightly alkaline, soil particles become negatively charged, and the mineral surface repels negatively charged organic molecules/colloids containing P, limiting the complexation of these P compounds on the soil mineral surface. Peng et al. [90] found that acidification tends to promote soil phosphorus loss into water bodies.

4.2. Influence of Soil Matter

Soil inorganic oxides have a certain effect on the adsorption capacity of soil for phosphorus. Mng'Ong'o et al. [91] reported that when the metal oxides content of aluminum, iron, and calcium were in the wide range of 234.56–3789.36 mg/kg, 456.78–2980.23 mg/kg, and 234.67–973.34 mg/kg, the adsorption capacity of the soil was higher than the average value under a high state, and the risk of phosphorus loss to water was lower. Therefore, some soils had deficient adsorption capacity, creating a risk of P loss to the water environment.

The soil organic matter is considered an important factor controlling the movement of P. An increase in organic matter provides a large amount of carbon and nutrients for microorganisms, promoting soil organic phosphate mineralization as the microbial population expands [92]. Organic acids produced by the decomposition of organic matter can decrease the precipitation of Ca-P minerals by increasing H^+, thereby increasing the concentration of soluble P in calcareous soil. Some of its active functional groups, such as the carboxyl group and phenolic hydroxyl group, can be complicated with metal ions (such as iron and aluminum) and reduce the available adsorption sites for P, promoting the release of soil phosphorus into the water.

4.3. Influence of Water Condition

Rainfall is the driving force of soil runoff and the main factor affecting phosphorus runoff into the water. Rainwater flows into farmland, and part of it is absorbed by crops and soil. When the moisture in soil reaches saturation, the excess rainwater gradually seeps down. With the continuous increase of rainfall, the loss of soil phosphorus to water increases. Yang et al. [93] reported that the runoff loss from soil P to water is not only related to rainfall but also closely related to rainfall intensity. When rainfall intensity increases, soil erosion will increase, and nutrients will be easily lost.

The flooding conditions will promote the reloss of P from soil to water. Under hydraulic erosion, soil phosphorus released by hydrolysis enters adjacent water along with surface runoff and soil erosion [90]. Xu et al. [94] demonstrated that residual P (H_2SO_4–H_2O_2 digestion) decreased by 18–27% in flooded paddy soil compared with aerobic soil. Zhang et al. [95] reported that in acidic soils, the effect of flooding conditions on P release was stronger, increasing by approximately 70%. Shaheen et al. [96] indicated flooding conditions increase soil pH and promote the dissolution of Fe and Al-P compounds in acidic soils, thereby increasing the risk of P loss.

5. Conclusions

With the rapid development of the economy, the application of P fertilizer has increased dramatically, and soil P nonpoint source pollution has caused serious water eutrophication. Currently, concerns over poor management of P nonpoint source pollution from soil to water and ongoing efforts to achieve environmental sustainability, especially the water environment, have piqued interest in soil restoration. Many studies have been conducted on the application of amendment materials to P nonpoint source pollution. This article reviews (1) different management strategies for P pollution from soil to water, including the knowledge of P forms and accumulation in soil, and P loss from soil to water bodies and management strategies; (2) the application of amendment materials to P nonpoint source pollution from soil to water, including recent advances in inorganic amendment materials, organic amendment materials, and composite amendment materials; and (3) the soil P retention reloss to water environmental risk. The theoretical investigations have provided insight into the research trends of soil P nonpoint source pollution to improve the understanding of using amendment materials to solve water pollution control.

6. Limitations of Existing Studies

The management of phosphorus pollution in water bodies has taken many results in the past, such as the removal of phosphorus from polluted water bodies by aquatic plants [97], the removal of phosphorus from polluted water bodies by amendment materials, and the modeling of the sources of total phosphorus pollution from nonpoint sources. However, due to the very complex composition of water bodies and the fluctuation of phosphorus concentration in water bodies due to climate and drastic human activities, phosphorus is still the primary form of pollution. There are still many problems to be improved in the management of phosphorus pollution in lakes and watersheds.

6.1. Watershed Phosphorus Pollution Management

Due to the complexity of the process of nonpoint source phosphorus pollution at the watershed scale, the many influencing factors and the late start of research on legacy effects, there are still many shortcomings and difficulties in process mechanisms and quantitative modeling.

In terms of process mechanisms, the understanding of phosphorus transport processes (especially subsurface runoff processes) in different hydrological pathways at the watershed scale is still relatively limited. Previous studies on nonpoint source phosphorus pollution processes in watersheds have focused on physical processes, such as soil erosion, with limited consideration of phosphorus biogeochemical processes (especially coupled with hydrological and biogeochemical regulation mechanisms). There is a lack of understanding of different media such as soil, groundwater, and sediment at the watershed or regional scale. The lack of research on phosphorus accumulation and its spatial distribution in different media such as soil, groundwater, and sediment at the watershed or regional scale has hindered the assessment and management of water environment pollution risks of phosphorus left in watersheds.

In terms of modeling studies, there is a lack of process models to describe the legacy effects of nonpoint source phosphorus pollution in watersheds. The understanding of the process mechanism of the legacy effects of nonpoint source phosphorus pollution is still unclear, and the current process models are still unable to comprehensively express the formation mechanism and action process of the legacy effects. The existing quantitative studies on the legacy effect of nonpoint source phosphorus pollution in watersheds are mainly on an interannual scale and basinwide analysis, and there is a lack of quantitative studies on the seasonal loss and spatial distribution of legacy phosphorus in different environmental media [98].

6.2. Management of Phosphorus Pollution in Lakes

At present, studies on lake phosphorus patterns are mainly focused on their important reservoir sediments or substrates, and on the differences between single lakes or different lakes in the same lake area; comparative studies on different types of lakes are relatively lacking.

Moreover, the management of eutrophication in phosphorus-controlled lakes remains a technical challenge, as there are no mature engineering techniques and experiences to draw on. Due to the influence of climate and intense human activities, phosphorus concentrations in lakes fluctuate greatly and even rebound in treatment, making it difficult to achieve phosphorus-control goals in lakes.

There are currently three forms of phosphorus in lake waters: water-soluble, particulate, and PH_3, of which the amount of PH_3 produced is small, and the relevant formation and transformation mechanisms are still unclear. The form of phosphorus in lake waters is more studied in the sediment-water transformation of inorganic phosphorus, while the transformation mechanisms of organic phosphorus between sediment and water bodies are yet to be further clarified. Most of the relevant lake water phosphorus management techniques are aimed at the reduction of total phosphorus concentrations, such as physical, chemical, and bioecological methods. The physical method, which has quicker results and obvious restoration effects, is not long-lasting enough, and the research focuses on the efficiency of phosphorus removal by adsorbent materials, while less attention is paid to the mechanism of phosphorus adsorption. The chemical method, which is easy to operate but has high chemical costs and secondary ecological risks, can be used as an auxiliary or emergency control technology. The bioecological method is a comprehensive technology, which is the mainstream method of lake restoration at present, with low economic costs, good landscape effects, and stable restoration effects. Most of the relevant studies focus on the effect and mechanism of functional organisms (plants, animals, and microorganisms) and their grouping on the removal of total phosphorus, while the ecological safety of engineered bacteria and the risk of phosphorus removal mechanisms of aquatic animals still need to be further strengthened [99].

Author Contributions: X.G.: Conceptualization, methodology, validation, formal analysis, investigation, data curation, writing—original draft preparation; X.C.: Conceptualization, methodology, software, validation, investigation, data curation, writing—original draft preparation; M.L.: Conceptualization, methodology, validation, investigation, data curation, writing—original draft preparation; Y.Z.: writing—review and editing; C.W.: writing—review and editing; Y.W.: writing—review and editing; H.-T.T.: writing—review and editing; S.J.: writing—review and editing; T.Z.: Writing—review and editing, visualization, supervision, project administration, funding acquisition. All authors have read and agreed to the published version of the manuscript.

Funding: The research was sustained by the grant from the Undergraduate Research Program of China Agricultural University, the National Key Technology Research and Development Program of China [Grant number 2022YFE012907], the Government Purchase Service Project of Ministry of Agriculture and Rural Affairs of China [Grant number 202205510310600], the National Natural Science Foundation of China [Grant Number 31401944].

Institutional Review Board Statement: Not applicable.

Data Availability Statement: Not applicable.

Conflicts of Interest: The authors declare no conflict of interest.

References

1. Springmann, M.; Clark, M.; Croz, D.M.D.; Wiebe, K.; Bodirsky, B.L.; Lassaletta, L.; de Vries, W.; Vermeulen, S.J.; Herrero, M.; Carlson, K.M.; et al. Options for keeping the food system within environmental limits. *Nature* **2018**, *562*, 519–525. [CrossRef] [PubMed]
2. Li, H.H.; Zhang, T.; Shaheen, S.M.; Abdelrahman, H.; Ali, E.F.; Bolan, N.S.; Li, G.X.; Rinklebe, J. Microbial inoculants and struvite improved organic matter humification and stabilized phosphorus during swine manure composting: Multivariate and mul-tiscale investigations. *Bioresour. Technol.* **2022**, *351*, 126976. [CrossRef]
3. Liu, G.F.; Dai, Z.M.; Liu, X.M.; Dahlgren, R.A.; Xu, J.M. Modification of agricultural wastes to improve sorption capacities for pol-lutant removal from water—A review. *Carbon Res.* **2022**, *1*, 24. [CrossRef]
4. Wu, F.C.; Li, F.B.; Zhao, X.L.; Bolan, N.S.; Fu, P.Q.; Lam, S.S.; Masek, O.; Ong, H.C.; Pan, B.; Qiu, X.Q.; et al. Meet the challenges in the "Carbon Age". *Carbon Res.* **2022**, *1*, 1. [CrossRef]
5. Deng, Y.X.; Zhang, T.; Clark, J.; Aminabhavi, T.; Kruse, A.; Tsang DC, W.; Sharma, B.K.; Zhang, F.S.; Ren, H.Q. Mechanisms and modelling of phosphorus solid–liquid transformation during the hydrothermal processing of swine manure. *Green Chem.* **2020**, *22*, 5628–5638. [CrossRef]
6. Xie, S.Y.; Zhang, T.; Mishra, A.; Tiwari, A.; Bolan, N.S. Assessment of catalytic thermal hydrolysis of swine manure slurry as liquid fertilizer: Insights into nutrients and metals. *Front. Environ. Sci.* **2022**, *10*, 1005290. [CrossRef]
7. Yu, J.X.; Xie, S.Y.; Zhang, T. Influences of hydrothermal carbonization on phosphorus availability of swine manure-derived hydrochar: Insights into reaction time and temperature. *Mater. Sci. Energy Technol.* **2022**, *5*, 416–423. [CrossRef]
8. Ge, X.F.; Zhang, T. Changes in inorganic and organic matters in processed water from hydrothermal-treated biogas slurry. *Mater. Sci. Energy Technol.* **2023**, *6*, 145–157. [CrossRef]
9. Guan, Q.; Zeng, G.; Song, J.; Liu, C.; Wang, Z.; Wu, S. Ultrasonic power combined with seed materials for recovery of phosphorus from swine wastewater via struvite crystallization process. *J. Environ. Manag.* **2021**, *293*, 112961. [CrossRef]
10. Godfray, H.C.J.; Beddington, J.R.; Crute, I.R.; Haddad, L.; Lawrence, D.; Muir, J.F.; Pretty, J.; Robinson, S.; Thomas, S.M.; Toulmin, C. Food security: The challenge of feeding 9 billion people. *Science* **2010**, *327*, 812–818. [CrossRef]
11. Vermeulen, S.J.; Campbell, B.M.; Ingram JS, I. Climate change and food systems. *Annu. Rev. Environ. Resour.* **2012**, *37*, 195–222. [CrossRef]

12. Chen, X.P.; Cui, Z.L.; Fan, M.S. Producing more grain with lower environmental costs. *Prod. More Grain Low. Environ. Costs Nat.* **2014**, *514*, 486–489. [CrossRef]
13. Chen, M.P.; Graedel, T.E. A half-century of global phosphorus flows, stocks, production, consumption, recycling, and environmental impacts. *Glob. Environ. Change* **2016**, *36*, 139–152. [CrossRef]
14. He, X.Y.; Wang, Y.K.; Zhang, Y.Y.; Wang, C.S.; Yu, J.X.; Ohtake, H.; Zhang, T. The potential for livestock manure valorization and phos-phorus recovery by hydrothermal technology—A critical review. *Mater. Sci. Energy Technol.* **2023**, *6*, 94–104.
15. Cordell, D.; White, S. Life's bottleneck: Sustaining the world's phosphorus for a food secure future. *Annu. Rev. Environ. Resour.* **2014**, *39*, 161–188. [CrossRef]
16. Wang, Q.M.; Zhang, T.; He, X.Y.; Jiang, R.F. Assessment of phosphorus recovery from swine wastewater in Beijing, China. *Sustainability* **2017**, *9*, 1845. [CrossRef]
17. Zou, L.; Liu, Y.; Wang, Y.; Hu, X. Assessment and analysis of agricultural non-point source pollution loads in China: 1978–2017. *J. Environ. Manag.* **2020**, *263*, 110400. [CrossRef]
18. Hu, X.; Zhou, Y.; Zhou, L.; Zhang, Y.; Wu, L.; Xu, H.; Zhu, G.; Jang, K.; Spencer, R.; Jeppesen, E.; et al. Urban and agricultural land use regulates the molecular composition and bio-lability of fluvial dissolved organic matter in human-impacted southeastern China. *Carbon Res.* **2022**, *1*, 19. [CrossRef]
19. Pradhan, S.N.; Ghosh, A.K.; Seema Ram, S.; Pal, Y.; Pradhan, C. Changes in degree of phosphorus saturation and risk of P loss upon twelve years of manuring and reduced tillage. *Geoderma* **2021**, *404*, 115277. [CrossRef]
20. Li, H.; Huang, G.; Meng, Q.; Ma, L.; Yuan, L.; Wang, F.; Zhang, W.; Cui, Z.; Shen, J.; Chen, X.; et al. Integrated soil and plant phosphorus management for crop and environment in China. A review. *Plant Soil* **2011**, *349*, 157–167. [CrossRef]
21. Sun, H.; Liang, W. Research on phosphorus pollution in China's water bodies and its monitoring and management technology. *J. Zhejiang Norm. Univ. Nat. Sci. Ed.* **2007**, *30*, 201–205.
22. Ma, E.P.; Cai, J.M.; Lin, J.; Liao, L.W.; Guo, H.; Han, Y. Pattern characteristics of nitrogen and phosphorus emissions from agricultural sources and water environment impacts in China in the last 30 years. *J. Nat. Resour.* **2021**, *36*, 752–770.
23. Tim, U.S.; Jolly, R. Evaluating agricultural nonpoint-source pollution using integrated geographic information systems and hydro-logic/water quality model. *J. Environ. Qual.* **1994**, *23*, 25–35. [CrossRef]
24. Sabo, R.D.; Clark, C.M.; Gibbs, D.A.; Metson, G.S.; Todd, M.J.; LeDuc, S.D.; Greiner, D.; Fry, M.M.; Polinsky, R.; Yang, Q.; et al. Phosphorus inventory for the conterminous United States (2002–2012). *J. Geophys. Res. Biogeosci.* **2021**, *126*, e2020JG005684. [CrossRef]
25. Kronvang, B.; Bruhn, A.J. Choice of sampling strategy and estimation method for calculating nitrogen and phosphorus transport in small lowland streams. *Hydrol. Process.* **1996**, *10*, 1483–1501. [CrossRef]
26. McDowell, R.W. The effectiveness of coal fly-ash to decrease phosphorus loss from grassland soils. *Soil Res.* **2005**, *43*, 853–860. [CrossRef]
27. Eduah, J.O.; Nartey, E.K.; Abekoe, M.K.; Breuning-Madsen, H.; Andersen, M.N. Phosphorus retention and availability in three contrasting soils amended with rice husk and corn cob biochar at varying pyrolysis temperatures. *Geoderma* **2019**, *341*, 10–17. [CrossRef]
28. Jiang, X.; Yan, B.; Chen, J.; Li, W.; Guan, Y. Transport and retention of phosphorus in soil with addition of Mg-Al layered double hydroxides: Effects of material dosage, flow velocity and pH. *Chem. Eng. J.* **2019**, *378*, 122154. [CrossRef]
29. Liang, Z.; Peng, X.; Luan, Z.; Li, W.; Zhao, Y. Reduction of phosphorus release from high phosphorus soil by red mud. *Environ. Earth Sci.* **2012**, *65*, 581–588. [CrossRef]
30. Fan, B.; Fenton, O.; Daly, K.; Ding, J.; Chen, S.; Chen, Q. Alum split applications strengthened phosphorus fixation and phosphate sorption in high legacy phosphorus calcareous soil. *J. Environ. Sci.* **2021**, *101*, 87–97. [CrossRef]
31. Yan, C.; Zhan, H.; Yan, S.; Dong, S.; Ma, C.; Song, Q.; Gong, Z.; Barbie, M. Effects of straw retention and phosphorous fertilizer application on available phosphorus content in the soil solution during rice growth. *Paddy Water Environ.* **2016**, *14*, 61–69. [CrossRef]
32. Liu, L.; Zheng, X.; Wei, X.; Kai, Z.; Xu, Y. Excessive application of chemical fertilizer and organophosphorus pesticides induced total phosphorus loss from planting causing surface water eutrophication. *Sci. Rep.* **2021**, *11*, 23015. [CrossRef] [PubMed]
33. Mardamootoo, T.; Du Preez, C.; Barnard, J.H. Agricultural phosphorus management for environmental protection: A review. *J. Geosci. Environ. Prot.* **2021**, *9*, 48–81. [CrossRef]
34. McDowell, R.W.; Sharpley, A.N. Approximating phosphorus release from soils to surface runoff and subsurface drainage. *J. Environ. Qual.* **2001**, *30*, 508–520. [CrossRef]
35. Hesketh, N.; Brookes, P.C. Development of an indicator for risk of phosphorus leaching. *J. Environ. Qual.* **2000**, *29*, 105–110. [CrossRef]
36. McDowell, R.W.; Condron, L.M. Estimating phosphorus loss from New Zealand grassland soils. *N. Z. J. Agric. Res.* **2004**, *47*, 137–145. [CrossRef]
37. Khan, A.; Lu, G.; Ayaz, M.; Zhang, H.; Wang, R.; Lv, F.; Yang, X.; Sun, B.; Zhang, S. Phosphorus efficiency, soil phosphorus dynamics and critical phosphorus level under long-term fertilization for single and double cropping systems. *Agric. Ecosyst. Environ.* **2018**, *256*, 1–11. [CrossRef]
38. Weihrauch, C.; Weber, C.J. Comparative risk assessment of phosphorus loss from "deep phosphorus stocks" in floodplain subsoils to surface waters. *Sci. Total Environ.* **2021**, *796*, 149037. [CrossRef]

39. Casson, J.P.; Bennett, D.R.; Nolan, S.C.; Olson, B.M.; Ontkean, G.R. Degree of Phosphorus saturation thresholds in manure-amended soils of alberta. *J. Environ. Qual.* **2006**, *35*, 2212–2221. [CrossRef]
40. Zak, D.; Gelbrecht, J.; Steinberg, C. Phosphorus retention at the redox interface of peatlands adjacent to surface waters in northeast Germany. *Biogeochemistry* **2004**, *70*, 357–368. [CrossRef]
41. McDowell, R.; Dodd, R.; Pletnyakov, P.; Noble, A. The ability to reduce soil legacy phosphorus at a country scale. *Front. Environ. Sci.* **2020**, *8*, 6. [CrossRef]
42. Khattab, R.M.; Badr, H.A.; Abo-Almaged, H.H.; Sadek HE, H. Recycling of alum sludge for alpha Al2O3 production using different chemical treatments. *Desalination Water Treat.* **2018**, *113*, 148–159. [CrossRef]
43. Ujaczki, E.; Zimmermann, Y.S.; Gasser, C.A.; Molnar, M.; Feigl, V.; Lenz, M. Red mud as secondary source for critical raw materials—Extraction study. *J. Chem. Technol. Biotechnol.* **2017**, *92*, 2835–2844. [CrossRef]
44. Ma, P.C.; Li, X.; Wen, Z.Y.; Meng, F.H.; Li, Z. Research progress on activation and mechanism of fly ash. *Inorg. Chem. Ind.* **2021**, *53*, 28–35.
45. McDowell, R.; Sharpley, A. Availability of residual phosphorus in high phosphorus soils. *Commun. Soil Sci. Plant Anal.* **2002**, *33*, 1235–1246. [CrossRef]
46. Eslamian, F.; Qi, Z.; Tate, M.J.; Zhang, T.; Prasher, S.O. Phosphorus loss mitigation in leachate and surface runoff from clay loam soil using four lime-based materials. *Water Air Soil Pollut.* **2018**, *229*, 91–97. [CrossRef]
47. Torbert, H.; Watts, D. Impact of flue gas desulfurization gypsum application on water quality in a coastal plain soil. *J. Environ. Qual.* **2014**, *43*, 273–280. [CrossRef]
48. Fan, B.; Ding, J.; Fenton, O.; Daly, K.; Chen, Q. Understanding phosphate sorption characteristics of mineral amendments in relation to stabilising high legacy P calcareous soil. *Environ. Pollut.* **2020**, *261*, 114175. [CrossRef]
49. Miyittah, M.K.; Stanley, C.D.; Mackowiak, C.; Rhue, D.R.; Rechcigl, J.E. Developing a remediation strategy for phosphorus immobilization: Effect of co-blending, Al-residual and Ca-Mg amendments in a manure-impacted spodosol. *Soil Sediment Contam.* **2011**, *20*, 337–352. [CrossRef]
50. Fan, B.; Ding, J.; Fenton, O.; Daly, K.; Chen, S.; Zhang, S.; Chen, Q. Investigation of differential levels of phosphorus fixation in dolomite and calcium carbonate amended red soil. *J. Sci. Food Agric.* **2021**, *102*, 740–749. [CrossRef]
51. Liu, W.; Ji, H.; Kerr, P.; Wu, Y.; Fang, Y. The application of soil amendments benefits to the reduction of phosphorus depletion and the growth of cabbage and corn. *Environ. Sci. Pollut. Res.* **2015**, *22*, 16772–16780. [CrossRef]
52. Brennan, R.B.; Murnane, J.G.; Sharpley, A.N.; Herron, S.; Brye, K.R.; Simmons, T. Soil phosphorus dynamics following land application of unsaturated and partially saturated red mud and water treatment residuals. *J. Environ. Manag.* **2019**, *248*, 109296. [CrossRef] [PubMed]
53. Lin, J.; Kim, M.; Li, D.; Kim, H.; Huang, C. The removal of phosphate by thermally treated red mud from water: The effect of surface chemistry on phosphate immobilization. *Chemosphere* **2020**, *247*, 125867. [CrossRef] [PubMed]
54. Fan, B.; Wang, J.; Fenton, O.; Daly, K.; Ezzati, G.; Chen, Q. Strategic differences in phosphorus stabilization by alum and dolomite amendments in calcareous and red soils. *Environ. Sci. Pollut. Res.* **2019**, *26*, 4842–4854. [CrossRef] [PubMed]
55. Uusi-Kämppä, J.; Turtola, E.; Närvänen, A.; Jauhiainen, L.; Uusitalo, R. Phosphorus mitigation during springtime runoff by amendments applied to grassed soil. *J. Environ. Qual.* **2012**, *41*, 420–426. [CrossRef] [PubMed]
56. Callery, O.; Brennan, R.; Healy, M. Use of amendments in a peat soil to reduce phosphorus losses from forestry operations. *Ecol. Eng.* **2015**, *85*, 193–200. [CrossRef]
57. He, X.Y.; Zhang, T.; Niu, Y.Q.; Xue, Q.; Ali, E.F.; Shaheen, S.M.; Tsang DC, W.; Rinklebe, J. Impact of catalytic hydrothermal treatment and Ca/Al-modified hydrochar on lability, sorption, and speciation of phosphorus in swine manure: Microscopic and spectroscopic investigations. *Environ. Pollut.* **2022**, *299*, 118877. [CrossRef]
58. Bacelo, H.; Pintor, A.; Santos, S.; Boaventura, R.; Botelho, C. Performance and prospects of different adsorbents for phosphorus uptake and recovery from water. *Chem. Eng. J.* **2020**, *381*, 122566. [CrossRef]
59. Lim, S.; Lee, D.; Kwak, J.; Park, H.; Kim, H.; Choi, W. Fly ash and zeolite amendments increase soil nutrient retention but decrease paddy rice growth in a low fertility soil. *J. Soils Sediments* **2016**, *16*, 756–766. [CrossRef]
60. Li, X.; Xie, Q.; Chen, S.; Xing, M.; Guan, T.; Wu, D. Inactivation of phosphorus in the sediment of the Lake Taihu by lanthanum modified zeolite using laboratory studies. *Environ. Pollut.* **2019**, *247*, 9–17. [CrossRef]
61. Wu, H.J.; Wang, S.; Gao, L.M.; Zhang, L.; Yuan, Z.W.; Fan, T.Y.; Wei, K.P.; Huang, L. Nutrient-derived environmental impacts in Chinese agriculture during 1978–2015. *J. Environ. Manag.* **2018**, *217*, 762–774. [CrossRef] [PubMed]
62. Jalali, M.; Antoniadis, V. Impact of sewage sludge, nanoparticles, and clay minerals addition on cucumber growth, phosphorus uptake, soil phosphorus status, and potential risk of phosphorus loss. *Environ. Technol. Innov.* **2021**, *23*, 101702. [CrossRef]
63. Moharami, S.; Jalali, M. Effect of time on the sorption and distribution of phosphorus in treated soil with minerals and nanoparticles. *Environ. Earth Sci.* **2015**, *73*, 8599–8608. [CrossRef]
64. Irshad, M.; Hafeez, F.; Naseem, M.; Rizwan, M.; Al-Wabel, M. Effect of coal and wood ash on phosphorus immobilization in different textured soils. *Arab. J. Geosci.* **2018**, *11*, 536. [CrossRef]
65. Faridullah; Malik, N.; Fareed, I.; Irshad, M. Reducing the leachability of nitrate, phosphorus and heavy metals from soil using waste material. *Braz. J. Chem. Eng.* **2017**, *34*, 715–726. [CrossRef]
66. Udeigwe, T.; Wang, J.; Zhang, H. Effectiveness of bauxite residues in immobilizing contaminants in manure-amended soils. *Soil Sci.* **2009**, *174*, 676–686. [CrossRef]

67. Zhang, L.; Chen, J.; Chu, G. Legacy phosphorus in calcareous soil under 33 years of P fertilizer application: Implications for efficient P management in agriculture. *Soil Use Manag.* **2022**, *38*, 1380–1393. [CrossRef]
68. Fang, C.; Zhang, T.; Li, P.; Jiang, R.F.; Wu, S.B.; Nie, H.Y.; Wang, Y.C. Phosphorus recovery from biogas fermentation liquid by Ca-Mg loaded biochar. *J. Environ. Sci.* **2015**, *29*, 106–114. [CrossRef]
69. Zhang, T.; Xu, H.Y.; Li, H.H.; He, X.Y.; Shi, Y.J.; Kruse, A. Microwave digestion-assisted HFO/biochar adsorption to recover phosphorus from swine manure. *Sci. Total Environ.* **2018**, *621*, 1512–1526. [CrossRef]
70. Shaheen, S.M.; Natasha Mosa, A.; El-Naggar, A.; Hossain Md, F.; Abdelrahman, H.; Niazi, N.K.; Shahid, M.; Zhang, T.; Tsang, Y.F.; Trakal, L.; et al. Manganese oxide-modified biochar: Production, characterization and applications for the removal of pollutants from aqueous environments—A review. *Bioresour. Technol.* **2022**, *346*, 126581. [CrossRef]
71. Chen, H.; Gao, Y.; Li, J.; Fang, Z.; Bolan, N.; Bhatnagar, A.; Gao, B.; Hou, D.; Wang, S.; Song, H.; et al. Engineered biochar for environmental decontamination in aquatic and soil systems: A review. *Carbon Res.* **2022**, *1*, 4. [CrossRef]
72. Matin, N.H.; Jalali, M.; Antoniadis, V.; Shaheen, S.M.; Wang, J.; Zhang, T.; Wang, H.; Rinklebe, J. Almond and walnut shell-derived biochars affect sorption-desorption, fractionation, and release of phosphorus in two different soils. *Chemosphere* **2020**, *241*, 124888. [CrossRef]
73. Peng, Y.; Sun, Y.; Fan, B.; Zhang, S.; Tsang DC, W. Fe/Al (hydr)oxides engineered biochar for reducing phosphorus leaching from a fertile calcareous soil. *J. Clean. Prod.* **2021**, *279*, 123877. [CrossRef]
74. Xu, M.; Wu, J.; Yang, G.; Zhang, X.; Peng, H.; Yu, X.; Xiao, Y.; Qi, H. Biochar addition to soil highly increases P retention and decreases the risk of phosphate contamination of waters. *Environ. Chem. Lett.* **2019**, *17*, 533–541. [CrossRef]
75. Niyungeko, C.; Liang, X.; Shan, S.; Zhang, J.; Tiimub, B.; SeyedHamid, H.; Khan, S.; Li, F.; Tian, G. Synergistic effects of anionic polyacrylamide and gypsum to control phosphorus losses from biogas slurry applied soils. *Chemosphere* **2019**, *234*, 953–961. [CrossRef] [PubMed]
76. Liang, X.Q.; Liu, Z.W.; Liu, J.; Chen, L.L.; Tian, G.M. Soil colloidal P release potentials under various polyacrylamide addition levels. *Land Degrad. Dev.* **2017**, *28*, 2245–2254. [CrossRef]
77. Cao, D.; Chen, W.; Yang, P.; Lan, Y.; Sun, D. Spatio-temporal variabilities of soil phosphorus pool and phosphorus uptake with maize stover biochar amendment for 5 years of maize. *Environ. Sci. Pollut. Res.* **2020**, *27*, 36350–36361. [CrossRef]
78. Li, F.; Jin, Y.; He, S.; Jin, J.; Wang, Z.; Khan, S.; Tian, G.; Liang, X. Use of polyacrylamide modified biochar coupled with organic and chemical fertilizers for reducing phosphorus loss under different cropping systems. *Agric. Ecosyst. Environ.* **2021**, *310*, 107306. [CrossRef]
79. Motaghian, H.; Hosseinpur, A.; Safian, M. The effects of sugarcane-derived biochar on phosphorus release characteristics in a calcareous soil. *J. Soil Sci. Plant Nutr.* **2020**, *20*, 66–74. [CrossRef]
80. Zhang, Q.; Song, Y.; Wu, Z.; Yan, X.; Gunina, A.; Kuzyakov, Y.; Xiong, Z. Effects of six-year biochar amendment on soil aggregation, crop growth, and nitrogen and phosphorus use efficiencies in a rice-wheat rotation. *J. Clean. Prod.* **2020**, *242*, 118435. [CrossRef]
81. Mukherjee, S.; Mavi, M.; Singh, J. Differential response of biochar derived from rice-residue waste on phosphorus availability in soils with dissimilar pH. *Int. J. Environ. Sci. Technol.* **2020**, *17*, 3065–3074. [CrossRef]
82. Wang, Y.; Zhang, Y.; Zhao, H.; Hu, W.; Zhang, H.; Zhou, X.; Luo, G. The effectiveness of reed-biochar in mitigating phosphorus losses and enhancing microbially-driven phosphorus dynamics in paddy soil. *J. Environ. Manag.* **2022**, *314*, 115087. [CrossRef] [PubMed]
83. Vanek, S.; Lehmann, J. Phosphorus availability to beans via interactions between mycorrhizas and biochar. *Plant Soil* **2015**, *395*, 105–123. [CrossRef]
84. Fang, C.; Zhang, T.; Li, P.; Jiang, R.F.; Wang, Y.C. Application of magnesium modified corn biochar for phosphorus removal and recovery from Swine wastewater. *Int. J. Environ. Res. Public Health* **2014**, *11*, 9217–9237. [CrossRef] [PubMed]
85. He, X.M.; Zhang, T.; Ren, H.Q.; Li, G.X.; Ding, L.L.; Pawlowski, L. Phosphorus recovery from biogas slurry by ultrasound/H2O2 digestion coupled with HFO/biochar adsorption process. *Waste Manag.* **2017**, *60*, 219–229. [CrossRef] [PubMed]
86. Deng, Y.X.; Zhang, T.; Sharma, B.K.; Nie, H.Y. Optimization and mechanism studies on cell disruption and phosphorus recovery from microalgae with magnesium modified hydrochar in assisted hydrothermal system. *Sci. Total Environ.* **2019**, *646*, 1140–1154. [CrossRef]
87. Chen, Q.; Qin, J.; Cheng, Z.; Huang, L.; Sun, P.; Shen, G. Synthesis of a stable magnesium-impregnated biochar and its reduction of phosphorus leaching from soil. *Chemosphere* **2018**, *199*, 402–408. [CrossRef]
88. Zhao, D.; Luo, Y.; Feng, Y.; He, Q.; Zhang, L.; Zhang, K.; Wang, F. Enhanced adsorption of phosphorus in soil by lanthanum-modified biochar: Improving phosphorus retention and storage capacity. *Environ. Sci. Pollut. Res.* **2021**, *28*, 68982–68995. [CrossRef]
89. Feng, Y.; Lu, H.; Liu, Y.; Xue, L.; Dionysiou, D.; Yang, L.; Xing, B. Nano-cerium oxide functionalized biochar for phosphate retention: Preparation, optimization and rice paddy application. *Chemosphere* **2017**, *185*, 816–825. [CrossRef]
90. Peng, L.; Xue, X.; Tang, Q.; Zhu, Y.; Xiao, L.; Yang, Y.; Lin, Q. Phosphorus retention and loss in three types of soils with implications for geographical pattern of eutrophication in china. *Water Environ. J.* **2020**, *34*, 9–18. [CrossRef]
91. Mng'Ong'o, M.; Munishi, L.K.; Blake, W.; Ndakidemi, P.A.; Comber, S.; Hutchinson, T.H. Characterization of soil phosphate status, sorption and saturation in paddy wetlands in usangu basin-tanzania. *Chemosphere* **2021**, *278*, 130466. [CrossRef] [PubMed]

92. Sun, Q.; Hu, Y.; Chen, X.; Wei, X.; Shen, J.; Ge, T.; Su, Y. Flooding and straw returning regulates the partitioning of soil phosphorus fractions and phoD-harboring bacterial community in paddy soils. *Appl. Microbiol. Biotechnol.* **2021**, *105*, 9343–9357. [CrossRef] [PubMed]
93. Yang, W.; Xia, Y.; Jiang, X.; Yan, X. The influence factors and loss of total phosphorus runoff loss are estimated. *J. Agric. Environ. Sci.* **2015**, *4*, 319–325.
94. Xu, X.; Mao, X.; Van Zwieten, L.; Niazi, N.K.; Wang, H.L. Wetting-drying cycles during a rice-wheat crop rotation rapidly (im)mobilize recalcitrant soil phosphorus. *J. Soils Sediments* **2020**, *20*, 3921–3930. [CrossRef]
95. Zhang, S.; Yang, X.; Hsu, L.; Liu, Y.; Wang, S.; White, J.R.; Shaheen, S.M.; Chen, Q.; Rinklebe, J. Soil acidification enhances the mobilization of phosphorus under anoxic conditions in an agricultural soil: Investigating the potential for loss of phosphorus to water and the associated environmental risk. *Sci. Total Environ.* **2021**, *793*, 148531. [CrossRef]
96. Shaheen, S.M.; Wang, J.; Baumann, K.; Wang, S.L.; Rinklebe, J. Redox-induced mobilization of phosphorus in groundwater affected arable soil profiles. *Chemosphere* **2021**, *275*, 129928. [CrossRef]
97. Zhu, S.J.; Gui, S.B.; Xu, W.; Xiang, P.; Meng, W.; Zhu, J.; Li, K.K. Experimental study on the tolerance of four species of water-holding plants to high concentrations of nitrogen and phosphorus pollution. *China Rural Water Hydropower* **2022**, *2*, 34–38.
98. Wu, H.; Chen, D.J. Progress in the study of legacy effects of non-point source phosphorus pollution in watersheds. *J. Agric. Environ. Sci.* **2022**, *41*, 2352–2364.
99. Cui, J.; Du, Y.; Ding, C.C.; Li, J.F.; Gao, F.S.; Chang, Y.J.; Zhang, J.B.; Liu, X.J.; Yao, D.R. Fugitive patterns of phosphorus in Chinese lake waters and progress of pollution control measures. *J. Ecol. Environ.* **2022**, *31*, 621–633.

water

MDPI

Review

The Promotion of Anaerobic Digestion Technology Upgrades in Waste Stream Treatment Plants for Circular Economy in the Context of "Dual Carbon": Global Status, Development Trend, and Future Challenges

Xinjia Huang

Bartlett School of Sustainable Construction, University College London, London WC1E 6BT, UK; ucbvxh2@ucl.ac.uk

Abstract: This review provides a comprehensive overview of the advancements and challenges of anaerobic digestion technology in waste stream treatment plants under the framework of the circular economy, emphasizing its role in achieving "dual carbon" goals. As climate change intensifies, with waste stream treatment contributing significantly to global emissions, there is a pressing need to optimize energy efficiency and reduce carbon outputs in this sector. Anaerobic digestion is highlighted as a solution for converting organic waste into renewable biogas and digestate, enabling energy self-sufficiency and reducing greenhouse gasses. The study highlights that anaerobic digestion enables the conversion of organic waste into renewable biogas and nutrient-rich digestate, facilitating energy self-sufficiency and significant reductions in GHG emissions. Successful implementations, such as in Weifang, China, demonstrate the feasibility of upgrading biogas into biomethane for local energy use. Advanced technologies like bioelectrochemical methanation and membrane bioreactors enhance biogas production efficiency, while co-digestion proves effective even in challenging conditions. Despite these advancements, the review identifies critical challenges, including high investment costs, technical inefficiencies, and regulatory barriers, particularly in developing countries. This study provides insights into integrating anaerobic digestion with circular economy principles and offers a foundation for future policies and research aimed at achieving carbon neutrality and sustainable waste management.

Keywords: bioenergy recovery; sustainable waste management; carbon neutrality; energy efficiency; green technologies

check for updates

Citation: Huang, X. The Promotion of Anaerobic Digestion Technology Upgrades in Waste Stream Treatment Plants for Circular Economy in the Context of "Dual Carbon": Global Status, Development Trend, and Future Challenges. *Water* **2024**, *16*, 3718. https://doi.org/10.3390/w16243718

Academic Editor: Fanying Kong

Received: 16 November 2024
Revised: 7 December 2024
Accepted: 10 December 2024
Published: 23 December 2024

1. Introduction

In recent years, the increase in greenhouse gas (GHG) emissions has exacerbated global climate change, leading to widespread adverse effects worldwide. The frequent occurrence of extreme weather events, such as hurricanes, floods, and droughts, poses significant threats to human life and economic development, while rising sea levels force mass migrations from coastal areas and further contribute to international instability [1]. In response to these challenges, the United Nations adopted 17 Sustainable Development Goals in 2015, urging countries to take immediate action to combat climate change. Later that year, the Paris Agreement was established during the Paris Climate Conference, requiring nations to reduce GHG emissions and limit global temperature rise to well below 2 °C, with efforts to cap the increase at 1.5 °C [2]. As one of the largest global emitters of carbon dioxide, China has also accelerated its transition towards a green economy. In 2020, the country announced its commitment to peaking carbon dioxide emissions before 2030 and achieving carbon neutrality before 2060 [3]. Subsequently, the 2021 Action Plan for Peaking CO_2 Emissions by 2030 emphasized industrial restructuring, improving energy efficiency, and promoting green technologies [4].

Despite significant progress across several sectors, waste stream treatment has often been overlooked due to its perception as a low-carbon industry. However, wastewater treatment involved in the waste stream treatment is an energy-intensive process, contributing nearly 5% of global non-carbon dioxide emissions, which is projected to rise to 22% by 2030 [5]. Additionally, electricity consumption in the sewage sector is expected to increase by 80% over the next 25 years [6]. Therefore, advancing technological solutions to achieve energy self-sufficiency and reducing carbon emissions in this sector has become a critical priority.

Anaerobic digestion technology is a well-established and widely applied method in sewage treatment, capable of breaking down organic wet waste and sewage sludge under anaerobic conditions. Through microbial activity, these materials are converted into biogas and digestate, producing both renewable energy and recyclable fertilizers [7–9]. This process not only reduces reliance on fossil fuels but also significantly curtails GHG emissions. Despite these benefits, the current global economy remains heavily dependent on a linear economic model, which is based on extracting raw materials from nature, converting them into products, and ultimately disposing of them as waste [10]. Over the past decades, this linear approach has led to severe pollution and resource depletion, exacerbating both the climate and environmental crises. In contrast, circular economic strategies—focused on the efficient use of industrial materials and the recycling of resources to generate renewable energy—offer a sustainable alternative. These strategies can significantly reduce global greenhouse gas emissions while enhancing resilience to extreme weather and natural disasters [11–13].

However, it is important to note that while recent advancements in digitization and intelligent applications have enhanced the ability of anaerobic digestion systems to convert organic waste into biogas, several technical and economic bottlenecks still hinder their large-scale deployment. These challenges include efficiency limitations that are highly sensitive to external factors, technological gaps related to scalability, as well as high initial investment costs and extended payback periods.

This review aims to examine how anaerobic digestion technology can facilitate the transition of waste stream treatment towards a circular economy. By analyzing the current global development of anaerobic digestion technology in waste stream treatment, the study will explore the key drivers and challenges, including technological advancements, policy frameworks, and market demand, while anticipating future trends. Finally, the review will synthesize existing research and practices, highlighting the essential role of anaerobic digestion technology in advancing circular economic principles. This will provide a scientific foundation and valuable insights to guide future research and policymaking.

2. Overview of Anaerobic Digestion Techniques

2.1. Basic Principles and Techniques of Anaerobic Digestion

Anaerobic digestion is a biological process in which microorganisms break down organic matter in an oxygen-free environment. Common organic feedstocks include animal manure, food waste, agricultural residues, and wastewater solids [8,9,14]. Through this process, organic matter is converted into methane and carbon dioxide, primarily to produce biogas. The material remaining after digestion, known as digestate, is nutrient-rich and can be separated into solid and liquid fractions, both of which are commonly used as fertilizers for crops. In wastewater treatment, the by-products of digestion are often referred to as biosolids [15].

Notably, while digestate is a nutrient-rich by-product of anaerobic digestion, containing high levels of nitrogen, phosphorus, and potassium that are valuable for improving soil fertility and promoting sustainable agricultural practices, it may not always be suitable for direct agricultural use. Factors such as high salinity, heavy metal contamination, and the presence of pathogens or organic pollutants can negatively impact crop health and soil ecosystems [16]. Furthermore, logistical challenges, including high transportation costs and regulatory restrictions, may further constrain its agricultural application [17]. In some instances, alternative uses, such as bioplastic production, conversion into biochar, or energy recovery through incineration, may present more viable and efficient options [18].

As shown in Figure 1, the anaerobic digestion process for converting organic waste into biogas involves four main stages: hydrolysis, acidification, acetogenesis, and methanogenesis [19]. In the hydrolysis stage, hydrolytic bacteria decompose complex organic compounds into simpler molecules, breaking down carbohydrates, proteins, and fats into amino acids, sugars, and long-chain fatty acids. In the subsequent acidogenesis phase, fermenting bacteria converts these monomers into volatile fatty acids, organic acids, and alcohols, with the potential release of ammonia and hydrogen sulfide. During acetogenesis, acetic acid-forming bacteria convert the volatile fatty acids into acetic acid, carbon dioxide, and hydrogen. Finally, in the methanogenesis stage, methanogens convert acetic acid and hydrogen into a mixture of methane and carbon dioxide.

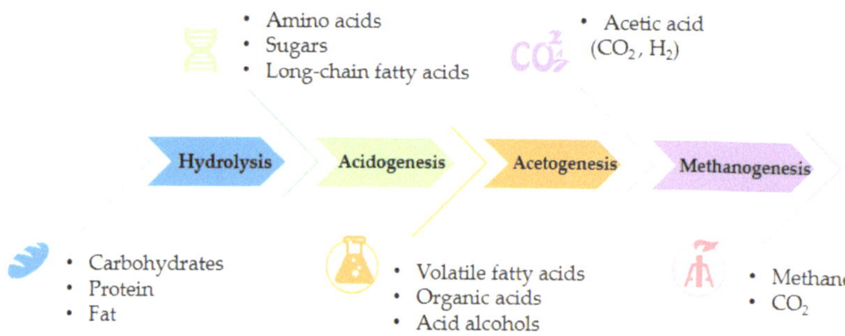

Figure 1. The four main stages of anaerobic digestion of organic waste.

As shown in Figure 2, anaerobic digestion technology can be categorized based on factors such as moisture content of the raw material, feed frequency, and mixing method. First, digesters are classified as either wet or dry systems. Wet digesters are suitable for substrates like sewage sludge and animal manure with a moisture content above 85%, while dry digesters are ideal for biosolid waste and food waste with a solids content exceeding 15% [20]. Second, based on feed frequency, digesters are categorized into batch and continuous systems; batch digesters produce biogas periodically, whereas continuous digesters generate biogas at a relatively stable rate [21]. In practice, many digesters operate in a semi-batch or semi-continuous mode, balancing biogas production continuity with routine maintenance needs. Finally, mixing methods for digesters vary between non-fixing and fully fixing systems. Fully fixing digesters, utilizing mechanical agitation and biogas recirculation, provide a uniform digestion environment and prevent temperature gradients or "dead zones" [22]. Although fully mixed digesters incur higher operating costs, these are partially offset by enhanced biogas yields.

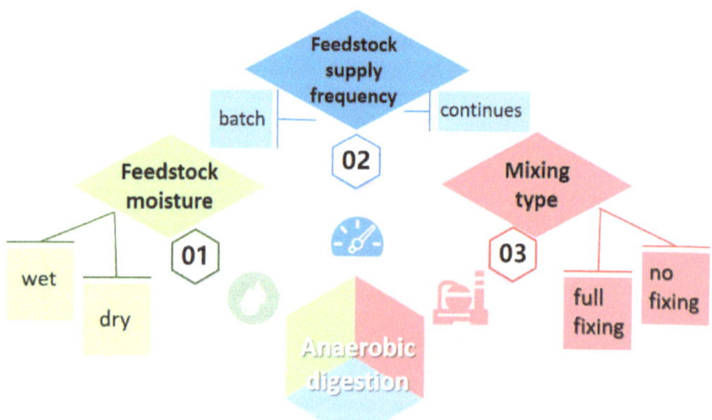

Figure 2. The main design parameters of anaerobic digestion technology.

2.2. Application Status of Anaerobic Digestion Technology in Wastewater Treatment

Anaerobic digestion technology has garnered significant attention due to its broad applications in climate and environmental protection, as well as its potential economic benefits. As an efficient waste management strategy, anaerobic digestion is widely applied in sewage sludge treatment, effectively reducing treatment costs and energy consumption. It also produces high-calorific biogas, which can be used for power generation, enabling plants to achieve energy self-sufficiency [23]. Additionally, the integration of cogeneration technology allows facilities not only to fulfill their own heating needs but also to supply thermal energy to surrounding areas, further enhancing economic benefits [20].

The application of anaerobic digestion technology varies significantly across regions. In Europe, stringent environmental regulations and renewable energy policies have driven the widespread adoption of this technology, particularly in Germany and the Nordic countries, where a mature industrial chain has formed. In Germany, the Renewable Energy Act (EEG) of 2000 spurred the development of biogas power, leading to the establishment of over 1000 biogas plants within 14 years [24]. Today, Germany is one of the world's largest biogas producers, and wastewater treatment plants often co-digest food and agricultural waste with sludge, enhancing anaerobic digestion efficiency [25]. In contrast, the development of anaerobic digestion technology in the United States has been propelled by technological innovations and advancements in pretreatment methods, which have significantly improved sludge decomposition rates and biogas production [26].

Compared to developed countries in Europe and the United States, developing nations face technical, economic, and policy-related challenges in adopting anaerobic digestion technology. In Malaysia, for instance, wastewater treatment largely relies on traditional methods despite technical potential due to resource limitations. Although European co-digestion methods have been applied to improve efficiency, results have been limited by inadequate technical support [27]. Similarly, African countries face barriers such as insufficient funding, technology, and policy frameworks, further complicated by inadequate infrastructure. In countries like Kenya and Uganda, the government has initiated biogas programs, but a lack of systematic support has hindered project sustainability [28]. To overcome these challenges, targeted strategies are needed in these regions, including enhanced technical training, increased public awareness, improved policy support, and strengthened international collaboration and financial assistance.

2.3. Economic and Environmental Benefits of Anaerobic Digestion Technology

The primary advantages of anaerobic digestion technology in the sewage treatment industry are twofold: economic benefits and environmental gains. First, as shown in Table 1, anaerobic digestion offers significant economic benefits by reducing fossil fuel costs associated with sludge incineration in wastewater treatment plants through biogas production and utilization. Additional revenue is generated through cogeneration and by-product sales, such as biogas electricity and reduced fertilizer input costs [29]. While large systems entail substantial capital investment, proper management and government incentives can enhance their economic feasibility.

From an environmental perspective, anaerobic digestion minimizes the combustion and emissions of organic waste, effectively reducing greenhouse gas emissions and supporting sustainable development [30]. In Japan, the anaerobic digestion project in Minamisanriku Town reduces waste disposal costs and increases local revenue by lowering incineration and landfill requirements. This project annually produces 2150 tons of liquid fertilizer and 255 tons of high-value rice, resulting in a 216.5% increase in net income when operational efficiency reaches 100%. Concurrently, greenhouse gas emissions are reduced to 126 kg CO_2-eq per ton of waste, marking an approximate 23% decrease [31]. In Russia's Sverdlovsk region, a sewage treatment plant generates approximately 9.9 million cubic meters of biogas annually, which is converted into 20.8 million kWh of electricity and 24.78 million kWh of heat, addressing regional energy demands. This €65.18 million investment promotes green infrastructure and reduces annual CO_2 emissions by 119,649 tonnes, accounting for 0.45% of the region's total wastewater treatment emissions [32]. Anaerobic digestion technology not only transforms sludge into biofertilizers and energy, aligning with circular economy principles, but also mitigates the risks associated with sludge treatment and alleviates environmental burdens.

Based on the analysis of various countries presented in Table 1, the application of anaerobic digestion technology generally yields substantial economic returns, primarily reflected in cost savings, increased net income, and enhanced energy efficiency, particularly in energy production and technological advancements. In terms of environmental benefits, most countries report significant reductions in greenhouse gas emissions, highlighting notable progress in sustainable development. Furthermore, energy efficiency and infrastructure investment are closely interconnected, with many countries benefiting from increased energy capacity and efficiency, especially those facing high energy demand. Factors influencing these outcomes include the economic context, environmental policies, energy demand, and technological innovation. For instance, wealthier countries and those with established energy industries are more likely to invest heavily in anaerobic digestion technologies, while environmental policies and technological innovations contribute to more substantial environmental benefits.

Table 1. Economic and environmental benefits of anaerobic digestion technology by region.

Region	Economic Benefit	Environmental Benefit	Energy Benefit	Infrastructure Investment	Reference
Japan	Net income increase by 216.5%	GHG reduction by approx. 23%	Production of 2150 tons of liquid fertilizer	Annual investment required: 5375 GBP	[31]
Russia	Green investment return of USD 651.8 million	Annual emission reduction of 11,964 tons, 0.45% to total emissions	Production of 2080 MWh of electricity and 2478 MWh of heat	Nationwide facility investment required: USD 651.8 million	[32]
Iran	Net income increases by 45.2%	Reduction in global warming potential by 79.16–92.65%	Total cumulative generation over 20 years: 4571.7 GWh	Average LCOE: USD 0.0335–0.0426 per kWh	[33]
China	Annual net profit: USD 7.85–11.67 million	Annual emission reduction of 785.4–3301.6 Mt CO_2-eq	Annual gas production: 17,090 m^3	Construction cost: USD 30–75 million	[20]
Germany	65% return on recycling and high technology reuse	20,000 tons GHG reduction	Annual production capacity: 980 million m^3	Upgrade investment required: USD 65 million	[34]
UK	Annual sales revenue: GBP 38.73 million	3–10% annual emission reduction	Annual production of approx. 1200 MWh of electricity	Construction cost: USD 10 million	[35]
Argentina	Energy cost savings of 35%	Better Acidification potential emission reduction effect	Methane purity after concentration: 65%	Average facility cost: USD 2 million	[36]
Spain	Income of EUR 973.59 million	79–80% reduction in GHG emissions	Annual capacity of 666.88 MWh of energy	Payback period: 3–5 years	[37]
USA	10-year lifecycle cost savings of USD 20.5 million	79.16–92.65% reduction in GHG emissions	CHP systems raise efficiency up to 80%	Capital investment: USD 12.3–16.4 million	[38]
Turkey	Savings of USD 1.3 billion in fossil fuel imports	12.2 Mt CO_2-eq reduction in emissions	Annual capacity of approx. 95 MWe of electricity	Initial investment: USD 21.71 million; payback period: 5–8 years	[39]

3. Status of Anaerobic Digestion Technology Upgrade in Global Wastewater Treatment Plants

3.1. Upgrade and Progress of Anaerobic Digestion Technology

As shown in Table 2, wastewater treatment plants worldwide are increasingly recognizing the potential of anaerobic digestion technology and actively pursuing technological upgrades. In terms of capacity enhancement, efficiency has been notably improved by optimizing operational conditions and implementing two-stage or multi-stage anaerobic digestion systems. These advanced configurations enhance the production of high value-added products, such as volatile fatty acids (VFAs) and hydrogen, by carefully adjusting pH, temperature, and the residence time of organic waste [40]. Integrating continuous stirred-tank reactors (CSTR) with upflow anaerobic sludge blanket (UASB) reactors has further boosted biogas yields, while two-stage anaerobic digestion has achieved 10% to 30% increases in methane production [41]. However, large-scale applications are constrained by high capital investment and operational complexity. Future developments should focus on large-scale testing, optimizing solid–liquid separation, and improving energy efficiency.

Second, smaller wastewater treatment plants with limited infrastructure have enhanced the production efficiency of methane and biofertilizers by employing photosynthetic biogas upgrading and bioelectrochemical methanation technologies. Photosynthetic biogas upgrading utilizes microalgae photosynthesis to fix CO_2 in the biogas, thus purifying it [42], while bioelectrochemical methanation applies voltage to facilitate CO_2 reduction and methane production [43]. However, due to the operational instability of these technologies, their scalabilities face significant challenges and the application remains limited to laboratory and small-scale settings, with industrial implementation requiring further research. Limitations include low current densities during large-scale applications, high capital and operational costs due to complex electrode and reactor designs, and sensitivity to operational parameters like pH and temperature. Additionally, issues such as biofouling, reduced efficiency in real wastewater environments, and difficulty maintaining electroactive microbial communities further constrain their industrial feasibility [44].

In wastewater treatment, anaerobic membrane bioreactors (AnMBRs), which integrate anaerobic digestion with membrane separation technology, overcome the limitations of traditional anaerobic reactors. This approach enables efficient treatment of urban sewage under low-temperature and low-organic-load conditions, achieving methane recovery rates as high as 99.5% [42]. Despite advances that have reduced the cost of membrane separation, the relatively low economic value of methane has driven wastewater treatment plants to explore its potential for producing high-value volatile fatty acids (VFAs). Current advancements in nanofiltration (NF), reverse osmosis (RO), and pervaporation (PV) membrane technologies have improved both the transmission rate and selectivity of VFAs [45]. Although AnMBRs can yield high-quality effluent by controlling hydraulic power and solid retention time [46], challenges remain, including membrane fouling, temperature fluctuations, and sulfate-induced corrosion, all of which require further resolution.

During anaerobic digestion, improper substrate composition, loading rates, or process parameters can lead to reaction interruptions, reduced biogas yields, or even complete process failure, resulting in economic losses for wastewater treatment plants. Studies indicate that adding substances such as ash, biochar, or zero-valent iron can help stabilize pH, create microbial habitats, and remove contaminants, thus optimizing reaction conditions and enhancing methane production [20]. Among these, biochar not only adsorbs inhibitory compounds like NH_3 and H_2S but also serves as a hydrogen transfer medium, promoting syntrophic metabolism and improving anaerobic digestion performance [47]. Biochar shows promising potential for application in small to medium-sized wastewater treatment plants, though its effectiveness on a larger scale requires further validation.

Hydrogen addition technology also holds promising applications. By adding H_2, hydrogenotrophic methanogens can convert H_2 and CO_2 into methane, thereby increasing the methane content in biogas [48]. Researchers at Aarhus University in Denmark developed a GasMix injection system to introduce H_2 into a full-scale thermophilic anaerobic reactor, achieving approximately a 45% increase in hydrogen conversion [49]. Although hydrogen conversion remains constrained by gas–liquid mass transfer, the successful operation of the GasMix system offers an effective solution for efficient and cost-effective hydrogen mass transfer in large-scale anaerobic digestion. Future studies can further optimize injector design and operational parameters to enhance performance.

Additionally, anaerobic digestion of single substrates tends to perform poorly in cold regions. To address this, wastewater treatment plants employ co-digestion technology, which improves the carbon-to-nitrogen ratio by mixing high-carbon substrates, such as crop residues, with high-nitrogen substrates in organic wastewater. This approach enhances the system's buffering capacity and helps prevent pH fluctuations. For example, in Tibet, China, studies have demonstrated that co-digesting wheat straw with wastewater sludge at a 9:1 ratio significantly increases biogas production at a low temperature of 15 °C, compared to the digestion of a single substrate [50]. Pretreatment techniques also play a crucial role in low-temperature anaerobic digestion. Methods such as using ultrasonic waves to break down organic particles or adjusting pH and moisture levels can improve substrate-microorganism contact, enhancing the digestion rate and methane yield [51]. In cold regions like Canada, the northern United States, and Northern Europe, various pretreatment technologies and reactor designs have been developed to sustain microbial activity under low temperatures, mitigating the adverse effects of cold on anaerobic digestion.

Table 2. The upgrading and progress of various anaerobic digestion technologies.

Upgrade Technology	Features	Advantages	Limitations	Application Prospects	Reference
Capacity Improvement	Optimization of pH, temperature, and retention time	Increased biogas yield, VFAs, and ammonia production	High capital investment; complex operation	Medium to small scale; experimental facilities	[40]
Reactor Combination	CSTR combined with UASB	10–30% increase in methane yield	Complex operation; higher costs	Small to medium scale; wastewater treatment plants	[41]
Phototrophic Bioreactor Upgrade	Microalgae-assisted purification	Increased methane purity	Difficult to scale up; high land area requirement	Small scale, decentralized; wastewater treatment	[42]
Bioelectrochemical Methanation	Generation of external voltage	Suitable for small-scale devices; current in experimental phase	System instability under high current densities	Laboratories and small scale applications	[43]
Membrane Bioreactor	Coupled membrane separation technology	Methane purity rate up to 99.5%	Membrane contamination; Temperature fluctuation effect	Medium to large scale; wastewater treatment	[42]
Membrane Application	Nanofiltration, reverse osmosis, and pervaporation	Increased VFAs production; mitigates membrane fouling issues	High maintenance costs	Small to medium scale wastewater treatment	[45,46]
Additives Technology	Addition of ash, biochar, etc.	pH regulation; inhibition of by-products; enhanced digestibility	Unproven on a large scale	Medium to small scale; experimental facilities	[20,47]
Ammonia Stripping	Injection of air and gas recirculation	Increased ammonia removal efficiency; improved mass transfer	Gas–liquid limitations; sensitive to operating conditions	Medium to large scale	[49]
Co-Substrate Technology	High-protein biomass co-digestion with organic substrates	Improved system stability and methane yield	Limited substrates in cold climates	Cold-climate regions; wastewater treatment plants	[50,52]

3.2. *Case Studies of the Current Situation in Various Regions of the World*

In Europe, stringent environmental regulations and supportive policies have driven the widespread adoption and advancement of anaerobic digestion technology, particularly in the treatment of sewage sludge and agricultural waste. As shown in Table 3, integrated anaerobic digestion and composting (IADC) is commonly used for organic waste treatment, with a focus on optimizing biogas productivity [53]. The EU's circular economy framework promotes legislation for waste recycling, encouraging the use of anaerobic digestion and composting technologies. While traditional wet anaerobic digestion still dominates large-scale organic waste treatment, small-scale anaerobic digestion (SSAD) has demonstrated significant potential for processing agricultural waste and organic sewage sludge. In Germany, for example, more than 560 small biogas plants have been established since 2016, generating an annual output of 271.3 GWh of electricity. This success is largely due to the Renewable Energy Act's Feed-in Tariff (FIT) policy, which provides subsidies for electricity produced by small biogas facilities [54]. In France, garage-type digesters are widely used for processing food waste and municipal wastewater in urban and suburban areas, while in

Italy, policy incentives support the development of biogas facilities to provide sustainable energy solutions for rural and local communities [55,56].

In the United States, upgrades to anaerobic digestion technology have focused on transforming wastewater treatment plants into water resource recovery facilities (WRRFs). Implementing real-time control for a 35% reduction in N_2O emissions, combined with electrolytic hydrogen production and pre-filtration technology, has enhanced organic matter collection efficiency and biogas production, thereby reducing reliance on fossil fuels [57]. Anaerobic digestion technology has also been effectively applied in large farms and slaughterhouses, leading to significant reductions in greenhouse gas emissions. For instance, large dairy farms have cut emissions by 2.45 to 3.52 million tonnes of CO_2-eq annually, with additional reductions achieved through cogeneration [58].

In Asia, particularly in China, anaerobic digestion technology has rapidly advanced under the impetus of the "dual carbon" policy. A successful example is seen in Weifang, Shandong Province, where biogas upgrading technology enables the conversion of 86,000 cubic meters out of 150,000 cubic meters of daily biogas production into biomethane for local use [59]. This biomethane, with chemical properties similar to natural gas, can be directly injected into the natural gas network, providing a pathway for the seamless integration of renewable energy with existing energy infrastructure. This approach not only reduces dependence on fossil fuels but also enhances energy security and supports decarbonization efforts. In the Nan'an community of Wuwei, Gansu Province, the central gas supply system maintains anaerobic digestion temperatures through biomass combustion, achieving a thermal efficiency of 86.5% [60]. In China, anaerobic digestion technology is transitioning from small-scale household applications to large-scale industrial systems, with significant contributions from large biogas plants expected to drive national biogas production by 2020 [61]. In this context, leveraging existing natural gas distribution networks for biomethane injection can facilitate efficient energy distribution, reduce greenhouse gas emissions, and establish a circular economy model that maximizes the value of organic waste. Simultaneously, it supports national energy and environmental goals, promoting sustainable development and resource efficiency.

Moreover, in other Asian countries like India, Nepal, and Vietnam, governments and NGOs are actively promoting household anaerobic digestion systems. In India, financial incentives and loans from the government enabled rural residents to install over 90,000 home-grade biogas digesters between 2017 and 2021, significantly improving rural sanitation [62]. In Indonesia, the implementation of decentralized wastewater treatment systems (DEWATS) combined with anaerobic digestion technology has successfully addressed weak wastewater treatment infrastructure, providing effective wastewater treatment and biogas production solutions to more than 15,000 communities [63].

Table 3. Case studies of anaerobic digestion technology development in different regions.

Region	Policy Support	Upgrade Status	Technology Application	Application Scenarios	References
Germany	Renewable Energy Law, small-scale electricity subsidy	Small-scale anaerobic digestion technology (SSAD)	Applied in agricultural waste treatment, supports agricultural economy	Agricultural waste treatment	[54]
France	Policy for small-scale biogas plant construction	Vehicle-scale digesters	Applied in urban organic waste for electricity and heat production	Urban organic waste treatment	[55]
Italy	2030 Biofuel Target	Biofuel technology	Processes organic waste, supports rural circular economy	Agricultural waste treatment	[56]
USA	WRRF facility transformation support	Transition to resource recovery facilities	Increases biogas yield, reduces GHG emissions	Large wastewater plants, landfill sites	[57,58,64]
China	Dual carbon policy	Industrial-scale development from small to large	Widespread biomass heating systems, increased gas contribution	Wastewater treatment, livestock waste management	[59–61]

Table 3. *Cont.*

Region	Policy Support	Upgrade Status	Technology Application	Application Scenarios	References
India	Government incentives and loans	Household-scale biogas systems	Household biogas plants improve rural sanitation	Agricultural and household waste treatment	[62]
Indonesia	Local government and NGO promotion	Decentralized wastewater treatment (DEWATS)	Reduction in organic pollutants, daily gas production over 20 m^3	Agricultural and residential areas	[63]
Nepal, Vietnam	NGO promotion of household biogas systems	Household-scale anaerobic digestion	Improved sanitation, supports agricultural resource recycling	Agricultural and household waste treatment	[62]
Brazil	Renewable Energy Law, National Solid Waste Policy	Large-scale landfill and co-digestion facilities	67% recycling rate with installed capacity of 835 MW, meeting 2.8% of national demand	Landfills, agricultural waste treatment	[65]
Chile	Policy support for traditional energy alternatives	Solid waste treatment technology	Reduce urban landfill emissions, supports energy recovery	Urban and agricultural waste treatment	[66]
Argentina	Decentralized renewable energy regulation	Co-digestion facilities	Annual generation of 6387 MWh, GHG emission reduction by 44.1–70.5%	Urban regions, electricity production	[67]

4. The Promoting Role of Anaerobic Digestion Technology in Circular Economy

4.1. Potential for Energy Self-Sufficiency and Carbon Reduction

Anaerobic digestion technology plays a crucial role in advancing the circular economy. As shown in Table 4, anaerobic digestion enables the conversion of sewage sludge into biogas, supports energy self-sufficiency through cogeneration, reduces dependence on fossil fuels, and facilitates the transformation of wastewater treatment plants into resource recovery facilities, thus lowering their carbon footprint. In Europe, policy support and technological advancements have positioned anaerobic digestion as a global leader in energy self-sufficiency and carbon reduction within wastewater treatment. For instance, Austria's Wolfgangsee-Ischl and Strass wastewater treatment plants produce 3 GWh of biogas annually using highly efficient combined heat and power (CHP) equipment, achieving complete energy self-sufficiency by 2008, with surplus electricity available for sale [68]. In Germany, the Grevesmühlen and Kohlbrandhoft wastewater treatment plants employed co-digestion technology to increase electricity production by 15% and generate a 20% energy surplus [69]. Additionally, Dutch dairy farms convert livestock manure and sewage sludge into energy through co-digestion and CHP systems, cutting their carbon footprint by 88–92% and reducing operational costs by 26% [70].

In the Americas, particularly in the United States, anaerobic digestion technology enables wastewater treatment plants to achieve energy self-sufficiency and recycle waste. For example, the East Bay Municipal Utility District (EBMUD) generates over $2 million annually from excess electricity produced through anaerobic digestion. At the Des Moines wastewater treatment plant, co-digestion technology has increased biogas production by 40%–50%, allowing for profitable biogas sales [71]. In Latin America, anaerobic digestion is widely applied on dairy farms, where it reduces greenhouse gas emissions by an estimated 32.8 million tonnes of CO_2-eq per year [72]. In Mexico, municipal wastewater treatment plants achieve 50%–60% energy self-sufficiency through anaerobic digestion, generating an annual output of 34 GWh of biogas [73].

In Asia, anaerobic digestion technology is aiding agriculture and livestock sectors in reducing environmental impacts and advancing the shift toward a circular economy. For example, in BB Nagar district, India, anaerobic digestion converts livestock and poultry waste into 10,637 kW of electricity, reducing CO_2 emissions by approximately 51,956.4 tonnes annually [74]. In China, 31 wastewater treatment plants in Guangzhou have achieved regional energy self-sufficiency, cutting electricity costs by 80%, reducing CO_2 emissions by about 37.5 tonnes per day, and generating substantial benefits in the carbon trading market. By 2050, carbon credit revenues in the region are projected to reach 30 million yuan [75].

Table 4. Potential of anaerobic digestion technology in energy self-sufficiency and carbon reduction.

Region	Application Field	Energy Self-Sufficiency Potential	Carbon Reduction Capacity	Representative Case	Reference
Europe	Wastewater Treatment Plant	100% power self-sufficiency, over 10% surplus energy	Reduces CO_2 by over 10%	Wolfgangsee-Ischl WWTP, Austria	[68]
	Wastewater Treatment Plant	50.5% average energy recovery; 1.54 kWh per m^3 biogas	Reduces methane and N_2O emissions	Krosno WWTP, Poland	[76]
	Co-digestion	Electricity increases by 15%; 20% energy surplus	Reduces CO_2 emissions by 83–92%	Grevenmuehlen and Kohlbrand WWTP, Germany	[69]
	Dairy Farm	CHP systems achieve 100% energy self-sufficiency	Environmental impact reduced by 83–87%	Dutch dairy farm Annual cost savings of 26%	[70]
North America	Wastewater Treatment Plant	Surplus power generation generates 200 million USD annually	Significant carbon reduction	EBMUD WWTP, USA	[71]
	Co-digestion	Biogas yield increased by 40%; 50% biogas sold	Reduces methane and other GHGs	Des Moines WWTP, USA	[77]
	Dairy Farm	Energy self-sufficiency through bioenergy and heat recovery	60% reduction in methane and nitrous oxide emissions	Latin American dairy farm Annual reduction of 32.8 m tons of CO_2-eq	[72]
	Manure Processing	Over 20% energy surplus	Annual CO_2 reduction: 124 Mt	Annual reduction in Brazil agricultural waste about 12% of total emissions	[78]
	Municipal Wastewater Treatment Plant	Annual biogas production: 340 GWh	Reduces methane and CO_2 emissions	Mexico WWTP 50–60% energy self-sufficiency rate	[73]
Asia	Manure Processing	Power conversion over 1063 kW	Annual CO_2 reduction: 51,596 tons	BB Nagar, India Reduce dependence on fossil fuels	[74]
	Wastewater Treatment Plant	Daily biogas production: 390 kWh	Reduces GHG emissions from energy and fuel usage	Anaerobic digestion and photovoltaic system combination in South Korea Jeju Island WWTP	[79]
	Municipal Wastewater Treatment Plant	Electricity Cost reduction by 8%	Reduces 37.5 tons of CO_2-eq About 12.56% of total emissions	31 WWTP in Guangzhou, China Carbon markets generate revenue	[75]

4.2. Economic Role in Resource Recovery and Waste Management

As shown in Table 5, the contribution of anaerobic digestion technology to a circular economy is evident in its effective waste management and resource recycling capabilities [8,9,12]. By converting agricultural and livestock waste as well as sewage sludge into biogas and organic fertilizers [80–83], anaerobic digestion technology not only reduces the cost of traditional fertilizers but also enhances soil and environmental quality, supporting resource reuse [84]. Furthermore, processing digestate allows for the production of high-value materials, such as carbon-based compounds, nanofibers, and bio-oils, which find applications in energy storage and pollutant adsorption [85–87]. These materials not only improve economic efficiency but also foster sustainable development of the circular economy in sewage treatment and waste management.

In emerging economies, anaerobic digestion technology has shown substantial economic benefits, driving sustainable green economic growth. For instance, in Poland, approximately 34% and 30% of sludge are managed through agricultural applications and thermal conversion technologies, respectively [88]. Research indicates that sludge gasification technology offers significant economic advantages within the circular economy, providing a net benefit of up to $3.5 million for small, decentralized wastewater treatment plants in Poland and reducing landfill costs. Additionally, the bio-oil generated through rapid sludge pyrolysis and the hydrogen yield during gasification, reaching 72.8–82.9%, further enhance energy recovery efficiency [89]. At the ZGO Gac plant in Poland, over 100,000 tonnes of solid waste are processed annually using anaerobic digestion and combined heat and power (CHP) systems, generating 6000 MWh of electricity—one-third of which is sold—resulting in a net surplus of €180,000. This demonstrates the critical role of anaerobic digestion technology in supporting the circular economy [90].

Since Croatia joined the European Union in 2013, it has introduced anaerobic digestion technology to treat municipal solid waste in order to meet EU waste management targets. The city of Zagreb collects about 270,000 tons of solid waste every year, of which 13.46% is converted into biogas through anaerobic digestion for power generation and heating, with an investment of 710,000 euros and an annual power generation income of about 1 million euros. At the same time, the cost of fertilizer is saved by about 500,000 euros per year [91]. In addition, the technology reduces greenhouse gas emissions by 43.59 tons of carbon dioxide equivalent per year [92], bringing Croatia a double advantage in terms of environmental and economic benefits. In Africa, despite slower overall economic and technological development, several rapidly urbanizing countries and regions have successfully implemented anaerobic digestion technology. For example, in South Africa's Western Cape Province, anaerobic digestion produces 100–150 m^3 of biogas per tonne of waste treated, reduces CO_2 emissions by 1 tonne, and enhances energy self-sufficiency [93]. In eThekwini, an anaerobic digestion project set to launch in 2027 aims to treat 13.1% of the city's total waste by 2035, potentially saving over R3.8 billion in treatment costs [94] and further advancing the circular economy.

In general, based on the analysis of anaerobic digestion technology adopted by various countries in Table 5, most countries achieve economic benefits through energy savings, waste conversion into electricity or biogas, and reductions in waste management costs. Additionally, many countries have successfully reduced GHG emissions and pollutants, contributing to sustainable development. In terms of resource recovery, the technology effectively transforms waste into valuable resources, thereby fostering a circular economy. The factors influencing these outcomes include the country's economic context, environmental policies, technological innovations, and energy demand. Developed countries and those with advanced infrastructure tend to reap higher economic and environmental benefits, which are further enhanced by supportive policies and technological advancements.

Table 5. The economic role of anaerobic digestion technology in resource recovery and waste management.

Region	Technology Application Type	Resource Recovery	Economic Benefits	Environmental Impact	Reference
Poland	Municipal Sludge Anaerobic Digestion	Hydrogen concentration 72.8–82.9%	Net income: USD 350,000 for small wastewater plants	Reduces sludge volume; 0.55–1.5% phosphorus recovery	[88–90]
	Municipal solid waste	Generates 6000 MWh of electricity; one-third sold	Annual revenue: USD 1.8 million	Provide organic fertilizer for agriculture, Reduce fertilizer costs, Reduce methane emissions	[90,95]
Mexico City	Biological pretreatment MBP	Consumes approx. 387 MJ fossil/Mg of energy	Annual savings of USD 207.6 in waste management costs	Reduces GHG emissions by 148 kg-CO_2e/Mg	[96]
Brazil	Two-Stage Membrane Bioreactor	Organic matter removal rate: 96.9%; methane concentration: 70.1%	Production increase: 25.7%, COD removal efficiency	Reduced water and soil pollution	[97]
USA	Waste Heat Energy Recovery	Recovers 20–30% of energy from waste heat	Annual savings on heating costs 20–30%	Reduce demand for fossil fuels, Reduce greenhouse gas emissions by 30%	[98]

Table 5. *Cont.*

Region	Technology Application Type	Resource Recovery	Economic Benefits	Environmental Impact	Reference
China	Food Waste Treatment	Biogas production: 130 m^3/ton	Annual revenue of RMB 20 million	Reduces solid waste by 46.8 kg CO_2-eq per ton	[99]
Norway	Reduces solid waste by 46.8 kg CO_2-eq per ton	Acid treatment for methane production; 42–50% yield	Reduces processing costs by 30%	Improves nutrient separation and reduces pollution	[100]
Germany	Municipal Solid Waste	100 kW waste-to-energy plant; 27.1 m^3 waste treated annually	Significant reductions in energy spending from fossil fuels	Reduces fossil fuel consumption by replacing 620 kg of CO_2-eq emissions	[101]
Croatia	Municipal Solid Waste	Annual electricity revenue: approx. EUR 1 million	Annual fertilizer cost savings: EUR 500,000	Reduces 43.59 tons CO_2 emissions; decreases reliance on agricultural fertilizers	[91,92]
South Africa	Municipal Waste Treatment	100–150 million m^3 annual biogas production	Estimated savings of ZAR 3.8 billion by 2035	Reduces 1 ton of CO_2 emissions; decreases reliance on traditional energy sources	[93,94]

5. The Development Trend of Anaerobic Digestion Technology

5.1. Emerging Technologies and Intelligent Applications

Anaerobic digestion technology is evolving alongside advancements in science and technology. As shown in Table 6, new processes such as temperature-phased anaerobic digestion (TPAD) and autogenic high-pressure digestion (AHPD) significantly boost biogas production and methane concentration [102,103]. Additionally, the implementation of forward transport membrane (FTM) technology and microalgae CO_2 capture has proven effective in enhancing methane purity and yield [69,104]. Pretreatment technologies, including wet oxidation and steam explosion, greatly improve the biodegradability of lignocellulosic materials [105], while low-temperature condensation enables the co-production of high-purity methane and carbon dioxide [106]. Furthermore, integrating anaerobic digestion with biorefinery technologies allows for the conversion of organic biomass into a diverse range of green energy and chemical products [107].

Digital and intelligent technologies have significantly advanced wastewater treatment, particularly in anaerobic digestion, enhancing both process accuracy and efficiency [108]. Artificial neural networks (ANN) and machine learning algorithms, such as random forests and support vector machines (SVM), optimize biogas production and facilitate predictive maintenance, achieving efficiency gains of up to 15% [109,110]. Additionally, the use of fuzzy logic systems and Internet of Things (IoT) technologies optimizes feeding systems and equipment maintenance, further improving overall operational efficiency [111,112]. The integration of life cycle assessment (LCA) tools has led to a 20–30% reduction in greenhouse gas emissions from wastewater treatment plants [113], providing substantial support for sustainable development.

Overall, the integration of artificial intelligence (AI) and the Internet of Things (IoT) enhances decision-making capabilities through advanced data analysis and automated systems, enabling operators to dynamically adjust feeding strategies and proactively resolve inefficiencies. This integration also minimizes downtime and maintenance costs, significantly improving operational performance. Collectively, these technologies provide transformative potential for predictive maintenance and process optimization, enhancing the efficiency and reliability of anaerobic digestion systems while aligning with the objectives of the circular economy.

Table 6. Development of anaerobic digestion technology in emerging technologies and intelligent applications.

Application Field	Technology Type	Key Performance Indicators	Application Scenarios	Technical Challenges	Reference
Energy Production	Temperature Phased Anaerobic Digestion (TPAD)	Methane yield increase of 9.6%	Various organic waste	Complex temperature control, energy demand	[102]
	Self-pressurized Gasification	Methane purity up to 82%, low-grade heat utilization	Various gasification reactors	High equipment manufacturing cost, complex pressure control	[103]
Methane Enrichment	Membrane Transport Enhancement	Improved methane purity	Biogas upgrading systems	High cost of nanoporous membrane materials	[104]
Methane Yield Increase	Microalgae Cultivation with CO_2	Methane yield up to 92%, removes ammonia	Waste treatment facilities	Nutrient demand additional environmental control	[69]
Agricultural Waste Treatment	Steam Explosion Pretreatment	Enhanced methane production	Agricultural and organic fertilizers	High equipment cost, complex pretreatment process	[105]
Wastewater and Sludge Treatment	Low-Temperature Liquefaction	Methane purity up to 99%	Wastewater treatment plants	Requires large-scale cooling infrastructure	[106]
Biomass Resource Utilization	Bio-refining and Anaerobic Digestion Integration	Comprehensive production of ethanol, bio-oil, syngas	Agricultural waste (e.g., sugarcane)	Complex industrial processes, high energy input	[107]
Process Control and Monitoring	Artificial Neural Networks (ANN)	Production efficiency increase by 20%	Biogas production systems	High sensor demand, complex model training	[110]
Optimization Maintenance and Prediction	SVM and Random Forest	Biogas yield increase of 15%, reduced maintenance costs	Large biogas plants	Requires extensive historical data for training	[109]
Biogas Yield Prediction	Random Forest Algorithm	Optimized biogas production forecast	Large biogas plants	High computational complexity, increased reliability	[110]
Feed System Optimization	Fuzzy Logic	Biogas yield increase of 10%, reduced labor dependency	Various biogas production systems	Complex algorithm adjustments, high transmission accuracy requirement	[111]
Life Cycle Assessment	Integrated LCA Model	30% reduction in GHG emissions, improved resource utilization	Wastewater treatment plants	Complex data model construction, extensive data collection time	[113]
Remote Monitoring and Maintenance	Internet of Things IoT	Enhanced data centralization, increased efficiency and operational rate	Large biogas plants, remote facility management	High system setup cost, need for continuous maintenance updates	[112]

5.2. Policy and Market Drivers Factors

Figure 3 illustrates the main evolution of the European biogas industry from 2020 to 2050, highlighting a gradual shift toward biomethane production and carbon appreciation. This evolution underscores the rapid development of anaerobic digestion technology, propelled by both policy and market forces. On the policy front, governments have introduced incentives to encourage the adoption and expansion of this technology. For instance, the European Green Deal and the EU's Long-Term Climate Strategy emphasize enhancing resource efficiency and promoting a circular economy through anaerobic digestion [114]. EU waste management policies, including the Landfill Directive and the Waste Framework Directive, also support anaerobic digestion's role in reducing landfill use and pollutant emissions [115].

Specific national policies have further spurred market growth. In France, the Energy Transition Act and the Multi-Annual Energy Plan have strengthened the biogas sector, aiming for biogas to comprise 30% of natural gas consumption by 2030 [116]. In Sweden, the Climate Investment Program promotes the biogas market with the goal of making the transport system fossil-fuel-free by 2030 [117]. Similarly, in Brazil, the Biogas and Biomethane State Policy outlines the rights and obligations of stakeholders, fostering greater market integration [118].

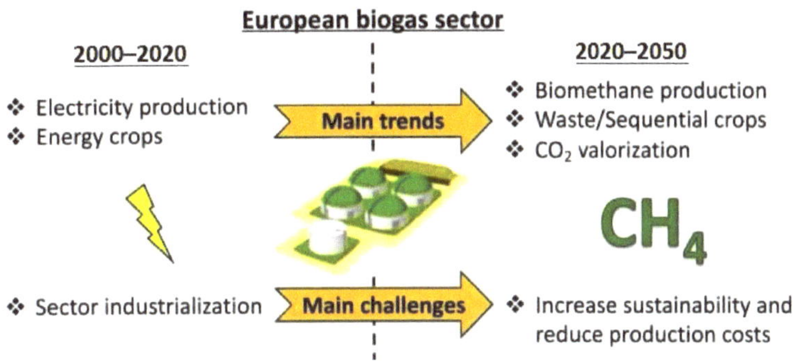

Figure 3. Development trend of European biogas industry 2000–2050 [114].

Market drivers are also accelerating the advancement of anaerobic digestion technologies. In Poland, rising global grain prices have led operators to shift towards using food industry and municipal waste as feedstocks, thereby enhancing economic efficiency and reducing competition with food resources [119]. In the Czech Republic, biofuel use is encouraged through subsidies and tax incentives [120]. In China, abundant supplies of livestock manure and crop residues serve as substantial feedstocks for anaerobic digestion, with large-scale biomethane and biogas projects expected to deliver significant economic benefits [43]. Together, these policy and market factors have fostered a supportive environment for the growth of anaerobic digestion technology, advancing resource recovery and the circular economy.

6. Challenges and Obstacles

6.1. Technical Bottlenecks and Economic Feasibility Challenges

Anaerobic digestion technology faces several technical and economic challenges in practical applications. As shown in Table 7, on the technical side, high ammonia nitrogen concentrations and volatile fatty acid accumulation can inhibit microbial activity, adversely affecting methane production [121]. While mature ammonia removal technologies exist, nitrogen and phosphorus recovery efficiencies remain below optimal levels, and the use of membrane technologies is constrained by clogging issues [122]. Additionally, untreated high concentrations of ammonia nitrogen and heavy metals can lead to environmental pollution, limiting the feasibility of agricultural applications [123]. Generally, the main technical challenge faced by anaerobic digestion technology in practical application is generally the environmental pollution and blockage caused by the immaturity of various technologies.

At the economic level, the high costs of nutrient recovery and raw material pretreatment add to operational expenses [123–127]. Although two-stage anaerobic digestion can enhance energy recovery, it requires additional equipment and energy input, increasing costs [128]. Furthermore, low market demand for biogas persists due to the competitive pricing of traditional fertilizers and fossil fuels, limiting commercial expansion [129]. Therefore, the main economic challenges encountered in the application process of anaerobic digestion technology include an increase in costs and a decrease in demand. Instability in methane production also directly affects economic viability, with reduced gas yields leading to lower energy recovery [130]. Existing subsidy policies primarily support project construction but lack operational subsidies, making it challenging for some biogas projects to maintain profitability [131], which is an additional significant concern that cannot be overlooked.

Additionally, the challenges encountered by different countries in the application of anaerobic digestion technology vary based on their technological development, economic conditions, and policy environment. Developed countries generally face fewer challenges related to equipment and technology maintenance; however, they still struggle with marketing difficulties and the lack of operational subsidies, largely due to the competitive pricing of fossil fuels and insufficient market demand for green energy. In contrast, developing countries face more severe technical challenges, such as membrane fouling and lower ammonia nitrogen recovery efficiency, while the high costs associated with pretreatment and dual-stage systems exacerbate the economic burden. For countries with significant agricultural needs, there is a heightened focus on mitigating the environmental issues caused by nitrogen and heavy metal pollution, which directly impact soil health and agricultural productivity. Overall, the performance of technological and economic challenges varies considerably across countries, shaped by a combination of national infrastructure, economic development levels, and policy support.

Table 7. The technical bottleneck and economic feasibility challenge of anaerobic digestion.

Field	Technical/Economic Category	Challenges	Impact	Solutions	Reference
Technical Challenges	Substrate Processing Efficiency, pH Value Control	Excessive ammonia concentration, accumulation of volatile fatty acids	Reduced gas yield, unstable fermentation process	Optimize substrate ratio, adjust pH of feedstock	[121]
	Nitrogen and phosphorus recovery technology	Membrane separation process Fouling problem	Reduced membrane lifespan, increased operational maintenance costs	Develop fouling-resistant membrane materials, enhance cleaning efficiency	[122]
	Ammonia recovery technology	Low maturity of membrane contactor technology	Increased operational and management complexity, higher system maintenance costs	Enhance membrane contactor, implement automatic control	[132]
	Environmental Impact Control	Untreated Nitrogen and Heavy Metals, Environmental and soil pollution	Soil degradation, impact on agriculture and environment	Improve nitrogen recovery efficiency, enhance heavy metal recovery technologies	[123]
Economic Challenges	Pretreatment Costs	High time and equipment investment in pretreatment process	Increased operational costs	Optimize pretreatment techniques, reduce equipment investment	[125]
	Dual-Stage System Costs	High costs for dual-stage gasification and energy system	System costs and energy consumption increase, Reduce in economic efficiency	Increase single-stage system efficiency, reduce equipment costs	[128]
	Methane Yield Instability	Fluctuating methane yield impacts economic benefits	Unstable system energy recovery, Reduced economic efficiency	Develop stable microbial agents, optimize operating conditions	[130]
	Market Promotion Difficulty	Immature bio-fertilizer and biofuel market; insufficient demand	Hindered commercial promotion of green fuels	Increase policy support, promote green fuel applications	[129]
	Lack of Operational Subsidies	Existing subsidy policies lack long-term support	Reduced economic viability, difficult to maintain operations	Government to increase long-term operational subsidies	[131]

6.2. Regulatory Challenges and Public Acceptance Issues

In terms of policy and regulation, regulatory differences in the application of anaerobic digestion technology in different countries limit the uniform implementation of the technology and affect the efficiency of cross-border trade and digestion management. While

Germany and the United Kingdom have developed standards (such as PAS-110) to regulate heavy metals and organic pollutants in digesters, many countries still lack similar regulations, which may limit the use of digesters as agricultural fertilizers [133]. In addition, the lack of global harmonized digestion regulations makes cross-border agricultural trade challenging. For example, EU restrictions on mechanically separated organic waste may not apply to other countries, affecting international trade in digestives [134]. Despite the potential environmental and economic benefits of anaerobic digestion in agriculture, the impact on the environment and human health needs to be supported by more research. Currently, the tools used to assess digestive risk (such as life cycle assessment and ecological risk assessment) are not widely available in many countries, which affects government and industry assessments of the quality and safety of disinfectants [36]. At the same time, as shown in Figure 4, as anaerobic digestion is an emerging technology, the EU has adopted the process of formulating policies through expert opinions, evidence review and scientific suggestions, which shows that the multi-layered regulatory system leads to poor information transmission, especially the lack of coordination among government departments, resulting in inefficient policy implementation [135].

In terms of public acceptance, limited awareness of anaerobic digestion technology often leads to concerns about the composition and safety of residual compounds [136]. Project size can also influence acceptance, with smaller facilities generally gaining easier support, whereas larger projects may encounter resistance due to external investment involvement and potential impacts on the local area [137]. Additionally, the "NIMBY" (Not in My Backyard) effect is particularly notable in the construction of large anaerobic digestion facilities. While the public may broadly support renewable energy, there is often opposition to facilities being located close to residential areas [138]. For instance, in Germany, the clustered construction of anaerobic digestion facilities has sometimes met with resident opposition. Conversely, in France, despite facility visibility, some residents perceive these installations as compatible with the rural landscape, making them more accepting.

In summary, the key to overcoming the "NIMBY" effect lies in enhancing community engagement and transparency by involving the public in decision-making from the early stages of the project and providing regular updates to build trust. Simultaneously, educational and outreach efforts should be undertaken to comprehensively communicate the environmental and economic benefits of anaerobic digestion facilities. Additionally, optimizing facility siting and design, such as incorporating greenery or adopting inconspicuous designs, can help minimize visual and environmental impacts. Providing economic incentives, such as sharing project revenues or offering free organic fertilizers to the local community, further ensures direct benefits to residents, fostering greater acceptance and support for the project.

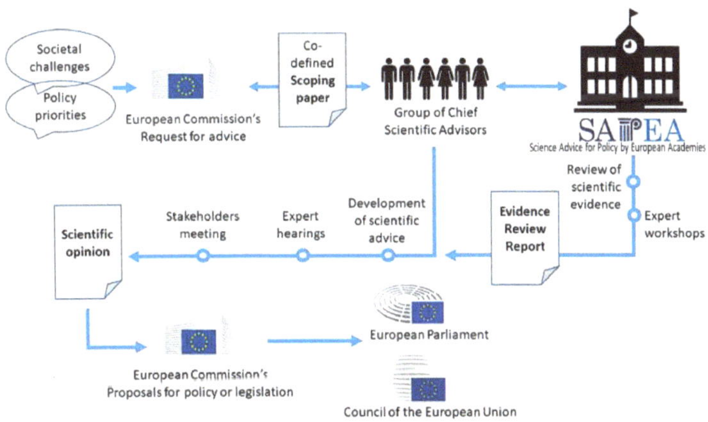

Figure 4. EU policy formulation and proposed adoption path [139].

7. Conclusions and Prospects

This review systematically examines the application of anaerobic digestion technology in wastewater treatment and its role in promoting the circular economy. As a sustainable approach for converting organic waste into biogas and organic fertilizers, anaerobic digestion holds significant potential to reduce emissions and energy consumption while enhancing energy self-sufficiency and resource recycling in wastewater treatment plants, thus becoming a key strategy in achieving the "dual carbon" goal. Although technology is relatively mature in developed regions, developing countries continue to face challenges related to limited financial and technical support; nonetheless, anaerobic digestion technology demonstrates broad prospects for growth.

Against the backdrop of the "dual carbon" goal, the future of anaerobic digestion technology presents both opportunities and challenges. Driven by climate policies and the imperative to reduce carbon emissions, advancements in co-digestion, multi-stage anaerobic digestion, and intelligent control methods promise to enhance efficiency and economic benefits. Policy incentives and market forces are also accelerating technological progress, with biogas markets and carbon trading systems improving economic viability. However, technical challenges remain, including digestate treatment, ammonia nitrogen inhibition, and volatile fatty acid accumulation, as well as the challenge of balancing capital investment with financial returns. Future research should prioritize improving technological stability, reducing costs, and refining policy support. Increasing public awareness of environmental protection and fostering social acceptance of anaerobic digestion projects are also crucial.

In conclusion, as a vital technology for promoting the circular economy and green development, anaerobic digestion will play an essential role in the global energy transition and in achieving carbon neutrality. Policymakers should advance supportive policies, standardization, and international cooperation, while the wastewater treatment industry must focus on technology integration and intelligent systems to achieve the Sustainable Development Goals.

Funding: This research received no external funding.

Data Availability Statement: Data sharing not applicable.

Conflicts of Interest: The author declares no conflicts of interest.

References

1. Moyer, J.; Hedden, S. Are we on the right path to achieve the sustainable development goals? *World Dev.* **2020**, *127*, 104749. [CrossRef]
2. Liu, W.; McKibbin, W.J.; Morris, A.C.; Wilcoxen, P.J. Global economic and environmental outcomes of the Paris Agreement. *Energy Econ.* **2020**, *90*, 104838. [CrossRef]
3. Zhou, S.; Tong, Q.; Pan, X.; Cao, M.; Wang, H.; Gao, J.; Ou, X. Research on low-carbon energy transformation of China necessary to achieve the Paris agreement goals: A global perspective. *Energy Econ.* **2021**, *95*, 105137. [CrossRef]
4. Zhang, S.; Chen, W. China's energy transition pathway in a carbon neutral vision. *Engineering* **2021**, *14*, 64–76. [CrossRef]
5. Maktabifard, M.; Hazmi, A.H.; Szulc, P.; Mousavizadegan, M.; Xu, X.; Zaborowska, E.; Li, X.; Mąkinia, J. Net-zero carbon condition in wastewater treatment plants: A systematic review of mitigation strategies and challenges. *Renew. Sustain. Energy Rev.* **2023**, *185*, 113638. [CrossRef]
6. Ren, Z.Y.; Schnoor, J.; Pagilla, K. *Toward a Net Zero Circular Water Economy*; IWA Publishing: London, UK, 2022; Volume 1, pp. 1–13.
7. Nabaterega, R.; Kumar, V.; Khoei, S.; Eskicioglu, C. A review on two-stage anaerobic digestion options for optimizing municipal wastewater sludge treatment process. *J. Environ. Chem. Eng.* **2021**, *9*, 105502. [CrossRef]
8. He, X.M.; Zhang, T.; Ren, H.Q.; Li, G.X.; Ding, L.L.; Pawlowski, L. Phosphorus recovery from biogas slurry by ultrasound/H_2O_2 digestion coupled with HFO/biochar adsorption process. *Waste Manag.* **2017**, *60*, 219–229. [CrossRef]
9. Zhang, T.; Wu, X.S.; Shaheen, S.M.; Zhao, Q.; Liu, X.J.; Rinklebe, J.; Ren, H.Q. Ammonium nitrogen recovery from digestate by hydrothermal pretreatment followed by activated hydrochar sorption. *Chem. Eng. J.* **2020**, *379*, 122254. [CrossRef]
10. Velenturf, A.; Purnell, P. Principles for a sustainable circular economy. *Sustain. Prod. Consum.* **2021**, *27*, 1437–1457. [CrossRef]
11. Barros, M.; Salvador, R.; Prado, F.G.; Francisco, C.A.; Piekarski, C.M. Circular economy as a driver to sustainable businesses. *Clean. Environ. Syst.* **2020**, *2*, 100006. [CrossRef]

12. Xu, Q.; Zhang, T.; Niu, Y.Q.; Mukherjee, S.; Elwafa, A.S.F.; Nguyen, N.S.H.; Aboud, A.N.M.; Wang, Y.K.; Pu, M.J.; Zhang, Y.R.; et al. A comprehensive review on agricultural waste utilization through sustainable conversion techniques, with a focus on the additives effect on the fate of phosphorus and toxic elements during composting process. *Sci. Total Environ.* **2024**, *942*, 173567. [CrossRef] [PubMed]
13. Liu, G.; Xu, Q.; Elwafa, A.S.F.; Alshehri, A.M.; Zhang, T. Hydrothermal carbonization technology for wastewater treatment under the "dual carbon" goals: Current status, trends, and challenges. *Water* **2024**, *16*, 1749. [CrossRef]
14. Ali, F.; Ali, N.; Ahmad, I.; Khurshid, R.; Javed, M.; Mulla, S.; Bilal, M.; Iqbal, H.M.N. *Sustainable Management of Municipal Solid Waste to Fuel: An Overview for a Better Tomorrow*; Elsevier: Amsterdam, The Netherlands, 2021; Volume 2, pp. 289–314.
15. Sanaye, S.; Mohammadi, M.; Yazdani, M.; Rashvanlou, R.B. Bio-gas augmentation and waste minimization by co-digestion process in anaerobic digestion system of a municipal waste water treatment plant. *Energy Convers. Manag.* **2022**, *268*, 115989. [CrossRef]
16. Baştabak, B.; Koçar, G. A review of the biogas digestate in agricultural framework. *J. Mater. Cycles Waste Manag.* **2020**, *22*, 1318–1327. [CrossRef]
17. Niemiec, M.; Sikora, J.; Sikora, S.A.; Szostak, G.Z.; Komorowska, M. Assessment of the Possibilities for the Use of Selected Waste in Terms of Biogas Yield and Further Use of Its Digestate in Agriculture. *J. Mater.* **2022**, *15*, 988. [CrossRef] [PubMed]
18. Barampouti, E.M.; Mai, S.; Malamis, D.; Moustakas, K.; Loizidou, M. Exploring technological alternatives of nutrient recovery from digestate as a secondary resource. *Renew. Sustain. Energy Rev.* **2020**, *134*, 110379. [CrossRef]
19. Ngo, T.; Ball, A.S.; Shahsavari, E. The current status, potential benefits and future prospects of the Australian biogas sector. *J. Sustain. Bioenergy Syst.* **2021**, *11*, 14–32. [CrossRef]
20. Uddin, M.; Wright, M. Anaerobic digestion fundamentals, challenges, and technological advances. *Phys. Sci. Rev.* **2022**, *8*, 2819–2837. [CrossRef]
21. Kunatsa, T.; Xia, X. A review on anaerobic digestion with focus on the role of biomass co-digestion, modelling and optimisation on biogas production and enhancement. *Bioresour. Technol.* **2021**, *344*, 126311. [CrossRef] [PubMed]
22. Roopnarain, A.; Rama, H.; Ndaba, B.; Bello-Akinosho, M.; Bamuza-Pemu, E.; Adeleke, R. Unravelling the anaerobic digestion 'black box': Biotechnological approaches for process optimization. *Renew. Sustain. Energy Rev.* **2021**, *152*, 111717. [CrossRef]
23. Ampese, L.; Sganzerla, W.; Ziero, D.D.H.; Mudhoo, A.; Martins, G.; Carneiro, F.T. Research progress, trends, and updates on anaerobic digestion technology: A bibliometric analysis. *J. Clean. Prod.* **2022**, *331*, 130004. [CrossRef]
24. Yang, X.; Liu, Y.; Thrän, D.; Bezama, A.; Wang, M. Effects of the German renewable energy sources act and environmental, social and economic factors on biogas plant adoption and agricultural land use change. *Energy Sustain. Soc.* **2021**, *11*, 6. [CrossRef]
25. Grando, L.R.; Antune, S.A.M.; Fonseca, F.V.; Sánchez, A.; Barrena, R.; Font, X. Technology overview of biogas production in anaerobic digestion plants: A European evaluation of research and development. *Renew. Sustain. Energy Rev.* **2017**, *80*, 44–53. [CrossRef]
26. Shrestha, B.; Hernandez, R.; Fortela, D.; Sharp, W.; Chistoserdov, A.; Gang, D.; Revellame, E.; Holmes, W.; Zappi, M. A review of pretreatment methods to enhance solids reduction during anaerobic digestion of municipal wastewater sludges and the resulting digester performance: Implications to future urban biorefineries. *Appl. Sci.* **2020**, *10*, 9141. [CrossRef]
27. Hanum, F.; Yuan, L.C.; Kamahara, H.; Aziz, H.A.; Atsuta, Y.; Yamada, T.; Daimon, H. Treatment of sewage sludge using anaerobic digestion in Malaysia: Current state and challenges. *Front. Energy Res.* **2019**, *7*, 19. [CrossRef]
28. Roopnarain, A.; Adeleke, R. Current status, hurdles and future prospects of biogas digestion technology in Africa. *Renew. Sustain. Energy Rev.* **2017**, *67*, 1162–1179. [CrossRef]
29. Correa, V.J.; Manandhar, A.; Shah, A. Economic Implications of Anaerobic Digestion for Bioenergy Production and Waste Management. Available online: https://ohioline.osu.edu/factsheet/fabe-6611 (accessed on 20 August 2024).
30. Subbarao, P.; D'Silva, T.; Adlak, K.; Kumar, S.; Chandra, R.; Vijay, V. Anaerobic digestion as a sustainable technology for efficiently utilizing biomass in the context of carbon neutrality and circular economy. *Environ. Res.* **2023**, *234*, 116286. [CrossRef]
31. Liu, C.; Moon, D.; Watabe, A. Assessing the economic and environmental impacts of anaerobic digestion for municipal organic waste: A case study of Minamisanriku Town, Japan. *Sustainability* **2024**, *16*, 6793. [CrossRef]
32. Kiselev, A.; Magaril, E.; Giurea, R. Environmental and economic forecast of the widespread use of anaerobic digestion techniques. *Recycling* **2024**, *9*, 62. [CrossRef]
33. Huang, W.; Fooladi, H. Economic and environmental estimated assessment of power production from municipal solid waste using anaerobic digestion and landfill gas technologies. *Energy Rep.* **2021**, *7*, 4460–4469. [CrossRef]
34. Li, Y.; Qi, C.; Zhang, Y.; Li, Y.; Wang, Y.; Li, G.; Luo, W. Anaerobic digestion of agricultural wastes from liquid to solid state: Performance and environ-economic comparison. *Bioresour. Technol.* **2021**, *332*, 125080. [CrossRef] [PubMed]
35. Chowdhury, T.H. Technical-economical analysis of anaerobic digestion process to produce clean energy. *Energy Rep.* **2021**, *7*, 247–253. [CrossRef]
36. Lamolinara, B.; Pérez-Martínez, A.; Guardado-Yordi, E.; Fiallos, C.G.; Diéguez-Santana, K.; Ruiz-Mercado, G.J. Anaerobic digestate management, environmental impacts, and techno-economic challenges. *Waste Manag.* **2022**, *140*, 14–30. [CrossRef]
37. Arias, A.; Feijoo, G.; Moreira, T.M. What is the best scale for implementing anaerobic digestion according to environmental and economic indicators? *J. Water Process Eng.* **2020**, *35*, 101235. [CrossRef]
38. Zhuang, Z.; Mohamed, B.A.; Li, L.Y.; Swei, O. An economic and global warming impact assessment of common sewage sludge treatment processes in North America. *J. Clean. Prod.* **2022**, *370*, 133539. [CrossRef]

39. Balcioglu, G.; Jeswani, H.K.; Azapagic, A. Evaluating the environmental and economic sustainability of energy from anaerobic digestion of different feedstocks in Turkey. *Sustain. Prod. Consum.* **2022**, *32*, 924–941. [CrossRef]
40. Patiño, A.T.; Sánchez, O.M.; Ariel, C. Prefeasibility analysis of different anaerobic digestion upgrading pathways using organic kitchen food waste as raw material. *J. Ferment.* **2024**, *10*, 300. [CrossRef]
41. Rajendran, K.; Mahapatra, D.; Venkatraman, A.; Muthuswamy, S.; Pugazhendhi, A. Advancing anaerobic digestion through two-stage processes: Current developments and future trends. *Renew. Sustain. Energy Rev.* **2020**, *123*, 109746. [CrossRef]
42. Iglesias, R.; Muñoz, R.; Polanco, M.; Díaz, I.; Susmozas, A.; Moreno, A.D.; Guirado, M.; Carreras, N.; Ballesteros, M. Biogas from anaerobic digestion as an energy vector: Current upgrading development. *Energies* **2021**, *14*, 2742. [CrossRef]
43. Zhao, J.; Li, Y.; Dong, R. Recent progress towards in-situ biogas upgrading technologies. *Sci. Total Environ.* **2021**, *800*, 149667. [CrossRef]
44. Nath, D.; Chakraborty, I.; Ghangrekar, M.M. Integrating microbial electrochemical technologies for methane-to-bioelectricity and water-splitting to impart self-sustainability to wastewater treatment plants. *Bioresour. Technol. Rep.* **2021**, *13*, 100644. [CrossRef]
45. Zhu, X.; Leininger, A.; Jassby, D.; Tsesmetzis, N.; Ren, J.Z.Y. Will membranes break barriers on volatile fatty acid recovery from anaerobic digestion? *ACS EST Eng.* **2020**, *1*, 141–153. [CrossRef]
46. Vinardell, S.; Astals, S.; Peces, M.; Cardete, M.; Fernández, I.; Alvarez, M.J.; Dosta, J. Advances in anaerobic membrane bioreactor technology for municipal wastewater treatment: A 2020 updated review. *Renew. Sustain. Energy Rev.* **2020**, *130*, 109936. [CrossRef]
47. Masebinu, S.; Akinlabi, E.; Muzenda, E.; Aboyade, A. A review of biochar properties and their roles in mitigating challenges with anaerobic digestion. *Renew. Sustain. Energy Rev.* **2019**, *103*, 291–307. [CrossRef]
48. D'Silva, C.T.; Isha, A.; Chandra, R.; Vijay, K.V.; Subbarao, P.M.V.; Kumar, R.; Chaudhary, P.V.; Singh, H.; Khan, A.; Tyagi, K.V.; et al. Enhancing methane production in anaerobic digestion through hydrogen assisted pathways—A state-of-the-art review. *Renew. Sustain. Energy Rev.* **2021**, *151*, 111536. [CrossRef]
49. Jensen, M.; Jensen, B.; Ottosen, L.; Kofoed, M. Integrating H_2 injection and reactor mixing for low-cost H_2 gas-liquid mass transfer in full-scale in situ biomethanation. *Biochem. Eng. J.* **2021**, *166*, 107869. [CrossRef]
50. Chen, D.; Yin, D.; Liu, W.; Lan, W.; Wang, Y. Anaerobic co-digestion scheme of biogas engineering based on feedstock and temperature. *BioResources* **2023**, *18*, 5777–5797. [CrossRef]
51. Yao, Y.; Huang, G.; An, C.; Chen, X.; Zhang, P.; Xin, X.; Shen, J.; Agnew, J. Anaerobic digestion of livestock manure in cold regions: Technological advancements and global impacts. *Renew. Sustain. Energy Rev.* **2020**, *119*, 109494. [CrossRef]
52. Karki, R.; Chuenchart, W.; Surendra, K.C.; Shrestha, S.; Raskin, L.; Sung, S.; Hashimoto, A.; Khanal, K.S. Anaerobic co-digestion: Current status and perspectives. *Bioresour. Technol.* **2021**, *330*, 125001. [CrossRef]
53. Cucina, M. Integrating anaerobic digestion and composting to boost energy and material recovery from organic wastes in the circular economy framework in Europe: A review. *Bioresour. Technol. Rep.* **2023**, *24*, 101642. [CrossRef]
54. O'Connor, S.; Ehimen, E.; Pillai, S.; Black, A.; Tormey, D.; Bartlett, J. Biogas production from small-scale anaerobic digestion plants on European farms. *Renew. Sustain. Energy Rev.* **2021**, *139*, 110580. [CrossRef]
55. Adami, L.; Schiavon, M.; Torretta, V.; Costa, L.; Rada, E. Evaluation of conventional and alternative anaerobic digestion technologies for applications to small and rural communities. *Waste Manag.* **2020**, *118*, 79–89. [CrossRef] [PubMed]
56. Hussain, Z.; Mishra, J.; Vanacore, E. Waste to energy and circular economy: The case of anaerobic digestion. *J. Enterp. Inf. Manag.* **2020**, *33*, 817–838. [CrossRef]
57. Faragò, M.; Damgaard, A.; Madsen, J.A.; Andersen, J.K.; Thornberg, D.; Andersen, M.H.; Rygaard, M. From wastewater treatment to water resource recovery: Environmental and economic impacts of full-scale implementation. *Water Res.* **2021**, *204*, 117554. [CrossRef]
58. Greene, J.M.; Wallace, J.; Williams, R.B.; Leytem, A.B.; Bock, B.R.; McCully, M.; Kaffka, S.R.; Rotz, A.C.; Quinn, J.C. National greenhouse gas emission reduction potential from adopting anaerobic digestion on large-scale dairy farms in the United States. *Environ. Sci. Technol.* **2024**, *58*, 12409–12419. [CrossRef] [PubMed]
59. Lu, J.; Gao, X. Biogas: Potential, challenges, and perspectives in a changing China. *Biomass Bioenergy* **2021**, *150*, 106127. [CrossRef]
60. Li, J.; Gan, C.; Zhou, J.; Novakovic, V. Performance analysis of biomass direct combustion heating and centralized biogas supply system for rural districts in China. *Energy Convers. Manag.* **2023**, *278*, 116730. [CrossRef]
61. Obileke, K.; Nwokolo, N.; Makaka, G.; Mukumba, P.; Onyeaka, H. Anaerobic digestion: Technology for biogas production as a source of renewable energy—A review. *Energy Environ.* **2020**, *32*, 191–225. [CrossRef]
62. Almansa, X.F.; Starostka, R.; Raskin, L.; Zeeman, G.; De Los Reyes, F., III; Waechter, J.; Yeh, D.; Radu, T. Anaerobic digestion as a core technology in addressing the global sanitation crisis: Challenges and opportunities. *Environ. Sci. Technol.* **2023**, *57*, 19078. [CrossRef] [PubMed]
63. Widyarani, D.R.; Hamidah, U.; Komarulzaman, A.; Rosmalina, R.T.; Sintawardani, N. Domestic wastewater in Indonesia: Generation, characteristics and treatment. *Environ. Sci. Pollut. Res.* **2022**, *29*, 32397–32414. [CrossRef]
64. Wang, S.; Sahoo, K.; Jena, U.; Dong, H.; Bergman, R.; Runge, T. Life-cycle assessment of treating slaughterhouse waste using anaerobic digestion systems. *J. Clean. Prod.* **2021**, *292*, 126038. [CrossRef]
65. Costa, J.M.; Beatriz, A.; Flavia, A.; Damasceno, B.; Duran, K.A.; Jerônimo, K.A.; Silva, M.M.; Cristina, T.; Rodriguez, R.P. Environmental aspects and perspectives of the Brazilian market for biogas and biomethane from anaerobic digestion: A review. *Bioenergy Res.* **2023**, *17*, 59–72. [CrossRef]

66. Vega, L.; Bautista, K.; Campos, H.; Daza, S.; Vargas, G. Biofuel production in Latin America: A review for Argentina, Brazil, Mexico, Chile, Costa Rica and Colombia. *Energy Rep.* **2024**, *11*, 28–38. [CrossRef]
67. Benito, A.; Verdezoto, C.P.; Burlot, A.; Arena, A. Hybrid power system for distributed energy deploying biogas from municipal solid waste and photovoltaic solar energy in Mendoza, Argentina. *E3S Web Conf.* **2024**, *532*, 01001. [CrossRef]
68. Tauber, J.; Ramsbacher, A.; Svardal, K.; Krampe, J. Energetic potential for biological methanation in anaerobic sewage sludge digesters in Austria. *Energies* **2021**, *14*, 6618. [CrossRef]
69. Nguyen, L.; Kumar, J.; Vu, M.; Mohammed, J.; Pathak, N.; Commault, A.; Sutherland, D.; Zdarta, J.; Tyagi, V.; Nghiem, L. Biomethane production from anaerobic co-digestion at wastewater treatment plants: A critical review on development and innovations in biogas upgrading techniques. *Sci. Total Environ.* **2020**, *765*, 142753. [CrossRef]
70. Middelburg, T. Achieving Energy Self-Sufficiency and Circularity in the Dutch Agricultural Sector, Using Anaerobic Digestion. Master's Thesis, University of Groningen, Groningen, The Netherlands, 2020.
71. Lima, D.; Appleby, G.; Li, L. A scoping review of options for increasing biogas production from sewage sludge: Challenges and opportunities for enhancing energy self-sufficiency in wastewater treatment plants. *Energies* **2023**, *16*, 2369. [CrossRef]
72. Schneider, V.J.; Isaksson, H.L.; Mainali, B.; Herrero, M.J.; Cardozo, E.; Malmquist, A.; Martin, A. Energy self-sufficiency and greenhouse gas emission reductions in Latin American dairy farms through massive implementation of biogas-based solutions. *Energy Convers. Manag.* **2022**, *261*, 115670. [CrossRef]
73. Paredes, M.; Güereca, L.; Molina, L.; Noyola, A. Methane emissions from anaerobic sludge digesters in Mexico: On-site determination vs. IPCC Tier 1 method. *Sci. Total Environ.* **2019**, *656*, 468–474. [CrossRef]
74. Vijay, V.; Subbarao, P.; Chandra, R. An evaluation on energy self–sufficiency model of a rural cluster through utilization of biomass residue resources: A case study in India. *Energy Clim. Chang.* **2021**, *2*, 100036. [CrossRef]
75. Ma, X.; Zhang, T.; Xie, Y.; Tan, Q. A simulation-assessment-optimization approach towards energy self-sufficiency and carbon reduction in regional-scale sewage systems. *Resour. Conserv. Recycl.* **2022**, *187*, 106595. [CrossRef]
76. Masłoń, A.; Czarnota, J.; Szczyrba, P.; Szaja, A.; Cieplak, S.J.; Łagód, G. Assessment of energy self-sufficiency of wastewater treatment plants—A case study from Poland. *Energies* **2024**, *17*, 1164. [CrossRef]
77. Dhara, S.; Das, P.P.; Uppaluri, R.; Purkait, K.M. *Biological Approach for Energy Self-Sufficiency of Municipal Wastewater Treatment Plants*; Elsevier: Amsterdam, The Netherlands, 2023; pp. 235–260.
78. Feng, Y.; Rosa, L. Global biomethane and carbon dioxide removal potential through anaerobic digestion of waste biomass. *Environ. Res. Lett.* **2024**, *19*, 024024. [CrossRef]
79. Woo, T.; Charmchi, A.; Ifaei, P.; Heo, S.; Nam, K.; Yoo, C. Three energy self-sufficient networks of wastewater treatment plants developed by nonlinear bi-level optimization models in Jeju Island. *J. Clean. Prod.* **2022**, *379*, 134465. [CrossRef]
80. Li, H.H.; Zhang, T.; Tsang, D.C.W.; Li, G.X. Effects of external additives biochar, bentonite, phosphate, on co-composting for swine manure and corn straw. *Chemosphere* **2020**, *248*, 125927. [CrossRef]
81. Li, H.H.; Zhang, T.; Shaheen, S.M.; Abdelrahman, H.; Ali, E.F.; Bolan, N.S.; Li, G.X.; Rinklebe, J. Microbial inoculants and struvite improved organic matter humification and stabilized phosphorus during swine manure composting: Multivariate and multiscale investigations. *Bioresour. Technol.* **2022**, *351*, 126976. [CrossRef]
82. Zhang, T.; Wu, X.S.; Shaheen, S.M.; Abdelrahman, H.; Ali, E.F.; Bolan, N.S.; Li, G.X.; Tsang, D.C.W.; Rinklebe, J. Improving the humification and phosphorus flow during swine manure composting: A trial for enhancing the beneficial applications of hazardous biowastes. *J. Hazard. Mater.* **2022**, *425*, 127906. [CrossRef]
83. Zhang, T.; Li, H.H.; Yan, T.; Shaheen, S.M.; Niu, Y.Q.; Zhang, Y.Y.; Abdelrahman, H.; Ali, E.F.; Bolan, N.S.; Rinklebe, J. Organic matter stabilization and phosphorus activation during vegetable waste composting: Multivariate and multiscale investigation. *Sci. Total Environ.* **2023**, *891*, 164608. [CrossRef] [PubMed]
84. Rekleitis, G.; Haralambous, K.; Loizidou, M.; Aravossis, K. Utilization of agricultural and livestock waste in anaerobic digestion: Applying the biorefinery concept in a circular economy. *Energies* **2020**, *13*, 4428. [CrossRef]
85. Sebastian, P.; Kalaiselvi, K.; Suganya, K.; Muthusamy, S.; Poornima, R.; Bush, R. Novel resources recovery from anaerobic digestates: Current trends and future perspectives. *Crit. Rev. Environ. Sci. Technol.* **2021**, *52*, 1915–1999.
86. He, X.Y.; Zhang, T.; Niu, Y.; Xue, Q.; Ali, E.F.; Shaheen, S.M.; Tsang, D.C.W.; Rinklebe, J. Impact of catalytic hydrothermal treatment and Ca/Al-modified hydrochar on lability, sorption, and speciation of phosphorus in swine manure: Microscopic and spectroscopic investigations. *Environ. Pollut.* **2022**, *299*, 118877. [CrossRef] [PubMed]
87. Su, X.H.; Zhang, T.; Zhao, J.Y.; Mukherjee, S.; Alotaibi, N.M.; Elwafa, A.S.F.; Tran, H.T.; Bolan, N.S. Phosphorus fraction in hydrochar from co-hydrothermal carbonization of swine manure and rice straw: An optimization analysis based on response surface methodology. *Water* **2024**, *16*, 2208. [CrossRef]
88. Przydatek, G.; Wota, A. Analysis of the comprehensive management of sewage sludge in Poland. *J. Mater. Cycles Waste Manag.* **2019**, *22*, 80–88. [CrossRef]
89. Bora, R.; Richardson, R.; You, F. Resource recovery and waste-to-energy from wastewater sludge via thermochemical conversion technologies in support of circular economy: A comprehensive review. *BMC Chem. Eng.* **2020**, *2*, 8. [CrossRef]
90. Seruga, P. The municipal solid waste management system with anaerobic digestion. *Energies* **2021**, *14*, 2067. [CrossRef]
91. Tomić, T.; Kremer, I.; Schneider, D. Economic efficiency of resource recovery—Analysis of time—Dependent changes on sustainability perception of waste management scenarios. *Clean Technol. Environ. Policy* **2021**, *24*, 543–562. [CrossRef]

92. Đurđević, D.; Žiković, S.; Blecich, P. Sustainable sewage sludge management technologies selection based on techno-economic-environmental criteria: Case study of Croatia. *Energies* **2022**, *15*, 3941. [CrossRef]
93. Adenuga, O.; Mpofu, K.; Modise, K. An approach for enhancing optimal resource recovery from different classes of waste in South Africa: Selection of appropriate waste to energy technology. *Sustain. Futur.* **2020**, *2*, 100033. [CrossRef]
94. Dell'Orto, A.; Trois, C. Double-stage anaerobic digestion for biohydrogen production: A strategy for organic waste diversion and emission reduction in a South African municipality. *Sustainability* **2024**, *16*, 7200. [CrossRef]
95. Demirtas, U.M.; Dalke, R.; Pagilla, K. Sludge management and utilization for decarbonization. In *Pathways to Water Sector Decarbonization, Carbon Capture and Utilization*; IWA Publishing: London, UK, 2022; Volume 10, p. 171.
96. Hernández, J.S. Energy, environmental, resource recovery, and economic dimensions of municipal solid waste management paths in Mexico city. *Waste Manag.* **2021**, *136*, 321–336. [CrossRef] [PubMed]
97. Silva, A.; Brasil, Y.; Koch, K.; Amaral, M. Resource recovery from sugarcane vinasse by anaerobic digestion–a review. *J. Environ. Manag.* **2021**, *295*, 113137. [CrossRef]
98. Zarei, M. Wastewater resources management for energy recovery from circular economy perspective. *Water-Energy Nexus* **2020**, *3*, 170–185. [CrossRef]
99. Jin, C.; Sun, S.; Yang, D.; Sheng, W.; Ma, Y.; He, W.; Li, G. Anaerobic digestion: An alternative resource treatment option for food waste in China. *Sci. Total Environ.* **2021**, *779*, 146397. [CrossRef]
100. Choudhury, A.; Lepine, C.; Witarsa, F.; Good, C. Anaerobic digestion challenges and resource recovery opportunities from land-based aquaculture waste and seafood processing byproducts: A review. *Bioresour. Technol.* **2022**, *354*, 127144. [CrossRef] [PubMed]
101. Shah, H.H.; Amin, M.; Pepe, F. Maximizing resource efficiency: Opportunities for energy recovery from municipal solid waste in Europe. *J. Mater. Cycles Waste Manag.* **2023**, *25*, 2766–2782. [CrossRef]
102. Amodeo, C.; Hattou, S.; Buffiere, P.; Benbelkacem, H. Temperature phased anaerobic digestion (TPAD) of organic fraction of municipal solid waste (OFMSW) and digested sludge (DS): Effect of different hydrolysis conditions. *Waste Manage.* **2021**, *126*, 21–29. [CrossRef] [PubMed]
103. Postawa, K.; Szczygieł, J.; Kułażyński, M. Innovations in anaerobic digestion: A model-based study. *Biotechnol. Biofuels* **2021**, *14*, 19. [CrossRef]
104. Archana, K.; Visckram, A.; Kumar, P.; Manikandan, S.; Saravanan, A.; Natrayan, L. A review on recent technological breakthroughs in anaerobic digestion of organic biowaste for biogas generation: Challenges towards sustainable development goals. *Fuel* **2024**, *358*, 130298. [CrossRef]
105. Loh, M.C.; Lues, R. Anaerobic digestion of lignocellulosic biomass: Substrate characteristics (challenge) and innovation. *Fermentation* **2023**, *9*, 755. [CrossRef]
106. Naquash, A.; Qyyum, M.; Haider, J.; Bokhari, A.; Lim, H.; Lee, M. State-of-the-art assessment of cryogenic technologies for biogas upgrading: Energy, economic, and environmental perspectives. *Renew. Sustain. Energy Rev.* **2022**, *154*, 111826. [CrossRef]
107. Sevillano, C.A.; Pesantes, A.A.; Carpio, E.P.; Martínez, E.J.; Gómez, X. Anaerobic digestion for producing renewable energy—The evolution of this technology in a new uncertain scenario. *Entropy* **2021**, *23*, 145. [CrossRef] [PubMed]
108. Zhang, T.; Pasha, A.M.K.; Sajadi, S.M.; Jasim, D.J.; Esfahani, N.N.; Maleki, H.; Salahshour, S.; Baghaei, S. Optimization of thermophysical properties of nanofluids using a hybrid procedure based on machine learning, multi-objective optimization, and multi-criteria decision-making. *Chem. Eng. J.* **2024**, *485*, 150059. [CrossRef]
109. Cinar, S.; Cinar, S.O.; Wieczorek, N.; Sohoo, I.; Kuchta, K. Integration of artificial intelligence into biogas plant operation. *Processes* **2021**, *9*, 85. [CrossRef]
110. Yildirim, O.; Ozkaya, B. Prediction of biogas production of industrial scale anaerobic digestion plant by machine learning algorithms. *Chemosphere* **2023**, *335*, 138976. [CrossRef]
111. Swami, S.; Suthar, S.; Singh, R.; Kumar, A.T.; Gupta, L.; Sikarwar, S.V. Integration of anaerobic digestion with artificial intelligence to optimise biogas plant operation. In *Environment, Development and Sustainability*; Springer: New Delhi, India, 2023; pp. 1–31.
112. Cheah, C.G.; Chia, W.Y.; Lai, S.F.; Chew, K.W.; Chia, S.R.; Show, P.L. Innovation designs of industry 4.0 based solid waste management: Machinery and digital circular economy. *Environ. Res.* **2022**, *213*, 113619. [CrossRef] [PubMed]
113. Sheik, G.A.; Krishna, S.; Patnaik, R.; Ambati, R.S.; Bux, F.; Kumari, S. Digitalization of phosphorous removal process in biological wastewater treatment systems: Challenges, and way forward. *Environ. Res.* **2024**, *252*, 119133. [CrossRef]
114. Brémond, U.; Bertrandias, A.; Steyer, J.P.; Bernet, N.; Carrere, H. A vision of European biogas sector development towards 2030: Trends and challenges. *J. Clean. Prod.* **2021**, *287*, 125065. [CrossRef]
115. Lybæk, R.; Christensen, T.B.; Thomsen, T.P. Enhancing policies for deployment of industrial symbiosis—What are the obstacles, drivers and future way forward? *J. Clean. Prod.* **2021**, *280*, 124351. [CrossRef]
116. Nevzorova, T.; Karakaya, E. Explaining the drivers of technological innovation systems: The case of biogas technologies in mature markets. *J. Clean. Prod.* **2020**, *259*, 120819. [CrossRef]
117. Gustafsson, M.; Anderberg, S. Great expectations—Future scenarios for production and use of biogas and digestate in Sweden. *Biofuels* **2022**, *14*, 93–107. [CrossRef]
118. Magnusson, T.; Zanatta, H.; Larsson, M.; Kanda, W.; Hjelm, O. Circular economy, varieties of capitalism and technology diffusion: Anaerobic digestion in Sweden and Paraná. *J. Clean. Prod.* **2022**, *335*, 130300. [CrossRef]

119. Miszczuk, C.J.; Martinát, S.; Horst, V.D.D. Changes in feedstocks of rural anaerobic digestion plants: External drivers towards a circular bioeconomy. *Renew. Sustain. Energy Rev.* **2021**, *148*, 111344. [CrossRef]
120. Martinát, S.; Chodkowska-Miszczuk, J.; Kulla, M.; Navrátil, J.; Klusáček, P.; Dvořák, P.; Novotný, L.; Krejčí, T.; Pregi, L.; Trojan, J.; et al. Best practice forever? Dynamics behind the perception of farm-fed anaerobic digestion plants in rural peripheries. *Energies* **2022**, *15*, 2533. [CrossRef]
121. Song, J.; Wang, Y.; Zhang, S.; Song, Y.; Xue, S.; Liu, L.; Lvy, X.; Wang, X.; Yang, G. Coupling biochar with anaerobic digestion in a circular economy perspective: A promising way to promote sustainable energy, environment and agriculture development in China. *Renew. Sustain. Energy Rev.* **2021**, *144*, 110973. [CrossRef]
122. Rizzioli, F.; Bertasini, D.; Bolzonella, D.; Frison, N.; Battista, F. A critical review on the techno-economic feasibility of nutrients recovery from anaerobic digestate in the agricultural sector. *Sep. Purif. Technol.* **2023**, *306*, 122690. [CrossRef]
123. Jacob, S.; Upadrasta, L.; Banerjee, R. Bottlenecks in biomethane production from agro-industrial wastes through anaerobic digestion. In *Practices and Perspectives in Sustainable Bioenergy*; Green Energy and Technology; Springer: New Delhi, India, 2020; pp. 75–104.
124. Deng, Y.X.; Zhang, T.; Clark, J.; Aminabhavi, T.; Kruse, A.; Tsang, D.C.W.; Sharma, B.K.; Zhang, F.S.; Ren, H.Q. Mechanisms and modelling of phosphorus solid-liquid transformation during the hydrothermal processing of swine manure. *Green Chem.* **2020**, *22*, 5628–5638. [CrossRef]
125. Shanmugam, S.; Mathimani, T.; Rene, R.E.; Geo, E.V.; Arun, A.; Brindhadevi, K.; Pugazhendhi, A. Biohythane production from organic waste: Recent advancements, technical bottlenecks and prospects. *Int. J. Hydrogen Energy* **2021**, *46*, 11201–11216. [CrossRef]
126. Wang, Y.; Zhang, T.; Westerhoff, P.; Jiang, R.F.; Ren, H.Q.; Yang, Y.; Li, Z. Microwave-assisted digestion and NaOH treatment of waste-activated sludge to recover phosphorus by crystallizing struvite. *Environ. Technol.* **2017**, *38*, 1211–1222. [CrossRef] [PubMed]
127. Xie, S.Y.; He, X.Y.; Alshehri, A.M.; Elwafa, A.S.F.; Zhang, T. Elevated effect of hydrothermal treatment on phosphorus transition between solid-liquid phase in swine manure. *Results Eng.* **2024**, *24*, 102887. [CrossRef]
128. Hoo, P.; Hashim, H.; Ho, W. Towards circular economy: Economic feasibility of waste to biomethane injection through proposed feed-in tariff. *J. Clean. Prod.* **2020**, *270*, 122160. [CrossRef]
129. Zieliński, M.; Kazimierowicz, J.; Dębowski, M. Advantages and limitations of anaerobic wastewater treatment—Technological basics, development directions, and technological innovations. *Energies* **2022**, *16*, 83. [CrossRef]
130. Mosquera, A.L.; Alvarado, A.A.; Arrieta, R.M.; Cattaneo, C.; Rene, E.; Depraect, G.O. Production of solid biofuels from organic waste in developing countries: A review from sustainability and economic feasibility perspectives. *Sci. Total Environ.* **2021**, *795*, 148816. [CrossRef] [PubMed]
131. Vinardell, S.; Astals, S.; Koch, K.; Alvarez, M.J.; Dosta, J. Co-digestion of sewage sludge and food waste in a wastewater treatment plant based on mainstream anaerobic membrane bioreactor technology: A techno-economic evaluation. *Bioresour. Technol.* **2021**, *330*, 124978. [CrossRef]
132. Kehrein, P.; Loosdrecht, V.M.; Osseweijer, P.; Garfí, M.; Dewulf, J.; Posada, J. A critical review of resource recovery from municipal wastewater treatment plants—Market supply potentials, technologies and bottlenecks. *Environ. Sci. Water Res. Technol.* **2020**, *6*, 877–910. [CrossRef]
133. Mateescu, C.; Dima, A.D. Critical analysis of key barriers and challenges to the growth of the biogas sector: A case study for Romania. *Biomass Convers. Biorefin.* **2020**, *12*, 5989–6002. [CrossRef]
134. Abanades, S.; Abbaspour, H.; Ahmadi, A.; Das, B.; Ehyaei, M.; Esmaeilion, F.; Assad, M.; Hajilounezhad, T.; Jamali, D.; Hmida, A.; et al. A critical review of biogas production and usage with legislations framework across the globe. *Int. J. Environ. Sci. Technol.* **2022**, *19*, 3377–3400. [CrossRef]
135. Ackrill, R.; Abdo, H. On-farm anaerobic digestion uptake barriers and required incentives: A case study of the UK East Midlands region. *J. Clean. Prod.* **2020**, *264*, 121727. [CrossRef]
136. Ahmed, N.; Qamar, S.; Jabeen, G.; Yan, Q.; Ahmad, M. Systematic analysis of factors affecting biogas technology acceptance: Insights from the diffusion of innovation. *Sustain. Energy Technol. Assess.* **2022**, *52*, 102122. [CrossRef]
137. O'Connor, S.; Ehimen, E.; Pillai, S.; Power, N.; Lyons, G.; Bartlett, J. An investigation of the potential adoption of anaerobic digestion for energy production in Irish farms. *Environments* **2021**, *8*, 8. [CrossRef]
138. Mancini, E.; Raggi, A. Out of sight, out of mind? The importance of local context and trust in understanding the social acceptance of biogas projects: A global scale review. *Energy Res. Soc. Sci.* **2022**, *91*, 102697. [CrossRef]
139. Duquennoi, C.; Martinez, J. European Union's policymaking on sustainable waste management and circularity in agroecosystems: The potential for innovative interactions between science and decision-making. *Front. Sustain. Food Syst.* **2022**, *6*, 937802. [CrossRef]

water

MDPI

Article

Technical–Economic Evaluation of Water Reuse at the WWTP El Salitre (Bogotá, Colombia): Example of Circular Economy

Michelle A. Urrea Vivas [1], Luis Seguí-Amórtegui [2,*], Cristina Tomás Pérez [3] and Hilda Guerrero-García Rojas [4]

1 Department of Civil and Environmental Engineering, Universitat Politècnica de Catalunya, 08034 Barcelona, Spain; michelle.atala.urrea@upc.edu
2 Strategy, Entrepreneurship and Sustainability Department, EAE Business School, Aragón, 55, 08015 Barcelona, Spain
3 Economics and Finance Department, EAE Business School, Aragón, 55, 08015 Barcelona, Spain; ctomas@eae.es
4 Department of Economics, Universidad Michoacana de San Nicolás de Hidalgo (UMSNH), Av. Fco J. Múgica, Morelia 58040, Michoacan, Mexico; hilda.guerrero@umich.mx
* Correspondence: lsegui@eae.es

Abstract: Water resource management should be conducted from a multidisciplinary perspective. In this sense, the objective of this work is to analyze, from the perspective of the circular economy, the technical–economic feasibility of implementing different alternatives for the regeneration of wastewater for its subsequent reuse in industrial and sports companies located in Bogotá, Colombia. The development of the methodology is carried out through the method of economic cost–benefit analysis (ACB) and the technique of net present value (NPV). These methodologies facilitate decision making based on the economic feasibility of recovering the initial investment costs and the operating costs during the useful life of the WWTP. Establishing the cost and price of reclaimed water is essential to the efficient management of water resources; so far, the studies carried out only focus on the economic viability of the internal costs of the system, while the private impacts and the externalities are excluded and relegated to unsubstantiated statements about the advantages of water reuse. The economic feasibility incorporating the analysis of externalities presents a total profit that ranges between 6.52 EUR/m^3 for the industrial sector and 2503 EUR/m^3 for the irrigation of golf courses. This analysis demonstrates the technical and economic feasibility of carrying out a circular economy where the water already used returns as a new source of supply.

Keywords: circular economy; externalities; opportunity costs; reuse of treated wastewater; supply and demand of reclaimed water

check for updates

Citation: Urrea Vivas, M.A.; Seguí-Amórtegui, L.; Tomás Pérez, C.; Guerrero-García Rojas, H. Technical–Economic Evaluation of Water Reuse at the WWTP El Salitre (Bogotá, Colombia): Example of Circular Economy. *Water* **2023**, *15*, 3374. https://doi.org/10.3390/w15193374

Academic Editors: Tao Zhang, Jing Yuan, Huu-Tuan Tran and Muhammad-Jamal Alhnidi

Received: 2 August 2023
Revised: 6 September 2023
Accepted: 18 September 2023
Published: 26 September 2023

1. Introduction

The "El Salitre" treatment plant is in the city of Bogotá, and it is the one in charge of treating the wastewater from the north of the town, with a current load capacity equivalent to 2 million people. The WWTP begins with a pretreatment followed by a primary treatment of settling tanks that are subsequently dumped into the Bogotá River. Due to the scarcity of water that is occurring, and the pollution generated by the bodies of water, a proposal is made that allows decisions to be made for a process of efficient regeneration and reuse of water. Resolution 1096 of 2000 of the Republic of Colombia emerged as a mandatory regulatory instrument within the territory to provide treatment to wastewater and include it again for reuse. The purpose of this study is to analyze technically and economically the feasibility of implementing an alternative that allows the recovery of the costs of water regeneration and reuse, obtaining private economic profit from the commercialization of this resource. In addition, the evaluation allows the analysis and evaluation of the externalities of the system, such as the decrease in contamination generated by surface water bodies.

The companies selected to supply them with reclaimed water are characterized by having processes with high water consumption and their proximity to the WWTP "El Salitre".

Considering this framework, the marginal costs of producing reclaimed water versus obtaining it from a conventional source and its total profits are established.

1.1. Wastewater Treatment and Reuse

Currently, the "El Salitre" treatment plant has a process that begins with a pre-treatment that allows the removal of those bulky and fine solids as well as removing sand, fats, and oils; this is organized so that it does not affect the other subsequent processes. The primary treatment is a sedimentation process that removes up to 40% of DBO_5 and 60% of total suspended solids (TSS). With this process, the effluent obtained is still of low quality to be reused in the different potential uses that exist. For this reason, the proposal is to analyze the use of a secondary treatment that involves a biological treatment process, either (1) activated sludge or (2) trickling filters; in both cases, these processes are accompanied by a settler. The secondary treatment achieves up to 95% removal of DBO_5 and suspended solids. To increase the quality of the regenerated water, two tertiary treatments are proposed through disinfection: (1) disinfection via chlorine or (2) UV radiation.

1.2. The Circular Economy as a Solution to the Problem of Scarcity

The existence of water as a non-renewable resource opens the door to the development of different techniques that make it possible to extend the useful life of this resource; even though water is a basic necessity for all living beings, it is wasted and contaminated indiscriminately, due to the industrial, commercial, agricultural, and residential activities carried out by man [1]. Many countries are seeing the need to adopt new strategies that allow them to make posthumous use of wastewater, both domestic and industrial, and studies have shown that the reuse of this is one of the most economical and efficient alternatives [2]. This reuse is essential for those areas that are arid, semi-arid, or that simply do not have a large quantity of the water resource. In the same way, it manages to solve pollution problems and allows an increase in the availability of a resource without the need to over-exploit the conventional sources (surface or underground waters).

The reclaiming of treated wastewater should be considered as a new source of unconventional resources, whose management must be included in the comprehensive planning of water resources, considering economic, social, and environmental issues [3,4]. Reclaimed water can be used in traditional processes that do not require high quality, releasing volumes of better quality for other and more demanding uses [5].

Currently, many cities are presenting limitations on the use of drinking water, which is why they have chosen to use systems that allow the reuse of wastewater that is subjected to different types of processes, which adapt its quality depending on what it will be used for. The circular economy concept is related to the efficient use of natural resources, thus seeking to develop two fundamental aspects: smart growth, which becomes inclusive and sustainable economically, socially, and environmentally. In this way, it is achieved that the behavior of exploitation and use of water is not linear; on the contrary, the life cycle is closed, making it a constant resource and maintaining it as long as possible [6].

The behavior of the circular economy revolves around the relationship that exists between the reclaimed water market and potential end users. This is based on the promotion of sustainable practices and the efficient use that should be made of water resources. This type of economic model not only ensures that the economy grows but at the same time allows for a reduction in the indiscriminate exploitation of water basins and aquifers, for which reason a linear model that generates pollution and does not allow for a fulfilling helpful life is not followed for this resource [7]. The transition that is made from the linear to the circular model is to encourage the most efficient use of water innovation, with economic incentives for those who could be potential end users; likewise, it improves the capacity of an economy to manage the demands of the growing imbalance between the supply and demand of water [8].

1.3. Economic Aspects of the Circular Model

From the economic perspective of applying the concept of circular economy, the reuse of wastewater is presented as a "win-to-win" relationship [9,10]. The specific purpose of the circular economy is to create a behavior that closes the cycles of the resources and extends their useful life by adapting their quality for future use. When talking about reuse, a sustainable alternative is provided to avoid overexploitation of watersheds or aquifers; unfortunately, as these resources are cheap, society will continue to overexploit the resource, causing constant waste without foreseeing it for the future. The fact that water is consecrated as a vital right implies that it is often free, or the price paid is very little; this is questioned, as there are more and more economic charges associated with exploitation. Therefore the economic convenience is reflected in the investment of infrastructure for regeneration plants. This implies evaluating associated costs incurred for investment, operation, and maintenance as well as the construction of networks for distribution to different end users [11]. Once these costs are available, the minimum sale price of reclaimed water must be established, which must guarantee the recovery of the costs incurred and generate an economic profit. Generally, the costs of wastewater regeneration vary depending on the final use, whether for potable use or not, as well as whether it includes aspects of quality, supply, and quantity requirements [12,13]. Generally, purchasing reclaimed water is better for potential end users since they are not only encouraged by discounts on the rates they currently pay, but they also benefit from a continuous supply due to the amount of water released.

Water and wastewater management is one of the biggest challenges for the CE as many kinds of industries depend on water [14], and limited access to clean water resources can limit both production capacity and profits.

2. Materials and Methods

2.1. Definition of Scope Study

Table 1 shows the main characteristics of the city of Bogotá. Currently, economic activities represent 23.6% of the primary sector, secondary 40%, and tertiary 36.3%. It is important to note that, in Bogotá, the impact on the consumption of water resources by industry is quite high, and agriculture is the same. Therefore, the reclaimed water supply will be focused on this area.

Table 1. General aspects of Bogotá. (Own elaboration from the cited sources.)

General Aspects of Bogotá	Amount	Units
Urban area [1]	307	km^2
Territorial extension [1]	1776	km^2
Population [1]	8,080,734	inhabitants
	Agricultural: 2320	ha
Land use [1]	Ranch: 1684	ha
	Industrial: 2980	ha
Drinking water network coverage [2]	97.5	%
Sewerage network coverage [2]	90.2	%
Plants available for wastewater treatment [2]	1	

Note: [1] [15]; [2] [16].

Table 2 stipulates the water flows demanded and the alternatives of the possible users that would make the reuse of the reclaimed water, always starting from the quality characteristics stipulated by the Colombian legal framework.

Table 2. Alternatives for companies to reuse treated water. (Own elaboration from the cited sources.)

Wastewater Treatment Plant	Effluent Flow Rate	Alternatives	Reuse	End User	Estimated Consumption
Wastewater treatment plant "El Salitre"	118,890,720 m^3/year (3.77 m^3/s)	1	Recreational	Golf club and park La Florida [1]	18,023 (m^3/year)–43 (m^3/day)
		2	Industrial	Papeles primavera (paper and school supplies company) [2]	21,250,000 (m^3/year)–58,219 (m^3/day)
		3	Industrial	Sellopack (plastics production company) [3]	10,489,600 (m^3/year)–28,740 (m^3/day)
		4	Industrial	Licorera Cundinamarca [4]	3,488,040 (m^3/year)–9689 (m^3/day)

Note: [1] [17]; [2] [18]; [3] [19]; and [4] [20].

The wastewater generated in the city of Bogotá is currently treated in a primary treatment system within the "El Salitre" plant (currently under expansion and will begin operations in mid-2021); from this effluent, a proposal of different alternatives that are adapted to the already existing infrastructure was developed. The criteria stipulated to be able to use reclaimed water are established by Resolution 1207/2014 of the Colombian government [21]. The exclusive use of reclaimed water can only be used for agricultural, industrial, or recreational use. Likewise, it starts from the guidelines recommended by the World Health Organization [22], demanding the elimination of any pathogen that may directly or indirectly harm human beings. Table 3 shows in detail the information previously mentioned.

Table 3. Quality criteria for reuse alternatives. Source: [21].

		Effluent Water Quality				
Parameter	**Unit of Measure**	**Quality Criterion for Reuse**				
			Industrial			
		Agriculture (Irrigation)	**Heat Exchange in Cooling Towers and Boilers**	**Discharge of Sanitary Equipment**	**Mechanical Cleaning of Roads and Irrigation of Roads to Control Particulate Matter**	**Firefighting Network Systems**
pH	Dimensionless	6.0–9.0	6.0–9.0	6.0–9.0	6.0–9.0	6.0–9.0
Thermo-tolerant coliforms	NMP/100 mL	1.0	1.0	1.0	1.0	1.0
Fecal Enterococci	NMP/100 mL	1.0	-	-	-	-
Helminths parasites in humans	eggs of larvae/L	1.0	0.1	1.0	1.0	0.1
Protozoa human parasites	cyst/L	1.0	0	1.0	1.0	1.0
Salmonella sp	NMP/100 mL	1.0	1.0	1.0	1.0	1.0
BOD_5	mg/L	0	1.0	1.0	1.0	1.0
Total phenols	mg/L	1.5	-	-	-	-
Free cyanide	mg CN/L	0.2	-	-	-	-
Chloride	mg Cl/L	300	-	-	-	-
Sulphate	mg SO_4/L	500	-	-	-	-
Mercury	mg Hg/L	0.002	0.001	-	0.001	-

Table 3. *Cont.*

Parameter	Unit of Measure	Effluent Water Quality				
		Quality Criterion for Reuse				
			Industrial			
		Agriculture (Irrigation)	Heat Exchange in Cooling Towers and Boilers	Discharge of Sanitary Equipment	Mechanical Cleaning of Roads and Irrigation of Roads to Control Particulate Matter	Firefighting Network Systems
Floating material	present or absent	Total absence	Total absence	-	-	-
Suspended solids	mg/L	Total absence	Total absence	-	-	-
Oil and fats	mg/L	Total absence	Total absence	-	-	-
Total residual chlorine (with at least 30 min of contact)	$mgCl_2/L$	Less than 1	-	-	-	-
Nitrate (NO_3^-, N)	mg NO_3^-/L	5	-	-	-	-

The expansion and optimization of the Wastewater Treatment Plant, PTAR Salitre Phase II, includes the following process: activated sludge followed by a secondary decanter to increase the efficiency of discharge quality; with these new treatment units, an approximate elimination will be achieved of SS (60–99%), BOD (60–99%), Total Coliforms (60–99%), and Nutrients (10–55%). Regarding tertiary treatment (Asano, 2007), a filtration process with disinfection is proposed, and this allows a high quality of the final effluent, achieving removal of SS (>99%), BOD (>99%), Total Coliforms (>99.9%), and Nutrients (>90%). Through interviews and consultations with experts [23–25] the costs incurred for construction, operation, and maintenance were established. On the other hand, the externalities that could impact the project were identified, either positive (income or benefit) or negative (expense or disadvantage), and likewise, the opportunity cost is included, which refers to the cost generated by investing the funds destined for the plant in another activity that generates greater profitability. In this case, an investment analysis was made of the total cost of carrying out the project in the purchase of government bonds, which raises a gain of 6.72% over 10 years.

2.2. Economic Analysis

The economic analysis starts with establishing, analyzing, quantifying, and monetarily evaluating the impacts of the project for a given area. This analysis is because the cost that is being incurred is real, which means the project manages to make the treatment of wastewater at a minimum cost. Said identification starts from those already existing impacts in the "El Salitre" treatment plant and those derived from carrying out the treatment based on the advantages and costs of investment, operation, and maintenance. This also allows for the comparison of the marginal costs of producing reclaimed water versus the cost of obtaining a quantity of water from another source of supply. The valuation methodology analyzes the economic behavior of the treatment plant, this includes all the private costs and profits that are measurable in financial terms; therefore, the total profits are maximized, using private profits as a starting point, taking into account positive or negative externalities and opportunity costs [26,27]. The objective function is determined as follows (see Equation (1)):

$$P_T = \sum_{n=0}^{n}[(RAV_n * SV_n) - (IC_n + COM_n + FC_n + T_n) + (PE_n - NE_n) - OC_n] \quad (1)$$

where P_T = Total Profit; RAV = Annual Volume of Regenerated Water; SV = Sale Price of reclaimed water; IC = Investment Cost; FC = Financial Costs; T = Tax; PE = Positive Externalities; NE = Negative Externalities; OC = Opportunity Costs; COM = Cost of Operation and Maintenance; and n = Year.

2.3. *Aggregation of Costs and Incomes*

The private costs obtained in Tables 4 and 5 refer to the costs of each of the processes where the selection was based on a cost-efficiency analysis, and from this analysis, the selected technology was activated sludge followed by a tertiary treatment of disinfection and UV rays due to its high level of efficiency in pollutant removal.

Table 4. Investment, operation, maintenance, and water regeneration costs per m^3 for selected water line options.

Investment, Operation, Maintenance, and Regeneration Costs of Selected Options for the Water Line						
Design for Medium Flow (4 m^3/s) and Peak Flow (9.9 m^3/s)		Investment Cost Civil Works and Equipment (EUR)	Operation and Maintenance Costs (EUR/Month)	Operation and Maintenance Costs (EUR/Year)	Cost Obtaining Reference	Regeneration Cost (MSP EUR/m^3)
Preliminary	Coarse and fine grinding	77,019,207 [1]*	169,281 [1]*	2,031,378	Bogotá water, sewerage, and cleanliness (2018). Monthly operation report	0.1028
	De-sanding and degreasing					
Primary	Sedimentator or primary decanter					
Secondary	Activated sludge—aerobic + Secondary settlers.	462,348,180 [2]*	6,472,873 [2]*	77,674,487	Database "CONSTRUDATA" ciudad de Bogotá, Escuela Colombiana de ingenieros Julio Garavito	1.1195
Tertiary	1. Filtration 2. Disinfection	4,740,227 [3]*	7193 [3]*	86,323	Lexington, Massachusetts (2018), wastewater treatment performance and cost data.	0.0058
Total cost of water treatment						1.2283

Notes: [1] [28]; [2] [15]; and [3] [29,30]. * Update of data based on statistical estimates (2022): update of capitals = (N × (CPI) + 1).

Table 5. Final cost of sludge treatment (own elaboration).

Investment Cost, Maintenance, Operation, and Final Treatment of Sludge					
Treatment	Investment Cost Civil Works and Equipment (EUR)	Operation and Maintenance Costs (EUR/Month)	Cost (EUR/Year)	Cost Obtaining Reference	Cost of Treatment (MSP EUR/m^3)
Initial thickening	29,518,581 [1]*	7369	88,430	Bogotá water, sewerage, and cleanliness (2018). Monthly operation report	0.0321
Anaerobic digester (primary sludge)	4,727,168 [1]*	10,620	127,447		0.0062
Dehydration band treatment	26,511,451 [1]*	15,971	191,659		0.0315
Thermal drying treatment (rotating drum)	80,498,393 [2]*	4655	55,868	PESA company price (2018)	0.0906
Total cost of sludge treatment					0.1604

Notes: [1] [28] and [2] [29,30]. * Update of data based on statistical estimates (2022): update of capitals = (N × (CPI) + 1).

The private costs obtained in Tables 4 and 5 refer to the costs of each of the processes. The selection was based on a cost-efficiency analysis, and from this analysis, the selected technology was activated sludge followed by a tertiary treatment of disinfection and UV rays due to its high level of efficiency in pollutant removal.

With the information presented in Tables 4 and 5, the minimum sale price (MSP) of reclaimed water is calculated, and this price guarantees the recovery of the expected costs and profits, and this investment must be made under the net present value criterion, which indicates the profitability of the project. Therefore, the (MSP) is calculated from the net cash flow (Equation (2)), when the net present value is equal to zero (see Equation (3)).

$$NCF = MSP–C–T \tag{2}$$

where NCF = Net cash flow; C = Costs; MSP = Minimum sale price; and T = Tax.

$$NPV = -IC + \sum_0^n \frac{NCF_n}{(1+i)^n} = 0 \tag{3}$$

where NPV = Net Present Value; i = discount rate; IC = Investment Cost; n = annual discount; and FNE = net cash flow.

The following assumptions for economic analysis are made for the technical and fiscal characteristics:

- Project lifespan: 20 years;
- Tax depression 5.5%;
- There is a 19% tax lien;
- Inflation and uncertainty are not considered since the analysis is for one year (it is a static analysis);
- Opportunity costs are calculated from interest rates of 6.72% on state bonds.

The construction of networks for the distribution of reclaimed water to end users depends on the distance from the treatment plant and is stipulated in Table 6.

Table 6. Total investment cost in distribution networks and distribution cost of end users of reclaimed water.

Company	Supply (m^3/Year)	Driving Distance (km)	Total Investment in Distribution (EUR)	Distribution (EUR/m^3)
Sellopack [1]	10,490,100	7.78	341,736	0.0336
Papeles primavera [2]	21,249,935	9.68	425,194	0.0200
Licorera Cundinamarca [3]	3,536,485	3.06	134,410	0.0380
La Florida Golf Course [4]	15,695	3.42	150,223	1.0586
Total			1,051,564	1.1483

Note: [1] [19]; [2] [18]; [3] [20]; and [4] [17].

In Table 7, the results of the total costs obtained for each of the companies are presented. As can be seen, in the case of Sellopack, a distribution cost of 0.0336 EUR/m^3, is added to the minimum price of sale of 1.3886 EUR/m^3 (this cost is the result of the sum of the MSP for water treatment, 1.2282 EUR/m^3, and the MSP for sludge treatment, 0.1604 EUR/m^3), meaning that the treatment and distribution of reclaimed water for this company is at 1.3922 EUR/m^3; currently, they pay 6.39 EUR/m^3 for conventional supply.

However, with bilateral agreements, it would be possible to establish a discount equivalent to 15% on the price they currently pay; this indicates that they have to pay 5.43 EUR/m^3, which is reflected in Table 8.

Table 7. Total cost of producing and distributing reclaimed water.

Company	Distribution (EUR/m^3)	Production (EUR/m^3)	Total Cost of Water (EUR/m^3)
Sellopack	0.0336	1.3886	1.3922
Papeles primavera	0.0200	1.3886	1.3908
Licorera Cundinamarca	0.0380	1.3886	1.3928
La Florida Golf Course	1.0586	1.3886	2.4463

Table 8. Sale price of reclaimed water with a 15% discount.

Company	Conventional Source Water Price (EUR/m^3)	Sale Price of Regenerated Water (EUR/m^3)
Sellopack	6.39 [1]	5.43
Papeles primavera	6.39 [2]	5.43
Licorera Cundinamarca	6.39 [3]	5.43
La Florida Golf Course	4.99 [4]	4.15

Note: [1] [19]; [2] [18]; [3] [20]; and [4] [17].

Table 9 establishes the private profits generated by the wastewater regeneration process; these are based not only on the profit obtained by end users but also by the company providing the service, "Acueducto de Bogotá".

Table 9. Private profits.

Companies That Buy Reclaimed Water				Acueducto de Bogotá	
Company	Supply (Mm^3/Year)	Profit (EUR/m^3)	Profit (MEUR/Year)	Profit (EUR/m^3)	Profit (MEUR/Year)
Sellopack	10.49	0.96	10.0	4.04	42.3
Papeles primavera	21.25	0.96	20.3	4.04	85.8
Licorera Cundinamarca	3.5	0.96	3.3	4.04	14.2
La Florida Golf Course	0.0015	2.24	0.035	1.70	0.026

On the other hand, there are the costs of private profits that are not only generated for the provider company Acueducto de Bogotá but also for end users that not only have a discount percentage but also positions them as environmental and sustainable companies.

2.4. Total Profit

Now, from these costs, the total profit is calculated, which is also based on generating a maximization based on private profits, the economic valuation of negative and positive externalities, and the opportunity cost (Equation (4)). This allows us to validate the economic decision making for the regeneration of residual water and its subsequent reuse, and this decision is given by the entity in charge, which is Acueducto de Bogotá.

$$P_T = P_P + P_E - OC \tag{4}$$

where P_T = Total Profit (Total Income - Total Costs); P_P = Private Profit (Private income − Private costs); P_E = Profit from Externalities (Income externalities − Costs externalities); and OC = Opportunity Costs.

3. Results and Discussion

As can be seen in Table 10, the quantification and assessment of the impacts generated by the implementation of the proposal allow us to determine that the total profit obtained by each of the companies varies in favor of distance factors and costs for the sale of m^3 of reclaimed water for each of the established sectors. In the four companies, the feasibility of implementation is determined due to the recovery of the investment and exploitation and maintenance costs, but it also shows the greatest difference between the probable maximum income and the minimum sale price. By the above, the sectors that generate the greatest profit are those with industrial characteristics, with a profit oscillating among 6.52 EUR/m^3 of reclaimed and distributed water, in the case of the agriculture sector a lower profit is presented due to being subsidized by the state.

Table 10. Analysis of the economic quantification of the externalities of the project.

Impact Group	Impacts Involved	References in the Literature on the Assessment of This Impact	Sellopack		Papeles Primavera		Licorera Cundinamarca		La Florida Golf Course	
			Quantification		Quantification		Quantification		Quantification	
			Positive	Negative	Positive	Negative	Positive	Negative	Positive	Negative
			EUR/m^3	EUR/m^3	EUR/m^3	EUR/m^3	EUR/m^3	EUR/m^3	EUR/m^3	EUR/m^3
Hydraulic infrastructure	Wastewater treatment (EUR/year)	[31]		1.3922 [1]		1.3908 [1]		1.3928 [1]		2.4463 [1]
	Regeneration and reuse of wastewater	[32]								
	Networks adaptation for the transport of reclaimed water	[33]		0.0336		0.0200		0.0380		1.0586
	Opportunity cost of investment	[34]		0.00480		0.00480		0.00480		0.00480
Conditioning and reuse of by-products	Sludges (biosolids)	[35]	0.045		0.045		0.045		0.045	
	Energy and biogas		N.C.		N.C.		N.C.		N.C.	
Use of resource	Quantities to supply	[36]	5.43		5.43		5.43		4.15	
	Supply guarantees		N.C.		N.C.		N.C.		N.C.	
Environment	Environmental improvement of the river	[37]	2.49		2.49		2.49		2.49	
	Atmospheric pollution CO_2 emissions	[38]		0.016 [2]		0.016 [2]		0.016 [2]		0.016 [2]
Total EUR			7.956	1.446	7.965	1.443	7.965	1.451	6.685	3.525
Total Profit			6.518		6.532		6.514		3.159	

Note: [1] Tables 5–7; [2] [39].

The results obtained allow us to open the change path from a linear economy based on extraction, adaptation, distribution, use, and discharge to a circular economy that materializes in repeatedly using water, managing to establish the dynamics of the natural cycle. This achieves a balance between the efficacy of the different treatment systems, their economic viability, and their impact as sustainable behavior on a large scale.

As can be seen in Table 9, it can be stated that the profits represent a gain of approximately 4.04 EUR/m^3 for the industrial sector and 1.70 EUR/m^3 for the sports sector. This indicates that not only the cost incurred by regeneration is recovered but also that profits were obtained. The four selected companies are feasible as end users, obtaining a maximum profit for them, ranging from 0.96 EUR/m^3 and 2.24 EUR/m^3 for the irrigation of golf courses. Despite not having quantified all externalities due to the lack of information, the model shows that since they are only positive externalities, their calculation would only increase the profitability of the proposed project, so the viability will not be affected. The model provides an appropriate tool for the planning and management of water resources.

4. Conclusions

The water sector in Bogotá, Colombia is in a transitional phase with unique opportunities for water reuse to be implemented on a large scale as a sustainable practice within a framework of integrated water management. Water is crucial for economic development since it interacts with all the sectors. The circular economy has become a fundamental model for environmental management, especially in the water sector. The main approach is based mainly on the fact that water reuse can spread the water already used, increasing the availability of water resources. Consequently, reclaimed water can be used in traditional processes that do not require high quality, releasing volumes of better quality for other and more demanding uses [5,40].

The main contribution made by this study is the analysis of the economic viability of the proposed alternative, generating lines of support for decision making. This described methodology makes it possible to ensure that the investment of costs allows for obtaining a maximization of total profits. Likewise, a series of externalities are evaluated to be able to identify, quantify, and value economically their impact on implementation.

The minimum sale price is determined under the net present value (NPV) criterion; the selection of the four companies presents a maximization of total profit for the industrial sector of 5.43 EUR/m^3 and companies in the special sector of 4.15 EUR/m^3. For this reason, the proposed alternative and the selection of possible end users show the feasibility of the implementation. For these companies, the economic profit will be presented at a low price for reclaimed water with a constant supply instead of paying for the conventional source, which is more expensive and is subjected to supply cuts due to dry seasons. This alternative must follow the quality standards that are legally required to guarantee the safety of the reuse of reclaimed water implemented in a circular economy context; in addition, this work provides a scientific contribution that should facilitate a comprehensive evaluation of costs and benefits.

For future research, it is recommended to address limitations that were not taken into account in the development of this methodology, such as the analysis of the selected technologies and determining their reliability, performance, robustness, and resilience; in the economic part, it is recommended to evaluate the price difference in the price of fresh water vs. reclaimed water, since the difference is not notable, making treated and reclaimed water not competitive enough; and in the social part, it is necessary to face challenges for public acceptance of the use of treated and reclaimed water in connection with the industrial, agricultural, or recreational sectors, thus achieving a successful exchange of resources.

Author Contributions: M.A.U.V.: prepared the article including conceptualization, methodology, investigation, writing—original draft, and writing—review, and editing. C.T.P.: investigation, financial methodology, writing—original draft, and editing. H.G.-G.R.: investigation, writing—review and editing. L.S.-A.: prepared the article including conceptualization, methodology, investigation, writing—original draft, and writing—review and editing. All authors have read and agreed to the published version of the manuscript.

Funding: This research received no external funding.

Data Availability Statement: The authors declare that the data supporting the findings of this study are available within the paper.

Conflicts of Interest: The authors declare no conflict of interest.

References

1. Schestak, I.; Styles, D.; Black, K.; Williams, A.P. Circular use of feed by-products from alcohol production mitigates water scarcity. *Sustain. Prod. Consum.* **2022**, *30*, 158–170. [CrossRef]
2. EPA. *U.S. EPA Guidelines for Water Reuse: 2012*; EPA: Washington, DC, USA, 2012.
3. Aldaya, M.M.; Custodio, E.; Llamas, R.; Fernández, M.F.; García, J.; Ródenas, M.Á. An academic analysis with recommendations for water management and planning at the basin scale: A review of water planning in the Segura River Basin. *Sci. Total Environ.* **2019**, *662*, 755–768. [CrossRef] [PubMed]
4. Molina, A.; Melgarejo, J. Water policy in Spain: Seeking a balance between transfers, desalination and wastewater reuse. *Int. J. Water Resour. Dev.* **2015**, *32*, 781–798. [CrossRef]
5. Morote, Á.F.; Olcina, J.; Hernández, M. The Use of Non-Conventional Water Resources as a Means of Adaptation to Drought and Climate Change in Semi-Arid Regions: South-Eastern Spain. *Water* **2019**, *11*, 93. [CrossRef]
6. Mayer, A.; Haas, W.; Wiedenhofer, D.; Krausmann, F.; Nuss, P.; Blengini, G.A. Measuring Progress towards a Circular Economy: A Monitoring Framework for Economy-wide Material Loop Closing in the EU28. *J. Ind. Ecol.* **2019**, *23*, 62–76. [CrossRef]

7. Bagatin, R.; Klemeš, J.J.; Reverberi, A.P.; Huisingh, D. Conservation and improvements in water resource management: A global challenge. *J. Clean. Prod.* **2019**, *77*, 1–9. [CrossRef]
8. Chen, Z.; Ngo, H.H.; Guo, W. A Critical Review on the End Uses of Recycled Water. *Crit. Rev. Environ. Sci. Technol.* **2013**, *43*, 1446–1516. [CrossRef]
9. Voulvoulis, N. Water reuse from a circular economy perspective and potential risks from an unregulated approach. *Curr. Opin. Environ. Sci. Health* **2018**, *2*, 32–45. [CrossRef]
10. Bellver-Domingo, Á.; Hernández-Sancho, F. Circular economy and payment for ecosystem services: A framework proposal based on water reuse. *J. Environ. Manag.* **2022**, *305*, 114416. [CrossRef]
11. Berbel, J.; Expósito, A. Economic challenges for the EU Water Framework Directive reform and implementation. *Eur. Plan. Stud.* **2018**, *26*, 20–34. [CrossRef]
12. Seguí, L.; Alfranca, O.; García, J. Techno-economical evaluation of water reuse for wetland restoration: A case study in a natural park in Catalonia, Northeastern Spain. *Desalination* **2009**, *246*, 179–189. [CrossRef]
13. Mannina, G.; Gulhan, H.; Ni, B.J. Water reuse from wastewater treatment: The transition towards circular economy in the water sector. *Bioresour. Technol.* **2022**, *363*, 127951. [CrossRef]
14. Mauchauffee, S.; Denieul, M.P.; Coste, M. Industrial wastewater re-use: Closure of water cycle in the main water consuming industries—The example of paper mills. *Environ. Technol.* **2012**, *33*, 2257–2262. [CrossRef]
15. Camacho, A. Ministerio de Relaciones Exteriores. Personal Communication. 2018.
16. Bohorquez, D. 'Alcaldía de Bogotá'. Personal communication. 2018.
17. Romero, C. 'Golf club and park La Florida'. Personal communication. 2018.
18. Paez, O. 'Papeles primavera'. Personal communication. *Cundinamarca*, 2018.
19. Soza, W. 'Sellopack S.A.S.'. Personal communication. 2018.
20. Lopez, E. 'Licorera Cundinamarca'. Personal communication. 2018.
21. Ministerio de Ambiente y Desarrollo Sostenible. *Resolución Número 1207 de 2014 por la Cual se Adoptan Disposiciones Relacionadas con el uso de Aguas Residuales Tratadas*; Ministerio de Ambiente y Desarrollo Sostenible: Bogotá Colombia, 2014.
22. OMS. Informe 2015 del PCM: Datos Esenciales. 2015. Available online: https://www.who.int/water_sanitation_health/monitoring/jmp-2015-key-facts/es/# (accessed on 24 August 2021).
23. AMB. EDAR del Besòs. 2021. Available online: https://www.amb.cat/web/ecologia/aigua/instalacions-i-equipaments/detall/-/equipament/edar-del-besos/275728/11818 (accessed on 24 August 2021).
24. Ministerio de Medio Ambiente. *Precios y Costes de los Servicios del Agua en España. Informe Integrado de Recuperación de Costes de los Servicios de Agua en España. Artículo 5 y Anejo III de la Directiva Marco de Agua*; Ministerio de Medio Ambiente: Madrid, Spain, 2007.
25. OCWD. Water Reuse. 2021. Available online: https://www.ocwd.com/what-we-do/water-reuse/ (accessed on 24 August 2021).
26. Chen, R.; Wang, X.C. Cost–benefit evaluation of a decentralized water system for wastewater reuse and environmental protection. *Water Sci. Technol.* **2009**, *59*, 1515–1522. [CrossRef] [PubMed]
27. Segui-Amortegui, L.; Alfranca-Burriel, O.; Moeller-Chavez, G. Economic analysis of wastewater reuse projects: A methodology for private reuse and public reuse cases. *Tecnol. Cienc.* **2008**, 316–331.
28. Pulido, R. Ptar Salitre. Acueducto de Bogotá. 2018.
29. Chen, L.; Liao, Y.; Ma, X. Economic analysis on sewage sludge drying and its co-combustion in municipal solid waste power plant. *Waste Manag.* **2021**, *121*, 11–22. [CrossRef] [PubMed]
30. Szypulska, D.; Kokurewicz, Ł.; Zięba, B.; Miodoński, S.; Muszyński-Huhajło, M.; Jurga, A.; Janiak, K. Impact of the thermal drying of sludge on the nitrogen mass balance of a WWTP, and GHG emissions with classical and novel treatment approach—A full-scale case study. *J. Environ. Manag.* **2021**, *294*, 113049. [CrossRef]
31. Leverenz, H.L.; Tchobanoglous, G.; Asano, T. Direct potable reuse: A future imperative. *J. Water Reuse Desalination* **2011**, *1*, 2–10. [CrossRef]
32. Philip, L.; Ramprasad, C.; Krithika, D. Sustainable Wastewater Management Through Decentralized Systems: Case Studies. In *Water Scarcity and Ways to Reduce the Impact*; Springer: Cham, Switzerland, 2019; pp. 15–45. [CrossRef]
33. Yi, L.; Jiao, W.; Chen, X.; Chen, W. An overview of reclaimed water reuse in China. *J. Environ. Sci.* **2011**, *23*, 1585–1593. [CrossRef]
34. Seiple, T.E.; Skaggs, R.L.; Fillmore, L.; Coleman, A.M. Municipal wastewater sludge as a renewable, cost-effective feedstock for transportation biofuels using hydrothermal liquefaction. *J. Environ. Manag.* **2020**, *270*, 110852. [CrossRef]
35. Lee, E.; Oliveira, D.S.B.L.; Oliveira, L.S.B.L.; Jimenez, E.; Kim, Y.; Wang, M.; Ergas, S.J.; Zhang, Q. Comparative environmental and economic life cycle assessment of high solids anaerobic co-digestion for biosolids and organic waste management. *Water Res.* **2020**, *171*, 115443. [CrossRef] [PubMed]
36. Butler, E.; Howell, N.; Guerrero, B. Finding new sources of water for semi-arid cities in unlikely places. *Environ. Sci. Pollut. Res.* **2020**, *27*, 6112–6125. [CrossRef] [PubMed]
37. Bolinches, A.; De Stefano, L.; Paredes-Arquiola, J. Too expensive to be worth it? A methodology to identify disproportionate costs of environmental measures as applied to the Middle Tagus River, Spain. *J. Environ. Plan. Manag.* **2020**, *63*, 2402–2424. [CrossRef]
38. Ross, B.N.; Lancellotti, B.V.; Brannon, E.Q.; Loomis, G.W.; Amador, J.A. Greenhouse gas emissions from advanced nitrogen-removal onsite wastewater treatment systems. *Sci. Total Environ.* **2020**, *737*, 140399. [CrossRef]

39. Shahabadi, M.B.; Yerushalmi, L.; Haghighat, F. Estimation of greenhouse gas generation in wastewater treatment plants—Model development and application. *Chemosphere* **2010**, *78*, 1085–1092. [CrossRef]
40. Melgarejo, J.; Prats, D.; Molina, A.; Trapote, A. A case study of urban wastewater reclamation in Spain: Comparison of water quality produced by using alternative processes and related costs. *J. Water Reuse Desalination* **2016**, *6*, 72–81. [CrossRef]

water

MDPI

Article

Promoting Circular Economy in the Palm Oil Industry through Biogas Codigestion of Palm Oil Mill Effluent and Empty Fruit Bunch Pressed Wastewater

Chaisri Suksaroj [1], Kanokwan Jearat [2], Nutthayus Cherypiew [3], Cheerawit Rattanapan [4] and Thunwadee Tachapattaworakul Suksaroj [4,*]

[1] Research Center for Sustainable Development, Department of Irrigation Engineering, Faculty of Engineering at Kamphaeng Saen, Kasetsart University, Kamphaeng Saen Campus, Nakhon Pathom 73140, Thailand; chaisri.s@ku.th

[2] Faculty of Management Sciences, Nakhon Si Thammarat Rajabhat University, Tha Ngio, Muang Nakhon Si Thammarat 80280, Thailand; kanokwan.micro19@gmail.com

[3] Environmental Management Program, Faculty of Environmental Management, Prince of Songkla University, Songkla 90112, Thailand; nutthatus_pto@hotmail.com

[4] ASEAN Institute for Health Development, Mahidol University, Salaya, Putthamonthon, Nakhon Pathom 73170, Thailand; cheerawit.rat@mahidol.edu

* Correspondence: thunwadee.suk@mahidol.edu

Abstract: This research aimed to investigate the biogas production and circular economy perspective in the palm oil industry through codigestion of oil palm empty fruit bunch (EFB) pressing wastewater and palm oil mill effluent (POME). The EFB pressing method constitutes an alternative new technology used to extract the remaining oil, increasing palm oil product; however, it produces highly polluted wastewater. Batch experiments were carried out at 35 °C to investigate the optimal ratios of EFB wastewater, inoculums, and POME. The optimal condition was 45% POME + 50% seed + 5% EFB wastewater. This condition was then used in semicontinuous fermentation where the optimal hydraulic retention time (HRT) totaled 25 days. The accumulated biogas was 18,679 mL/L while the accumulated methane totaled 6778 mL/L. The methane content was 62%, and the COD removal efficiency was 67%. The sludge produced from the HRT 25-days digester complied with the organic compost standard which could be further used to nourish the soil. An economic analysis of the EFB pressing project revealed a higher internal rate ratio with shorter payback compared with the conventional process. These results provide information on the circular economic approach to promote sustainable palm oil processing.

Keywords: biogas; palm oil industry; circular economy; clean energy; sustainability

check for updates

Citation: Suksaroj, C.; Jearat, K.; Cherypiew, N.; Rattanapan, C.; Suksaroj, T.T. Promoting Circular Economy in the Palm Oil Industry through Biogas Codigestion of Palm Oil Mill Effluent and Empty Fruit Bunch Pressed Wastewater. *Water* **2023**, *15*, 2153. https://doi.org/10.3390/w15122153

Academic Editors: Tao Zhang, Jing Yuan, Huu-Tuan Tran and Muhammad-Jamal Alhnidi

Received: 10 May 2023
Revised: 31 May 2023
Accepted: 2 June 2023
Published: 7 June 2023

1. Introduction

The ASEAN (Association of Southeast Asian Nations) Region stands as the primary hub for palm oil production, given its position as the largest producer. The palm oil industry holds significant importance, especially in the Republic of Indonesia and Malaysia. These two countries have emerged as the top global primary producers and exporters, accounting for 56 and 30% of the world's palm oil supply, respectively. Additionally, Thailand secures its place as the fourth largest palm oil exporter globally. Although Thailand's contribution to global palm oil supply is approximately 2% [1], it remains a crucial maincrop that significantly contributes to the economic growth of both the country and the entire region. The typical process for extracting palm oil is a wet method using a large amount of water in the production process, resulting in generating palm oil effluent (POME) at a rate of about 0.7 to 0.9 cubic meters per ton of fresh palm fruit [2,3]. POME is a nonharmful waste; however, it will pose environmental issues because of the vast oxygen draining capacity in oceanic frameworks because of natural and supplemental substances. The wastes are in the form of high organic matter concentration, such as cellulosic wastes with a mixture of carbohydrates and oils. This effluent has a dark brown color and contains organic matter at a concentration of over 20,000 mg/L [4]. The discharge of untreated POME creates adverse impacts to the environment. Currently, anaerobic action is commonly used to treat POME, efficiently removing organic matter and producing biogas as a renewable energy source [5,6].

Additionally, aside from POME, palm oil processing generates various waste products, including oil palm trunks (OPT), oil palm fronds (OPF), empty fruit bunches (EFBs), palm pressed fibers (PPF), palm kernel shells, and less fibrous material, such as palm kernel cake [7]. Among these waste materials, a substantial amount and mass are attributed to fresh EFBs. For every 100 kg of oil palm fresh fruit bunches processed for oil production, approximately 30 to 60 kg of oil palm EFBs are discarded as waste. The improper disposal of EFBs on land leads to pollution in the surrounding areas, as these EFBs still contain oil that can contaminate the local environment. To address this issue, composting has emerged as a favored alternative for managing solid waste in Thailand and other countries. Due to the residual oil content in the EFB, palm oil factories in Thailand now employ compression techniques to extract the remaining oil, resulting in increased productivity and providing a higher yield of compostable fiber after the re-pressing process.

Wastewater obtained from EFB pressing exhibits higher levels of chemical oxygen demand (COD), biological oxygen demand (BOD), suspended solids (SS), and various substances compared with the typical wastewater produced during crude palm oil extraction processes. These characteristics make it a viable resource for biogas production and energy generation. However, if the EFB pressing wastewater is introduced directly in the fermentation process, it could pose significant toxicity issues for microorganisms, potentially leading to system failure. Thus, it becomes crucial to mitigate its toxicity by diluting it with general wastewater from the palm oil extraction plant. Furthermore, conducting a cost analysis of this alternative process is essential to determine its feasibility and encourage wider application. Hence, this research aimed to investigate the production of biogas from EFB pressing wastewater through codigestion with general wastewater from the palm oil extraction plant. This study focused on analyzing the composition of EFB pressing wastewater and evaluated the effectiveness of its codigestion with general wastewater in enhancing biogas production and methane content. The findings of this study can serve as a foundation for informed decision making regarding integrating EFB pressing in palm oil extraction and this wastewater feeding in biogas production. This approach has the potential to optimize wastewater management, minimize resource consumption in renewable energy production, and enhance the value of waste generated during the palm oil extraction process.

2. Materials and Methods

2.1. Substrates

The characteristics of POME and EFB pressed wastewater were collected from the receiving tanks of a palm oil factory using the grab sampling method. The inoculum used for the study was collected from the anaerobic wastewater treatment plant of the same factory and stored at a temperature of 4 °C until analysis. The collected samples were analyzed for their characteristics, including COD, BOD, pH, SS, total solids (TS), volatile solids (VS), alkalinity, volatile fatty acids (VFA), ammonia nitrogen (NH_3-N), total Kjeldahl nitrogen (TKN), mixed liquor SS (MLSS), and grease and oil (G&O), using the Standard Methods for the Examination of Water and Wastewater, 23rd Edition [8]. The characteristics of the pressing machine, EFB wastewater, and EFB before and after pressing are illustrated in Figure 1.

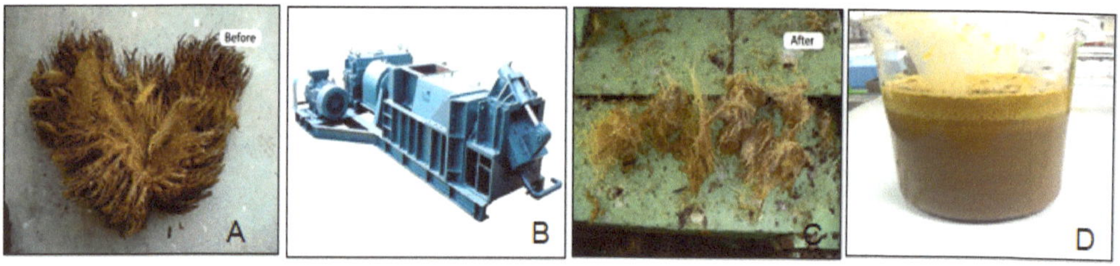

Figure 1. EFB before pressing (**A**), pressing machine (**B**), EFB after pressing (**C**), and EFB wastewater (**D**).

2.2. Laboratory-Scale Batch Reactor

The laboratory-scale reactor system consists of a glass bottle with a total volume of 1 L and a working volume of 0.5 L. The bottle was sealed with a rubber septum and covered with parafilm to ensure airtightness. The anaerobic fermentation process was conducted at a temperature of 35 °C with daily agitation by shaking the bottle once a day. The biogas produced in the bottle was displaced in a glass bottle filled with water to measure its volume daily. The gas composition was analyzed using gas chromatography (GC, Hewlett-Packard Model 6890, Palo Alto, CA, USA) every seven days. The percentage of methane analyzed was multiplied by the measured volume of total biogas to calculate the daily production of methane. The types and amounts of substrates used in the experiment are presented in Table 1.

Table 1. Types of substrates and the quantity of POME, seed inoculum, and EFB wastewater used in the batch-type biogas production experiment, totaling 500 mL.

Set No.	Type of Substrate	POME (mL)	Inoculum (mL)	EFB Press (mL)	C:N
1	65% POME + 35% seed	325.00	175.00	-	140:1
2	62.5% POME + 35% seed + 2.5% EFB ww	316.80	175.00	8.20	142:1
3	60% POME + 35% seed + 5% EFB ww	308.75	175.00	16.25	135:1
4	55% POME + 35% seed + 10% EFB ww	292.50	175.00	32.50	208:1
5	50% POME + 50% seed	250.00	250.00	-	63:1
6	47.5% POME + 50% seed + 2.5% EFB ww	243.75	250.00	6.25	116:1
7	45% POME + 50% seed + 5% EFB ww	237.50	250.00	12.50	126:1
8	40% POME + 50% seed + 10% EFB ww	225.00	250.00	25.00	147:1
9	25% POME + 75% seed	125.00	375.00	-	74:1
10	22.5% POME + 75% seed + 2.5% EFB ww	121.88	375.00	3.12	93:1
11	20% POME + 75% seed + 5% EFB ww	118.75	375.00	6.25	117:1
12	15% POME + 75% seed + 10% EFB ww	112.50	375.00	12.5	78:1

Note: ww—wastewater.

2.3. Semicontinuous Laboratory-Scale Reactor

The experimental setup for studying the production of biogas in a semicontinuous laboratory-scale reactor system consisted of a 4 L brown-colored glass bottle with a cylindrical shape as a fermenter, with a working volume of 3 L, and a 1 L gas storage system connected by a balloon tube to store the gas. The mixture was stirred using a small water pump to ensure adequate mixing inside the fermenter. The experiment was conducted in a semicontinuous completely stirred tank reactor (CSTR) system, controlling appropriate experimental conditions for anaerobic fermentation without air circulation, at a constant temperature of 35 ± 1 °C using a heater and a water bath to maintain the temperature level constant.

2.4. Codigestion Experiments

The study was conducted by codigesting normal wastewater from palm oil extraction and EFB wastewater to find the optimal ratio producing the highest methane yield. The experiment was conducted in a batch laboratory system, and the results were used to guide the following experiment, which was conducted using a semicontinuous system.

2.4.1. Batch Experiments

The batch experiments were conducted using the following steps: all three types of fermentable materials were brought to room temperature before conducting the experiment, the three types of fermentable materials were separated and placed in a container, the volume of each fermentable material was measured and combined according to the ratio calculation of the COD/N (chemical oxygen demand/nitrogen) ratio [9], using cofermentable materials in the ratios of 2.5, 5, and 10% v/v, and the fermentation material was added to the anaerobic digestion system as a one-time feeding. Then, the starting inoculum was used at 20, 35, 50, and 75% of the working volume of the fermentation material to determine the appropriate proportion of the microorganisms in gas production (Table 1) as recommended by [10]. They suggested that the amount of inoculum should be no less than 10% of the working volume, and then the pH was adjusted in the range of 7.0 to 7.2. The anaerobic fermentation system was created from a 1 L glass bottle with a working volume of 0.5 L. The control set, including main materials mixed with different ratios of inoculum, was used to establish a baseline to compare with experimental treatments, and the system was in batch mode.

Next, the fermentation bottle that had been filled with the fermentable material was placed in a hot water bath at a temperature of 35 ± 1 °C to measure the volume of biogas produced daily. Some of the water was replaced with gas to analyze the gas composition using GC. The parameters measured in wastewater samples are according to [11] (Table S1).

2.4.2. Semicontinuous Experiments

From the batch experiment, the optimum ratio and conditions efficient in producing the maximum methane yield were selected. This involves studying the optimum storage time and the efficiency in treating organic matter, as well as the rate of methane production in the system. In this study, a semicontinuous CSTR with an airless system was used to control the experimental conditions suitable for anaerobic fermentation at a constant temperature of 35 ± 1 °C. The composition of the gas was analyzed using GC every seven days, and the obtained methane percentage was multiplied by the volume of biogas produced to calculate the amount of methane produced each day. The diagram of the semicontinuous digester is presented in Figure 2.

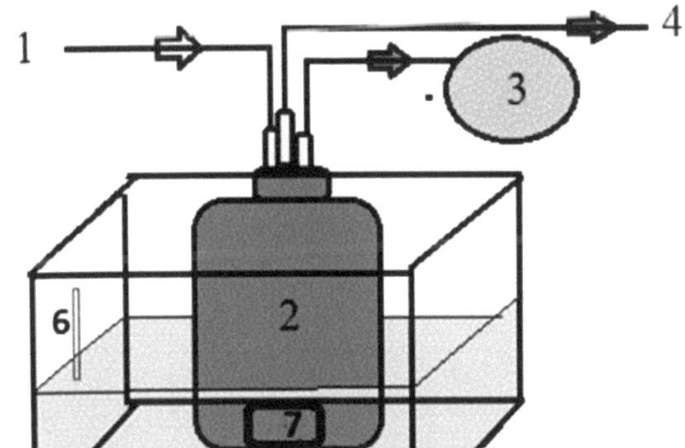

Figure 2. Diagram of semicontinuous digester including (1) inlet; (2) digester 4 L; (3) gas collector; (4) outlet; (5) heater bath control temperature for 35 ± 1 °C; (6) thermometer; and (7) stirrer.

The reactor performs experiments at different storage times or hydraulic retention times (HRTs) of 30, 25, 20, 14, and 7 days, consecutively. The longer HRT was preferred by the high buffer system. The short HRT is desirable in terms of minimizing the capital cost. The variation in storage times studied were in the range of the optimized anaerobic degradation rate from many studies [12–15]. A mixing rate of 100% means stirring for 24 h. The effluent discharge from the reactors was carried out once daily with volumes of 100, 120, 150, 214, and 428 mL, consecutively (Table 2). The volume of wastewater discharged depended on the liquid retention in the system. During the discharge, the wastewater was sucked through a feeding tube before refilling it with new wastewater. Continuous stirring was maintained during the discharge to ensure that the effluent had a consistent texture. Water samples were collected for chemical analysis (Table 3) to monitor the system performance by considering variable parameters, such as pH, alkalinity, COD, VFA, and MLSS. The biological gas produced was discharged in the balloon connected to the gas counter. The amount of biological gas produced daily was then measured. Gas samples were collected in a tube and placed in a vacuum test tube to analyze the gas composition. After measuring the gas, the gas line was closed. After completing the wastewater discharge, the effluent line was closed, and the wastewater line was opened to refill the system with the volume of discharged wastewater. The water inlet was then closed, and the balloon was opened to collect the biogas. The system was run until it reached a steady state, based on the volume and composition of biogas that varies within ±15%.

Table 2. Type of fermentation material, retention time, and volume of waste added and discharged daily in the semicontinuous system for biogas production experiments.

Set No.	Type of Substrate	Working Volume (mL)	HRT (Day)	Waste Added–Discharged (mL)
1	POME + 50% seed + 5% EFB ww	3000	7	428
2	POME + 50% seed + 5% EFB ww	3000	14	214
3	POME + 50% seed + 5% EFB ww	3000	20	150
4	POME + 50% seed + 5% EFB ww	3000	25	120
5	POME + 50% seed + 5% EFB ww	3000	30	100

Note: ww—wastewater.

Table 3. Measured parameters, analytical method, frequency sampling of anaerobic fermentation systems.

Parameter	Method	Frequency of Monitoring
Total solids	Gravimetric method	every 3 days
COD	Close reflux, Titrimetric	every 3 day
Temperature	Thermometer	every day
pH	pH meter	every day
Alkalinity	Direct titration method	every 6 days
VFA	Direct titration method	every 6 days
Biogas production	fluid displacement	every day
Biogas composition	GC-TCD	every week

2.4.3. Measuring the Volume of Biogas, Analyzing the Composition of the Gas and Liquid Samples

The biogas produced was measured in volumes using water displacement and collected for analysis. The biogas composition, including methane (CH_4) and carbon dioxide (CO_2), was analyzed every 12 days using GC with a thermal conductivity detector (TCD) sensor. The column used was an HP-plot/Q with a diameter of 1 mm and a length of 2 m, using helium gas as the carrier. The temperature in the injector, detector, and oven was maintained at 250 °C.

The characteristics of the influent and effluent of the anaerobic digestion system were analyzed according to Standard Methods for the Examination of Water and Wastewater, 22nd edition [11]. The parameters to be analyzed included total solids (TS), volatile solids (VS), volatile fatty acids (VFA), pH, alkalinity, and COD (Table 3).

2.4.4. Analysis of the Influent and Effluent Characteristics and Sludge from the Anaerobic Digestion System

This experiment used the residual sludge from the appropriate retention time of a semicontinuous anaerobic fermentation process. The analysis aimed to determine the amount of plant nutrients, including total nitrogen, total phosphorus, and potassium as well as the C/N ratio, moisture content, and organic carbon content. The results were used to apply the sludge as a soil conditioner material and were compared to the Thai Industrial Standard for organic compost set by the Industrial Product Standards Office, Ministry of Industry [16].

2.5. Circular Economy Capacity Analysis

The results obtained from this research experiment could be used as information to support decision making for entrepreneurs interested in producing electricity using the anaerobic wastewater treatment system of crude palm oil mills. The analysis focuses on identifying the costs, returns, and feasibility of using EFB wastewater in codigestion with POME to produce biogas as an energy source or fuel to generate electricity. Economic tools are used to evaluate the profitability and feasibility of the proposed system. This study refers to reliable secondary data and information obtained from supporting mills, which define the conditions for analysis as follows.

(1) Setting assumptions for project analysis.
  - The operational period of the biogas power plant is five years based on the minimum years of the power purchase agreement with the regional electricity authority, in a firm pricing scheme.
  - The return on investment starts from year 1 and ends at the project's completion.
  - No salvage value is considered at the end of the operational period.
  - A discount rate of 10% is used.

(2) Analyzing the cost of the biogas power plant can be categorized as follows.
  (2.1) The investment cost or fixed cost is the initial cost incurred from constructing the power plant and equipment, including the following:
    - The cost of the biogas system, consisting of the biogas digester tank, gas storage tank, piping system, and monitoring and control equipment.
    - The cost of the electricity production system, consisting of the biogas-powered electricity generator, control equipment, gas piping system, and electrical wiring.
  (2.2) The operating cost or variable cost is the expense for general management and production, including labor cost and operational and maintenance expenses, calculated on an annual basis.

(3) Project Return Analysis: The business will generate revenue from selling electricity produced and selling it back to the state through the Small Power Producer (SPP) and Very SPP (VSPP) policies. This includes the value of the building and land when the project is completed. Machinery and equipment are considered to have zero salvage value at the end of their useful life.

3. Results and Discussion

3.1. The Chemical Characteristics of Wastewater from Palm Oil Mill Extraction, Wastewater from EFB Pressing, and Sludge

Both types of wastewater have different physical characteristics due to their different sources in the production process. Additionally, both types of wastewater have high levels of pollution that can cause water contamination if released into the external environment, especially the wastewater from the EFB pressing. This waste is generated as an additional byproduct of the main production process. The process of EFB pressing uses heat to help release organic substances that remain in the EFB, resulting in a large amount of organic matter contaminating the wastewater.

Typically, the factory can treat the general wastewater from the crude palm oil extraction process directly with an anaerobic sequencing batch reactor and upflow anaerobic sludge blanket treatment systems producing biogas as a renewable energy source for electricity generation. However, the factory collects the wastewater from the EFB extraction for later treatment. The wastewater is fed into the treatment system slowly and gradually due to its high level of pollution, which could cause the treatment system to fail if too much wastewater is added at once.

The results of the chemical analysis of two types of wastewater from the palm oil extraction process and sludge showed high levels of pollution in both types, as indicated by their COD and BOD values. The general wastewater from the palm oil extraction process had COD and BOD values of 61,000 and 29,798 milligrams per liter, respectively, while the wastewater from the EFB extraction had COD and BOD values of 74,750 and 31,339 mg/L, respectively. When comparing the pollution levels of both types of wastewater, the wastewater from the EFB of oil palm had higher levels of pollution than the general wastewater from the palm oil extraction process (Table 4).

Table 4. Chemical characteristics of wastewater and sludge used in the experiment.

Parameter	POME	EFB Wastewater	Sludge
pH	4.6	4.9	-
COD (mg/L)	61,000	74,750	-
BOD (mg/L)	29,798	31,339	-
TKN (mg/L)	550	325	-
NH_3-N (mg/L)	2.75	5.25	-
Grease and oil (mg/L)	970	8590	-
TS (mg/L)	20,010	96,320	-
SS (mg/L)	16,250	91,240	-
VFA (mg/L)	5288	10,613	-
MLSS (mg/L)	18,000	-	7.63
BOD:COD	0.48	0.41	

Based on these properties, if methane production is considered, 1 g of degraded COD will produce 0.351 L of methane. This means that the wastewater from the palm oil extraction process can produce 21.41 L of methane, while the wastewater from the EFB extraction can produce 26.24 L of methane. This theory shows that the wastewater from EFB pressing can produce more methane than the general wastewater from the palm oil extraction process, despite both types of wastewater having high levels of organic matter. This is because both types of wastewater can be a sufficient source of methane when treated in anaerobic conditions due to their high levels of easily biodegradable organic matter, as evidenced by their BOD/COD ratios ranging from 0.4 to 0.5 [17,18] (Table 4).

Apart from high levels of organic compounds in the form of BOD or COD, both types of wastewater also contain high amounts of volatile fatty acids. These fatty acids are considered organic pollutants in wastewater. However, they can be converted to biogas, although the process of converting fat substrate into biogas is difficult [5]. Simply adding only empty palm oil fruit bunches pressing wastewater in the anaerobic digestion process to produce methane may result in system failure due to the high levels of impurities and volatile fatty acids. This can cause volatile fatty acids to accumulate, which can be toxic to the group of microorganisms producing methane, leading to inhibition of the methane production process. Therefore, codigestion with palm oil mill effluent can dilute the toxicity that may occur in the system and adjust the nutrient ratio to be suitable for efficient microbial work, resulting in increased biodegradation rates and biogas production [19].

The black-colored microorganisms in the sludge have a high SS concentration. Regarding the concentration of microorganisms in the reactor tank, the MLSS is 18,000 mg/L, and the pH value is 7.63. The sludge was analyzed and used as inoculum in the fermentation system. The inoculum was taken directly from the biogas production tank of the factory, making it suitable for use as an inoculum in the fermentation process, which was consistent with other research [5,6,20]. The potential for generating gas from palm oil mill effluent was studied using a 2 L CSTR reactor with a stirring speed of 100 rounds per minute, operating at a thermophilic state and temperature of 55 °C for 6 days. The study found that at an MLSS concentration of 14,000 mg/L, the COD removal efficiency reached 90%, with a methane content of 64%.

3.2. The Study of Codigestion of POME and EFB Wastewater at Various Ratios Using a Batch System

The results of a batch codigestion process, using different mixing ratios and lasting ten days until the end of the fermentation process, are presented in Figure 3. Experimental set 7, comprising a mixing ratio of 45% POME + 50% seed + 5% EFB wastewater, exhibited the highest cumulative biogas and methane yields of 396 ± 4.58 mL and 294 ± 3.51 mL, respectively, compared with the other 11 experimental sets.

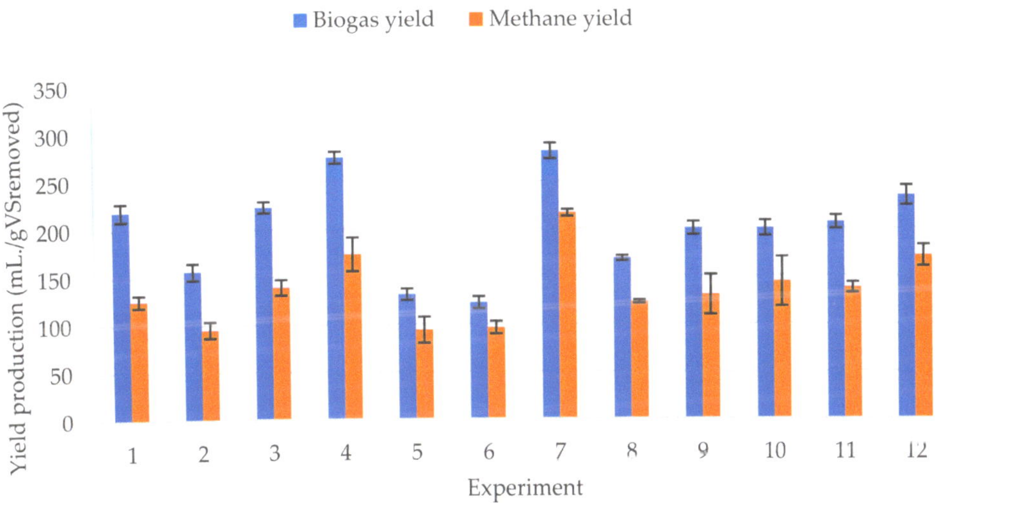

Figure 3. Biogas and methane yield for each experiment.

This resulted in the highest methane production efficiency of 0.016 L CH_4/g VS removed or 0.18 L CH_4/g VS added, increasing when cofermented with EFB wastewater, compared with the fermentation of POME and inoculum alone, as shown in Figure 4.

In Table 5, the biogas production is expressed in units of L CH_4/g VS removed to show the amount of methane produced from the organic matter decomposed in the substrate and in units of L CH_4/g VS added to show the total methane produced from all substrates, including the portion that cannot be decomposed by microorganisms.

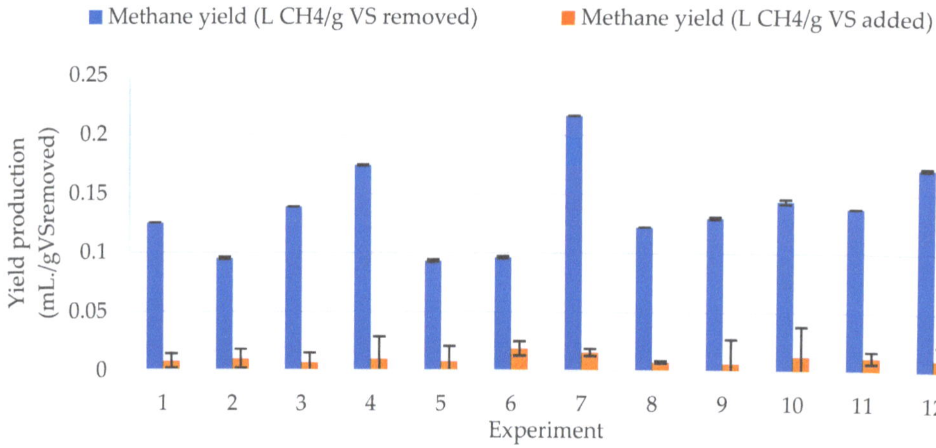

Figure 4. Methane yield by VS removal and VS added of each experiment.

Table 5. Biomethane potential of batch experiments.

No.	Condition (Total Volume 100%)	mL CH_4	VS Added (g/L)	VS Removed (g/L)	Methane Yield	
					L CH_4/gVS Removed	L CH_4/gVS Added
1	POME + 35% seed (control)	124 ± 6.43	16.12	0.99	0.125 ± 0.006	0.008 ± 0.000
2	POME + 35% seed + 2.5% EFB ww	211 ± 18.52	20.28	2.22	0.095 ± 0.008	0.010 ± 0.001
3	POME + 35% seed + 5% EFB ww	150 ± 9.02	22.41	1.08	0.139 ± 0.008	0.007 ± 0.000
4	POME + 35% seed + 10% EFB ww	170 ± 17.62	17.26	0.98	0.174 ± 0.018	0.010 ± 0.001
5	POME + 50% seed (control)	157 ± 21.36	15.98	1.67	0.093 ± 0.013	0.008 ± 0.001
6	POME + 50% seed + 2.5% EFB ww	233 ± 14.64	19.73	2.42	0.096 ± 0.006	0.019 ± 0.001
7	POME + 50% seed + 5% EFB ww	294 ± 3.51	18.46	1.36	0.216 ± 0.003	0.016 ± 0.000
8	POME + 50% seed + 10% EFB ww	140 ± 1.53	16.59	1.14	0.122 ± 0.001	0.008 ± 0.000
9	POME + 75% seed (control)	138 ± 15.72	18.99	1.06	0.130 ± 0.020	0.007 ± 0.001
10	POME + 75% seed + 2.5% EFB ww	229 ± 39.89	17.41	1.59	0.144 ± 0.025	0.013 ± 0.002
11	POME + 75% seed + 5% EFB ww	209 ± 8.33	17.61	1.52	0.138 ± 0.005	0.012 ± 0.000
12	POME + 75% seed + 10% EFB ww	212 ± 13.20	19.87	1.24	0.172 ± 0.011	0.011 ± 0.001

Note: ww—wastewater.

Experimental set 7 exhibited the highest methane production efficiency, occurring during codigestion of POME + 50% seed + 5% EFB wastewater, with a value of 0.216 ± 0.003 L CH4/g VS removed or 0.0159 ± 0.000 L CH_4/g VS added. This showed that codigestion with EFB wastewater could increase the methane production efficiency compared with the control set (experiment set 3), revealing codigestion of 50% POME + 50% seed without EFB wastewater, with a value of 0.093 ± 0.013 L CH_4/g VS removed or 0.0078 ± 0.001 L CH_4/g VS added. The results demonstrate that codigestion could increase the methane production efficiency up to twofold (Figures 3 and 4), which was consistent with other studies on codigestion to increase the methane production efficiency [9,19,21,22]

Upon analyzing the default parameters listed in Table 6, decreasing the volume of POME resulted in a reduced amount of initial volatile fatty acid (VFA) while increasing the quantity of inoculum. Typically, the recommended range for VFA is between 50 and 500 mg per liter of CH3COOH, with the maximum permissible value of 2000 mg per liter of CH3COOH in the system [23]. However, the initial VFA quantity in sets 1 to 4 exceeded the proposed theoretical value, rendering the conditions unfavorable for the gas production process. Moreover, the initial alkalinity levels in sets 2, 3, 4, and 7 exceeded 5000 mg per liter, whereas the acceptable range of general alkalinity is between 1000 and 5000 mg per liter as calcium carbonate [17]. Nonetheless, upon analyzing the ratio of VFA to bicarbonate alkalinity (VFA/HCO_3), all the experimental sets had values lower than 0.4, suggesting that the system was buffered.

Table 6. Parameters related to before (influent) and after (effluent) digestion processes, initial inoculum at 35%.

Parameter	65% POME + 35% Seed (Control)		62.5% POME + 35% Seed +2.5% EFB ww		60% POME + 35% Seed +5% EFB ww		55% POME + 35% Seed +10% EFB ww	
	Influent	Effluent	Influent	Effluent	Influent	Effluent	Influent	Effluent
pH	7.07	8.30	7.04	8.66	7.04	8.67	7.07	8.77
COD, mg/L	43,444	42,792	43,164	34,639	49,780	31,838	63,383	57,373
SS, mg/L	26,000	16,830	28,870	17,450	24,640	14,120	25,850	17,480
TS, mg/L	45,384	29,788	37,516	35,544	38,584	33,600	50,704	38,728
VS, mg/L	16,124	15,132	20,228	18,004	22,412	17,336	17,256	16,280
Alkalinity, mg/L	4800	5000	6400	8600	5600	7000	6600	8200
VFA, mg $CaCO_3$/L	1900	680	2160	400	2280	640	2300	1160
NH_3-N, mg/L	169	260	125	265	180	295	124	255
TKN, mg/L	309	465	304	464	368	445	305	432
C:N	140:1	92:1	142:1	74:1	135:1	71:1	208:1	133:1

Note: ww—wastewater.

From Tables 6–8, the effluents were lower than the influents, indicating that the organic matter in the system was being degraded into gas. Furthermore, the pH, alkalinity, NH_3-N, and TKN of the effluent increased. The fermentation process ended due to an increase in NH_4 during the final stage, leading to a pH value higher than 8, indicating an imbalance in the system's metabolism that can be toxic to the group of methane-producing bacteria. In general, the appropriate conditions for this group of bacteria should have a pH value within the range of 7 to 8 [24,25].

Table 7. Parameters related to before (influent) and after (effluent) digestion processes, initial inoculum at 50%.

Parameter	POME + 50% Seed (Control)		POME + 50% Seed +2.5% EFB ww		POME + 50% Seed +5% EFB ww		POME + 50% Seed +10% EFB ww	
	Influent	Effluent	Influent	Effluent	Influent	Effluent	Influent	Effluent
pH	7.05	8.54	7.07	8.68	7.06	8.55	7.06	8.55
COD, mg/L	50,245	28,038	62,110	33,630	63,608	31,600	65,919	30,588
SS, mg/L	29,120	24,060	29,360	24,620	27,820	37,310	55,010	49,720
TS, mg/L	16,388	32,412	48,120	44,284	44,968	39,408	39,516	36,228
VS, mg/L	15,180	14,296	19,728	17,312	18,460	17,620	16,588	15,444
Alkalinity, mg/L	4800	5500	4800	6200	5750	4750	4750	5500
VFA, mg $CaCO_3$/L	1260	250	1300	380	1200	325	1100	300
NH_3-N, mg/L	240	355	227	338	231	336	221	312
TKN, mg/L	794	1660	533	616	504	672	448	616
C:N	63:1	17:1	116:1	54:1	126:1	47:1	147:1	50:1

Note: ww—wastewater.

3.3. Study of Codigestion of POME and EFB Wastewater with Various Retention Times Using a Semicontinuous System

The semicontinuous fermentation process was carried out using a mixture ratio of 45% POME, 50% seed, and 5% EFB wastewater. This ratio was used for the initial fermentation and allowed to be retained for 10 to 30 days.

Table 8. Parameters related to before (influent) and after (effluent) digestion processes, initial inoculum at 75%.

Parameter	POME + 75% Seed (Control)		POME + 75% Seed +2.5% EFB ww		POME + 75% Seed +5% EFB ww		POME + 75% Seed +10% EFB ww	
	Influent	Effluent	Influent	Effluent	Influent	Effluent	Influent	Effluent
pH	7.09	8.79	7.09	8.68	7.05	8.64	7.07	8.59
COD, mg/L	58,082	25,034	62,340	22,813	65,741	27,478	74,903	36,947
SS, mg/L	37,520	32,370	41,960	35,330	59,260	28,490	66,050	32,470
TS, mg/L	48,792	44,280	51,476	44,528	49,660	45,980	49,720	44,524
VS, mg/L	18,992	16,648	17,008	16,500	17,212	16,496	19,872	18,836
Alkalinity, mg/L	3750	4750	4750	5000	4500	4500	4400	5200
VFA, mg $CaCO_3$/L	1425	450	850	200	525	250	500	240
NH_3-N, mg/L	252	266	289	390	305	395	315	379
TKN, mg/L	784	840	672	728	560	616	952	1568
C:N	74:1	30:1	93:1	31:1	117:1	46:1	78:1	23:1

Note: ww—wastewater.

3.3.1. Effect of the Retention Time on Methane Production Rate and Methane Yield

According to the study, the cofermentation of 45% POME, 50% seed, and 5% EFB wastewater resulted in the accumulation of biogas and methane. The accumulated biogas volumes after 7, 14, 20, 25, and 30 days of retention time were 33,963, 46,870, 45,841, 67,558, and 65,868 mL, respectively (as shown in Figure 5). The accumulated methane volumes after 7, 14, 20, 25, and 30 days of retention time were 1685, 9029, 10,549, 28,470, and 23,793 mL, respectively. At a retention time of 25 days, the organic loading rate (OLR) was 2.60 g COD/L·day, and the accumulated biogas and methane volumes were higher than those at other retention times during the entire 30-day experiment.

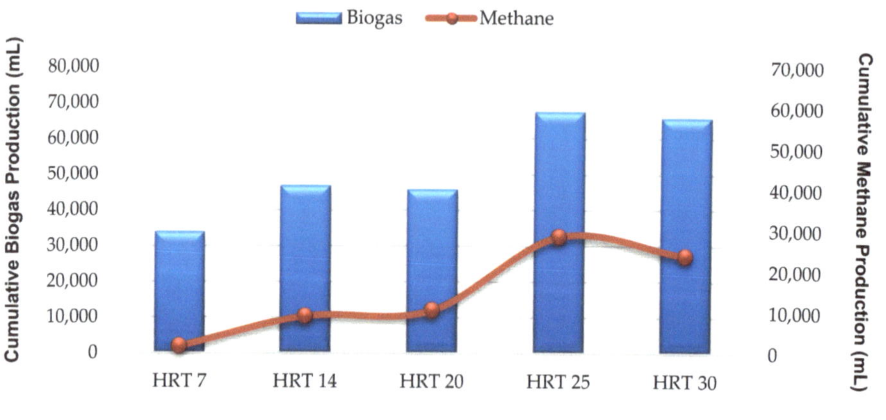

Figure 5. Accumulated biogas and methane volumes at the retention time of 30 days from the semicontinuous fermentation process.

Gas production with methane generated each day (illustrated in Figure 6) was found to have low methane production during the short retention times of 7 and 14 days, with an OLR of 9.27 g COD/L·day and 4.63 g COD/L·day, respectively. This was due to system failure caused by an increased amount of COD from adding wastewater to the system. Methane production stopped at 17 and 19 days, respectively. At a retention time of 20 days with an OLR of 3.25 g COD/L·day, methane production was observed between days 4 and 20 and stopped on day 21. Across those three retention times, the maximum methane production was only 7, 32, and 39%, respectively, indicating that these retention times were unsuitable for methane production. At retention times of 25 and 30 days, methane production was as high as 62 and 53%, respectively, indicating that these retention times were suitable for methane production. The system reached a steady state on days 25 and 22, respectively, and the fermentation process was completed after 30 and 27 days, respectively.

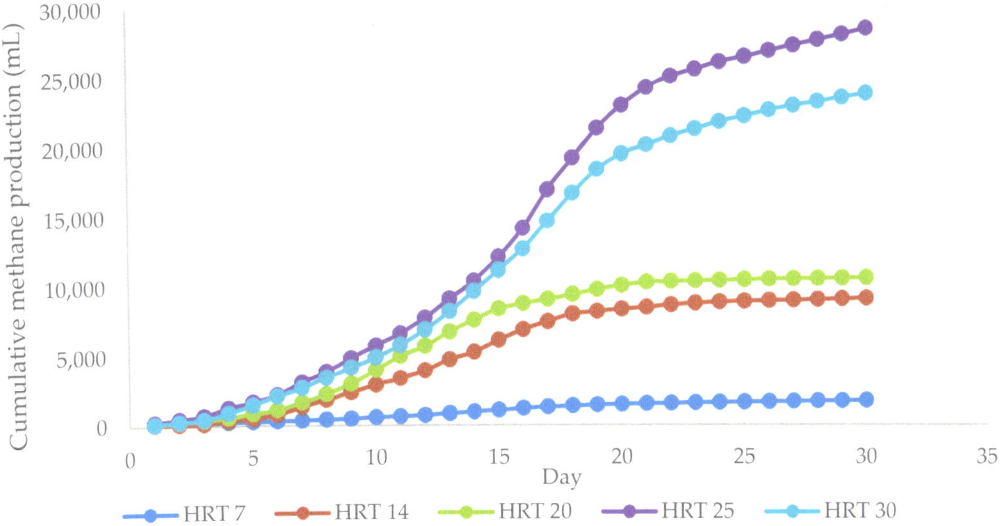

Figure 6. Amount of methane produced each day at different retention times using a semicontinuous system.

The results of the experiment are consistent with [26] study on the codigestion of wastewater and solid waste from an olive oil production plant in a mesophilic condition, with an organic loading rate ranging from 0.67 to 6.67 gCOD/L·day. The study found that biogas production was inhibited when the organic loading rate exceeded 4.67 gCOD/L·day.

Experiments conducted with a wide range of HRTs have proven beneficial for manufacturing operations in subsequent up-scale experiments in the biogas system. The variation in the HRT corresponds to changes in the organic loading rate (OLR), which can be traced back to variations in chemical oxygen demand (COD). Manufacturers can further refine the feeding conditions of the biogas system by calculating the OLR based on the instantaneous COD of the substrate, thus maximizing biogas production.

3.3.2. Changes in the System during Different Retention Times in Semicontinuous Anaerobic Fermentation Process

During the cofermentation process of POME + 50% seed + 5% EFB in a semicontinuous mode at different retention times, at a retention time of 7 days, the pH value (illustrated in Figure 7) decreased rapidly due to the accumulation of volatile fatty acids (VFAs) within the system. At retention times of 14 and 20 days, during the first 1 to 3 days of operation, the pH value decreased, but then gradually increased, even though it remained within the suitable range. However, the high organic loading and short retention time resulted in low methane production and a rapid end to the fermentation process.

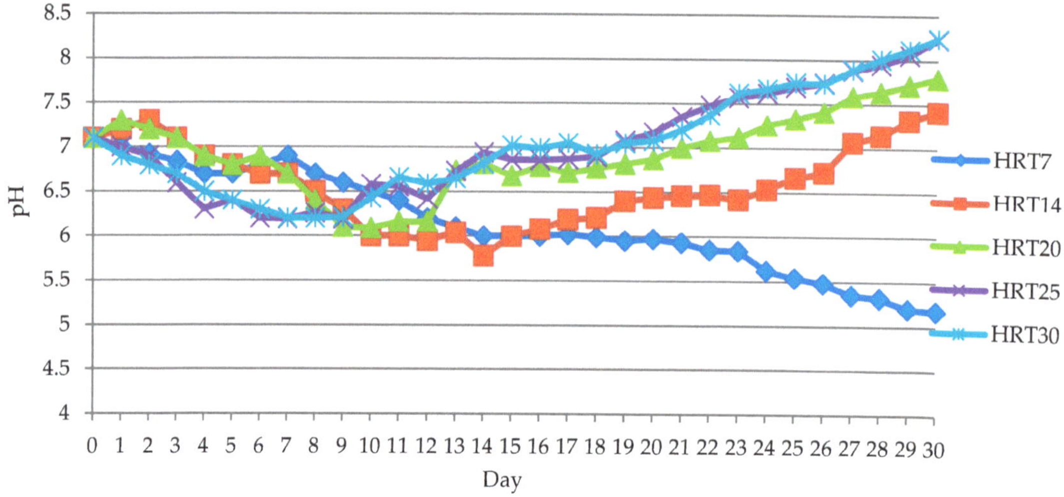

Figure 7. Changes in pH values at different retention times.

At the retention times of 25 and 30 days, the pH value continuously increased and remained within the suitable range, resulting in a higher methane production compared with other retention times. This indicated that a longer retention time was more efficient than a shorter one, especially for systems with high nutrient or COD concentrations. In the case of retention times of 25 and 30 days, the process could run up to 30 and 28 days, respectively. However, if the pH value exceeded 8.0, the fermentation process would end.

Nutrients containing fats and long-chain fatty acids (LCFAs) often lead to easy inhibition of the fermentation process. LCFAs can affect the methanogenesis process and other steps in anaerobic digestion. As shown in Figure 8, VFA accumulation increased at retention times of 7, 14, 20, and 25 days due to the degradation of organic matter under anaerobic conditions. However, at a retention time of 30 days, the VFA levels were slightly increased in the beginning and decreased later than 18 days, indicating the consumption of VFAs by microorganisms to produce methane gas and insufficient organic matter in the later phase (longer HRT provided lower OLR). The alkalinity values continuously decreased throughout all the retention times, affecting the buffer capacity of the reactor and ultimately leading to the end of the fermentation process.

3.3.3. COD Removal Efficiency

After the fermentation process experiment, at retention times of 25 and 30 days, the efficiency of the COD removal was up to 67 and 84%, respectively (Table 9). This was because longer retention times allow microorganisms to use nutrients for a longer period. In contrast, shorter retention times reduced the efficiency of COD removal. Specifically, at a retention time of 20 days, the efficiency of COD removal was only 45%, and at retention times of 7 and 14 days, no efficiency was observed in COD removal due to the short retention time. This was consistent with [26] research on the anaerobic codigestion of industrial wastewater and solid waste from an olive oil factory. The experiment was conducted at an average temperature, with retention times of 12, 24, and 36 days, and the wastewater had COD concentrations of 24, 56, and 80 gCOD/L. The highest organic removal efficiency was 89%, achieved with an organic loading rate of 0.67 gCOD/L/day (with a wastewater concentration of 24 gCOD/L) and a retention time of 36 days.

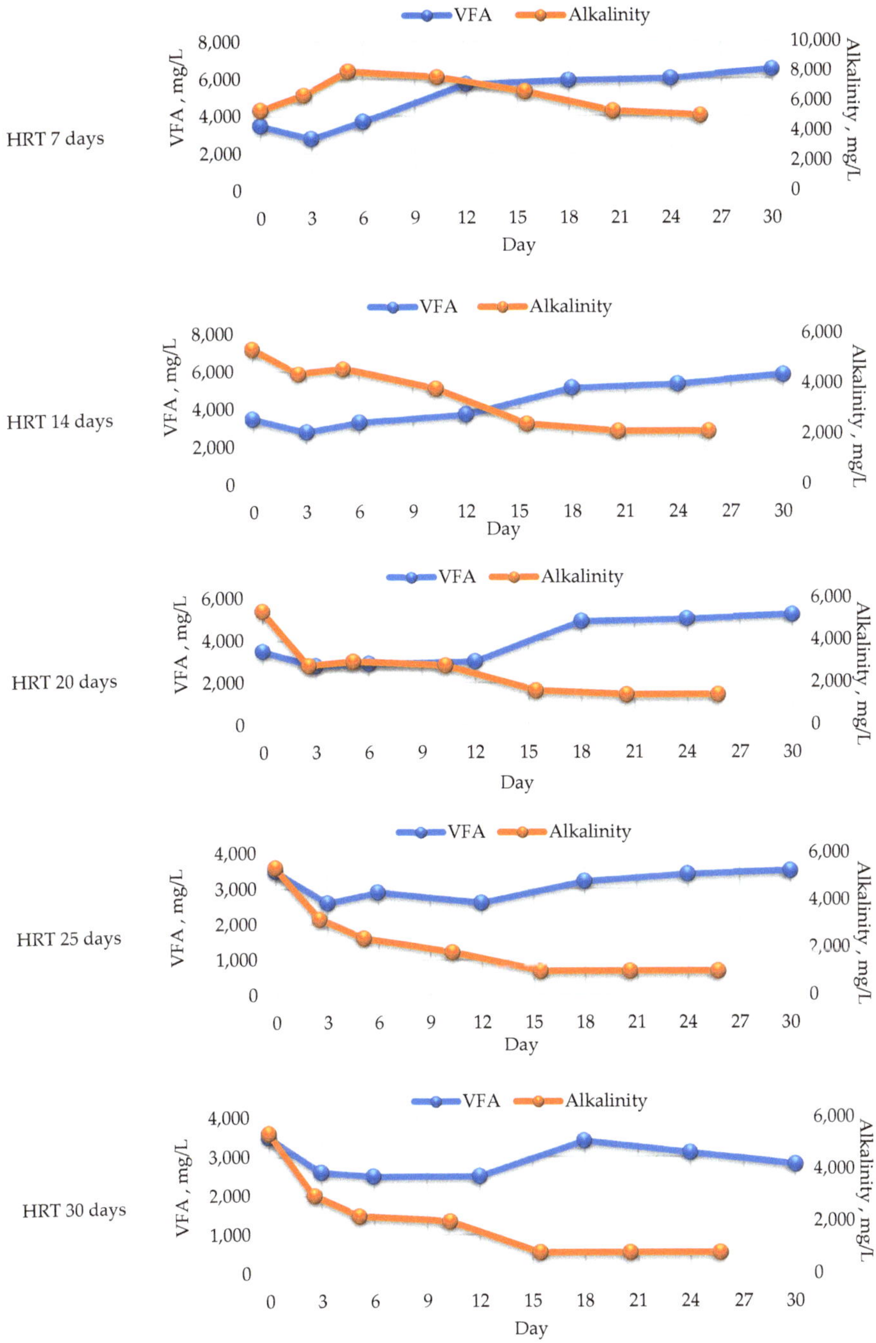

Figure 8. VFA and alkalinity of the experiments with different HRTs.

Table 9. Effluent characteristics of different retention time experiments.

Parameter	Influent	Effluent				
		HRT7	HRT14	HRT20	HRT25	HRT30
pH	7.1	5.18	7.42	7.80	8.24	8.25
COD (mg/L)	96,000	128,000	112,000	52,800	32,000	16,000
TKN (mg/L)	392	252	560	476	588	504
NH_3-N (mg/L)	252	280	224	280	392	392
TS (mg/L)	33,172	29,756	20,748	28,632	29,572	19,332
SS (mg/L)	28,305	17,703	17,790	11,616	20,686	19,508
VFA (mg/L)	3500	5900	5100	4900	3400	3100
Alkalinity (mg/L)	5400	6600	2400	1600	1000	800
C:N	131:1	106:1	92:1	102:1	75:1	49:1
COD removal efficiency (%)		−34	−17	45	67	84

When considering all the consensus results obtained from different HRTs, the 25 days was a suitable HRT with the highest cumulative biogas and methane production, the continuous VFA production during the experimental period contributing to methanogenesis continuously.

3.3.4. Characteristics of Remaining Sludge after Fermentation

An analysis of the characteristics of the sludge residue included the purpose of recycling waste for further use, such as soil conditioner, making fertilizer for plants, and so on. The sludge analyzed (Figure S1) was taken from the fermentation system with a suitable retention period of 25 days, providing the highest amount of nitrogen compared with other retention periods.

The properties of the sludge residue (Table 10) have an organic carbon value of 19.77%, total nitrogen value (N) of 2.62%, total phosphorus value (Total P_2O_5) of 2.80%, and total potassium value (K_2O) of 11.32%. The ratio of carbon to nitrogen was 7.0, whereas the moisture content value was 28.55%. These values fall within the standard range issued by the Department of Agriculture, Ministry of Agriculture and Cooperatives, Thailand [27].

Table 10. Sludge residue characteristics.

Parameter	% *w*/*w*	Standard
Organic carbon	19.77	Not lower than 20
Total N	2.62	Not lower than 1.00
Total P_2O_5	2.80	Not lower than 0.5
Total K_2O	11.32	Not lower than 0.5
C/N ratio	7.00	Not exceeding 20:1
Moisture content	28.55	

3.4. Circular Economy Perspective through Economic Analysis

Data from the factory supported research found that the factory had incurred expenses in setting up a biogas production system equal to 2,625,889 USD, including the construction cost, maintenance cost, electricity cost, and chemicals and materials (see Supplementary Materials Table S2).

During the production of crude palm oil, the factory generates EFBs at a rate of approximately 25% of the fresh fruit bunches. With a current production rate of 720 tons of fresh fruit bunches daily, this results in around 180 tons of EFBs daily. To increase production efficiency, a crude palm oil company has implemented a process of compressing the EFBs, as there is still residual oil in the EFBs that can be extracted and reused. This process increases oil production and helps to reduce oil pollution in the environment. With approximately 180 tons of EFBs daily, the EFB compression process generates wastewater at a rate of 54 cubic meters daily or 1.6×10^4 cubic meters yearly. Additionally, 1 kg of EFB can be converted to 0.00722 kg of oil, and the cost of adding the EFB compression process to the production process was 148,885 USD (Table S3).

From the codigestion experiment, the appropriate ratio of general wastewater from the palm oil extraction factory to EFBs was 45% POME + 50% seed + 5% EFB wastewater. When a working volume of 0.5 L was used with the above ratio, 0.294 L of methane gas was produced, which was greater than the control set (without EFB wastewater), producing only 0.157 L of methane gas. This showed the potential for methane gas production to increase up to twice when using EFB wastewater of 0.0125 L mixed with POME 0.238 L. When the EFB wastewater produced was 1.6×10^7 L/year, it required POME 3.0×10^8 L/year.

Therefore, the total volume of the two materials fermented together was 3.16×10^8 L/year. When the above ratio was tested in a semicontinuous CSTR reactor system, 0.95 L/day of methane gas was produced with an organic feed rate of 0.15 L/day. When the volume of the fermented material was equal to 3.16×10^8 L/year, methane gas could be produced at a rate of 2.0×10^6 m^3/year.

The results of the experiment showed that mixing EFB wastewater with POME could increase the production of biogas. If this project is implemented, the cost can be calculated based on the information presented in Table S4. Project A will produce biogas solely from the POME, requiring 2,625,889 USD, while Project B will produce biogas by codigestion of EFB wastewater with POME, requiring 2,804,431 USD.

3.5. Economic Feasibility Assessment

From the data provided by the supported factory, the yearly production of biogas was 3,600,000 cubic meters, which could be converted to 2.2 kWh of electricity per cubic meter of biogas. As a result, if the factory supplies this electricity to the power grid at a rate of 0.12 USD per unit, it could generate revenue of 939,502 USD yearly (Table S5). Moreover, by combining EFB wastewater with POME, the factory can double the amount of biogas produced. Therefore, the factory can generate electricity of 15.84×10^6 kWh per year, resulting in revenue of 1,879,004 USD yearly, while also increasing crude oil production of 388 tons/year or 345,907 USD/year. The preliminary data could be used to assist in making decisions about project construction using tools or criteria to evaluate the economic value. The economic viability of the project can be evaluated using the principles of a cost–benefit analysis, which consider the net present value and internal rate of return from Table S6. Both projects have constant revenue yearly and a project period of 5 years, with a minimum required return rate of 10%. From the table, it becomes evident that Project B, combining EFB wastewater with POME, had a higher net present value than Project A, with a net present value of 8,434,163 USD. When considering the internal rate of return of the projects, Project B had an internal rate of return of up to 73%, which was higher than Project A with an internal rate of return of 23%. Project A had a payback period of approximately 3 years, while Project B would have a payback period ranging from 1 to 2 years.

4. Conclusions

From the study results, the accumulation of biogas and methane occurred in cofermentation of the 45% POME + 50% seed + 5% EFB wastewater mixture during the retention periods of 30, 25, 20, 24, and 7 days, with cumulative biogas amounts of 15,024, 18,679, 11,896, 8120, and 6974 mL, respectively, and cumulative methane amounts of 4893, 6778, 2946, 1210, and 251 mL, respectively. The retention period of 25 days had higher cumulative biogas and methane amounts than the other retention periods throughout the entire experiment of 30 days. The characteristics of the sludge residue showed an organic carbon content of 19.17%, total nitrogen of 2.62%, total phosphorus of 2.80%, total potassium of 11.32%, a carbon-to-nitrogen ratio of 7.0, and a moisture content of 28.55%, showing its applicability for further use as organic compost. Based on the economic evaluation, it could be concluded that the codigestion of EFB wastewater with the 45% POME + 50% seed + 5% EFB wastewater mixture in the crude palm oil mill has a high economic value when considering the net present value and internal rate of return, which are 8,434,163 USD and 73%, respectively.

Supplementary Materials: The following supporting information can be downloaded at: https://www.mdpi.com/article/10.3390/w15122153/s1, Table S1: Parameters used to measure and the method of analyzing the wastewater samples in the batch experiment. Table S2: The Biogas system installation cost. Table S3: The additional cost from EFB pressing installation. Table S4: The cost of EFB pressing installation into the system. Table S5. Income from projects. Table S6: Net Present Value (NPV) and Internal Rate of Return (IRR) from Projects. Figure S1: Sludge residue after 25 days of digestion.

Author Contributions: Conceptualization, T.T.S.; Validation, C.S.; Resources, N.C.; Data curation, K.J.; Writing—original draft, C.S.; Writing—review & editing, C.R. and T.T.S.; Project administration, T.T.S.; Funding acquisition, T.T.S. All authors have read and agreed to the published version of the manuscript.

Funding: This research was funded by Thailand Science Research and Innovation (TSRI; Grant No. MDS56I0177).

Data Availability Statement: No new data were created or analyzed in this study. Data sharing is not applicable to this article.

Acknowledgments: The authors are grateful to Thailand Science Research and Innovation (TSRI; Grant No. MDS56I0177) for their financial support.

Conflicts of Interest: The authors declare they have no conflict of interest, and the funders had no role in the design of the study; in the collection, analyses, and interpretation of the data; in the writing of the manuscript; or in the decision to publish the results.

Abbreviations

ASBR	anaerobic sequencing batch reactor
BOD	biochemical oxygen demand
COD	chemical oxygen demand
CSTR	semicontinuous completely stirred tank reactor
EFB	empty fruit bunch
G&O	grease and oil
HRT	hydraulic retention time
MLSS	mixed liquor suspended solids
OLR	organic loading rate
OPF	oil palm fronds
OPT	oil palm trunks
PPF	palm pressed fibers
POME	palm oil mill effluent
SS	suspended solids
TKN	total Kjeldahl nitrogen
TS	total solids
UASB	upflow anaerobic sludge blanket
VS	volatile solids
VFA	volatile fatty acids

References

1. IPAD; USDA. Palm Oil 2022 World Exports. USDA; 2022. Available online: https://ipad.fas.usda.gov/cropexplorer/cropview/commodityView.aspx?cropid=4243000&sel_year=2022&rankby=Exports (accessed on 15 April 2023).
2. Prasertsan, S.; Prasertsan, P. Biomass residues from palm oil mills in Thailand: An overview on quantity and potential usage. *Biomass Bioenergy* **1996**, *11*, 387–395. [CrossRef]
3. Kaewmai, R.; H-Kittikun, A.; Suksaroj, C.; Musikavong, C. Alternative Technologies for the Reduction of Greenhouse Gas Emissions from Palm Oil Mills in Thailand. *Environ. Sci. Technol.* **2013**, *47*, 12417–12425. [CrossRef] [PubMed]
4. Lam, M.K.; Lee, K.T. Renewable and sustainable bioenergies production from palm oil mill effluent (POME): Win-win strategies toward better environmental protection. *Biotechnol. Adv.* **2011**, *29*, 124–141. [CrossRef] [PubMed]

5. Suksaroj, T.T.; Yaeed, S.; Suksaroj, C. The effect of POME ultrasonication on biogas production and reduction of greenhouse gases emissions from wastewater treatment units of palm oil mills. *Desalin. Water Treat.* **2020**, *202*, 86–94. [CrossRef]
6. Yaeed, S.; Suksaroj, T.T.; Suksaroj, C. Mechanical pretreatment processes for enhancement of biogas production from palm oil mill effluent (POME). *Desalin. Water Treat.* **2017**, *67*, 133–139. [CrossRef]
7. Waisundara, V. (Ed.) *Palm Oil*; InTech: London, UK, 2018. [CrossRef]
8. APHA. *Standard Methods for the Examination of Water and Wastewater*, 23rd ed.; APHA: Washington, DC, USA, 2015.
9. Ounsaneha, W.; Rattanapan, C.; Suksaroj, T.T.; Kantachote, D.; Klawech, W.; Rakkamon, T. Biogas production by co-digestion of municipal wastewater and food waste: Performance in semi-continuous and continuous operation. *Water Environ. Res.* **2021**, *93*, 306–315. [CrossRef] [PubMed]
10. Loehr, R.C. *Agricultural Waste Management: Problems, Processes, Approaches*; Academic Press: New York, NY, USA, 1974.
11. APHA. *Standard Methods for the Examination of Water and Wastewater*, 22nd ed.; APHA: Washington, DC, USA, 2012.
12. Li, C.; Champagne, P.; Anderson, B.C. Biogas production performance of mesophilic and thermophilic anaerobic co-digestion with fat, oil, and grease in semi-continuous flow digesters: Effects of temperature, hydraulic retention time, and organic loading rate. *Environ. Technol.* **2013**, *34*, 2125–2133. [CrossRef] [PubMed]
13. Onthong, U.; Juntarachat, N. Evaluation of Biogas Production Potential from Raw and Processed Agricultural Wastes. *Energy Procedia* **2017**, *138*, 205–210. [CrossRef]
14. Rubio, J.A.; Fdez-Güelfo, L.I.; Wilkie, A.C.; García-Morales, J.L. Mesophilic anaerobic co-digestion of two-phase olive-mill waste and cattle manure: Optimization of semi-continuous process. *Fuel* **2022**, *328*, 125354. [CrossRef]
15. Hobson, P.N.; Wheatley, A. *Anaerobic Digestion: Modern Theory and Practice*; Elsevier Applied Science: New York, NY, USA, 1993.
16. Industrial Product Standards Office, Ministry of Industry. *Thai Industrial Standard: Fertilizers*; Industrial Product Standards Office: Bangkok, Thailand, 2005.
17. Metcalf & Eddy Inc. *Wastewater Engineering Treatment and Reuse*, 4th ed.; McGraw-Hill: New York, NY, USA, 2004.
18. Deng, W.; Zheng, P.; Chen, Z. Anaerobic digestion and post-treatment of swine wastewater using IC–SBR process with bypass of raw wastewater. *Process Biochem.* **2006**, *41*, 965–969. [CrossRef]
19. Montusiewicz, A.; Lebiocka, M. Co-digestion of intermediate landfill leachate and sewage sludge as a method of leachate utilization. *Bioresour. Technol.* **2011**, *74*, 3–16. [CrossRef] [PubMed]
20. Poh, P.E.; Chong, M.F. Development of anaerobic digestion methods for palm oil mill effluent treatment (POME). *Bioresour. Technol.* **2010**, *100*, 1–9. [CrossRef] [PubMed]
21. Hallaji, S.M.; Kuroshkarim, M.; Moussavi, S.P. Enhancing methane production using anaerobic co-digestion of waste activated sludge with combined fruit waste and cheese whey. *BMC Biotechnol.* **2019**, *19*, 19. [CrossRef] [PubMed]
22. Gonzalez, R.; Pena, D.C.; Gomez, X. Anaerobic Co-Digestion of Wastes: Reviewing Current Status and Approaches for Enhancing Biogas Production. *Appl. Sci.* **2022**, *12*, 8884. [CrossRef]
23. Park, J.G.; Lee, B.; Jo, S.Y.; Lee, J.S.; Jun, H.B. Control of accumulated volatile fatty acids by recycling nitrified effluent. *J. Environ. Health Sci. Eng.* **2018**, *16*, 19–25. [CrossRef] [PubMed]
24. Dawei, Y.; Liu, J.; Sui, Q.W.; Wei, Y. Biogas-pH automation control strategy for optimizing organic loading rate of anaerobic membrane bioreactor treating high COD wastewater. *Bioresour. Technol.* **2016**, *203*, 62–70. [CrossRef]
25. Kouzi, A.; Puranen, M.; Kontro, M.H. Evaluation of the factors limiting biogas production in full-scale processes and increasing the biogas production efficiency. *Environ. Sci. Pollut. Res.* **2020**, *27*, 28155–28168. [CrossRef] [PubMed]
26. Fezzani, B.; BenCheikh, R. Optimisation of the mesophilic anaerobic co-digestion of olive mill wastewater with olive mill solid waste in a batch digestion. *Desalination* **2008**, *228*, 159–167. [CrossRef]
27. Department of Agriculture, Ministry of Agriculture and Cooperatives. *Thai Standard: Organic Fertilizers*; Department of Agriculture: Bangkok, Thailand, 2012.

water

Article

Utilizing a Novel Halotolerant *Bordetella* Bacterium Combined with Co-Metabolites to Boost the Degradation of P-Nitrophenol in High-Salinity Wastewater

Lei Qin [1], Haorui Li [1], Yingyu Tan [2], Xuenan Yan [1], Peng Tao [1], Zheng Fan [1,*], Tiejun Li [3,*], Jia Tan [4], Yiwei Wang [4] and Lei Jin [3]

[1] College of Chemical Engineering, State Key Lab Base of Green Chemical Synthesis Technology, Zhejiang University of Technology, Hangzhou 310014, China; qinlei0214@zjut.edu.cn (L.Q.); 17326023966@163.com (H.L.); yara_85@163.com (X.Y.); taopeng19550926@163.com (P.T.)
[2] Eco-Environmental Science Research & Design Institute of Zhejiang Province, Hangzhou 310007, China; ying_yutan@hotmail.com
[3] State Key Laboratory of Sustainable Utilization Technology Research of Marine Fishery Resources, Zhejiang Marine Fisheries Research Institute, Zhoushan 316021, China; jinlei2388@126.com
[4] Zhejiang Fenghe Detection Technology Co., Ltd., Jinhua 322000, China; 18875120633@163.com (J.T.); ywfhjc@163.com (Y.W.)
* Correspondence: fanzh@zjut.edu.cn (Z.F.); litiejun1982@126.com (T.L.)

Abstract: A novel strain capable of fully utilizing p-nitrophenol (PNP) as the sole carbon source under high-salinity conditions was isolated from the sediments of wastewater discharged from an aquaculture company. The identification of the strain as *Bordetella* sp. was confirmed by analyzing its morphological, physiological, and biochemical traits in conjunction with its 16S rDNA sequence. Furthermore, pantothenic acid, serving as a carbon source for co-metabolites, could significantly enhance the biodegradation process of the tricarboxylic acid (TCA) cycle. Under the optimal growth conditions at a temperature of 30 °C, pH of 8.0, aeration of 0.32 $m^3 \cdot (m^3 \cdot min)^{-1}$ and salinity of 3% (NaCl, w/v), the degradation rate of 350 $mg \cdot L^{-1}$ PNP increased from 60.8% to 85.9% within 72 h after adding 30 $mg \cdot L^{-1}$ of pantothenic acid to a 12-liter bioreactor. The intermediate products from the degradation process, analyzed via GC/MS, were determined to be hydroquinone, which suggests that the degradation pathway of the bacterium for PNP involves the breakdown of hydroquinone. Benefits have been derived from the microorganism's tolerance to high salinity and high PNP concentrations, coupled with its superior PNP degradation performance, offering new insights and a research basis for the efficient biological treatment of high-salinity PNP wastewater.

Keywords: *Bordetella* sp.; p-ntrophenol; PNP; cleavage pathway; co-metabolites; high salinity

check for updates

Citation: Qin, L.; Li, H.; Tan, Y.; Yan, X.; Tao, P.; Fan, Z.; Li, T.; Tan, J.; Wang, Y.; Jin, L. Utilizing a Novel Halotolerant *Bordetella* Bacterium Combined with Co-Metabolites to Boost the Degradation of P-Nitrophenol in High-Salinity Wastewater. *Water* **2024**, *16*, 3360. https://doi.org/10.3390/w16233360

Academic Editors: Christos S. Akratos and Alejandro Gonzalez-Martinez

Received: 10 October 2024
Revised: 3 November 2024
Accepted: 20 November 2024
Published: 22 November 2024

1. Introduction

P-nitrophenol (PNP), an essential chemical material, is used in a wide range of manufacturing applications, including pharmaceuticals, dyes, pigments, plastic, bactericides, industrial solvents, and phosphorus organic pesticides [1]. PNP shows continuous environmental toxicity due to its chemical stability. Moreover, it can be carried along with discharged industrial waste; thus, it can easily flow into the natural environment and penetrate through the soil to contaminate groundwater [2]. The combustion of biomass and hydrolysis of pesticides such as parathion contribute to its formation in the atmosphere [3]. Further, PNP poses a great threat to human health and ecosystems. For example, after a 24-hour exposure to 10-micromolar PNP, the viability of two mammalian testicular somatic cell lines, TM3 and TM4, reduced by ~10% and 25%, respectively. Moreover, apoptosis rates in these cells were significantly elevated, with TM3 cells showing a 6.3-fold increase and TM4 cells a 10-fold increase compared to controls [4]. PNP elicits oxidative stress, consequently causing impairment of energy metabolism and ultimately influencing the estrogen signaling pathway in Caenorhabditis elegans. Exposure to 8 ng L^{-1} and 8 μg L^{-1} for 48 h resulted in an approximate reduction of 8% and 17.8% in a larval population compared to the control; similarly, the ovulation rate experienced a decrease, reducing from the initial 8.94% to 8.31% and 7.48%, respectively [5]. Moreover, it was found that a concentration of 250 mg kg^{-1} PNP exerted a severe inhibitory effect on the biochemical properties of soil, with a dramatic decrease in the nitrification potential of soil to zero within a 15-day period [6]. In summary, it is essential to investigate efficient methods for removing PNP.

Adsorbent nanocomposites exhibit high efficiency toward the removal of PNP from water bodies through principles such as electrostatic interactions, π-π stacking, hydrophobic interactions, hydrogen bonding, and electron donor–acceptor interactions [7,8]. Biochar, known for its effective adsorption traits, is used for capturing pollutants such as ammoniacal nitrogen [9]. For example, according to a literature study, by utilizing coconut-shell-derived activated carbon—which is dominated by basic surface groups formed through physical activation—the full elimination of 50 mg/L^{-1} PNP was accomplished in a short period of 45 min in mono-component kinetic tests [10]. The covalent–organic framework (COF) SNW-1 has been utilized to produce COF-derived carbons (CDC(x)), with a variant named CDC (800) demonstrating an outstanding maximum PNP adsorption capacity of 1190 mg g^{-1} at pH 7. This capacity is six-fold greater than that of activated carbon and exceeds the highest adsorption capacities of all other documented materials [11]. A previous study reported that the photocatalytic properties of zinc oxide (ZnO) nanoparticles enabled the degradation of recalcitrant contaminants, including methyl Orange and PNP [12]. The reduction of PNP to a full extent was achieved by ZnO nanoparticles when subjected to sodium borohydride (NaBH4) reduction. Moreover, the catalytic performance of this material remained unimpaired across five reuse cycles [13]. The catalyst prepared from cobalt-diethylene glycol terephthalate aided in the effective degradation of PNP under visible light irradiation, achieving an impressive 99.74% efficiency within just 5 minutes of exposure; however, the process generated intermediate products such as p-benzoquinone, p-aminophenol, and ethyl isobutyrate, leading to secondary pollution [14]. Although the aforementioned methods are effective for PNP removal, the adsorbent materials cannot completely decompose PNP, and there is a possibility of desorption of PNP molecules back into the water body. Catalytic reduction can reduce PNP to p-aminophenol but thorough removal cannot be achieved, while catalytic oxidation may lead to the production of secondary pollutants [15,16].

Biodegradation, lauded for its safety and efficacy, has attracted significant attention and is now widely utilized in treating wastewater [17]. It offers a cost-effective alternative to adsorption and catalytic techniques, which reduces financial requirements for implementation [18]. However, PNP triggers stress responses in sludge, highlighting a considerable effect on cellular processes [19].

Notably, it is imperative to undertake broad-strain screenings to find strains capable of exhibiting heightened degradation activity and a strong ability to endure high PNP concentrations [20]. For instance, enterococcus gallinarum JT-02, isolated from poultry farm sludge, achieved a degradation rate of 98.21% for an initial concentration of 100 mg L^{-1} PNP under conditions of pH 7, a temperature of 30 °C, and over a 4-day reaction period [21]. Moreover, a PNP-degrading bacterial strain, exhibiting the greatest genetic sequence homology with Pseudomonas juntendi BML 3 and Pseudomonas inefficax JV 551 A3, was isolated from pesticide-affected soil in Lucknow. This strain demonstrated an exceptional degradation capability, reaching a 100% efficiency rate within 24 h when utilizing 0.5 mM PNP as its carbon source [22]. Derived from soil contaminated with industrial waste, *Nocardioides* sp. ZS2 possesses both denitrification capabilities and the ability to degrade PNP. In shake flask experiments, the concentration of NO_3^--N was reduced from 16.53 mg L^{-1} to 0.09 mg L^{-1} within 24 h, and that of PNP decreased from 2.01 mg L^{-1} to 0.05 mg L^{-1} over the initial 36 h [23]. Nevertheless, the aforementioned strains cannot effectively degrade high concentrations of PNP in high-salinity wastewater environments.

The objective of this research is to formulate a microbial agent for degrading PNP, thereby enabling efficient biodegradation of PNP at high concentrations in environments with high salinity. Herein, a new genus *Bordetella* sp. was obtained by isolation and acclimation. *Bordetella* sp. could effectively biodegrade high concentrations of PNP under high-salinity conditions. Furthermore, the influence of different conditions on bacterial growth and reproduction was also explored, and the effects of adding different co-metabolites as a carbon source on the metabolism of major pollutants were studied comprehensively. In addition, through technical characterization and analysis, the degradation pathway of PNP by *Bordetella* sp. was proposed herein.

2. Materials and Methods

2.1. Materials

The chromatographically pure PNP test standards and methanol (>98%) were purchased from Sigma-Aldrich Co., Ltd. (Shanghai, China) Other reagents were purchased from Hangzhou Huadong Pharmaceutical Co., Ltd. (Hangzhou, China), and the remaining solvents and chemicals were analytically pure.Minimal medium (MM): 0.8 g KH_2PO_4, 0.4 g K_2HPO_4, 0.1 g $MgSO_4$, 0.02 g $FeSO_4$, 0.3 g $(NH_4)_2SO_4$, 30.0 g NaCl, 0.05 g $CaCl_2$, 0.01 g $ZnSO_4$, 0.01 g $MnSO_4{\cdot}7H_2O$, 0.4 mg $(NH_4)_6Mo_7O_{24}{\cdot}4H_2O$, 0.4 mg $CoCl_2{\cdot}6H_2O$ per liter.

Enrichment medium (EM): the yeast extracts (1 g L^{-1}) and beef east (1 g L^{-1}) were added to the MM.

Plate screening medium (PSM): Agar (2%) and PNP (10 mg L^{-1}) were added to the MM. Medium with pH 7.5, NaOH or HCl (2 mol L^{-1}), and a volume of 1 L was prepared with sterile water. All media were sterilized at 120 °C for 20 min (sterilization conditions).

2.2. Analytical Methods

Determination of the PNP solution concentration in wastewater: The PNP concentration was ascertained by high-performance liquid chromatography (HPLC, Scincos-3100, Hanam-si, Republic of Korea). The chromatographic conditions were as follows: mobile phase (methanol–water = 30:70), stationary phase (C18 anti-phase non-polar column), and flow rate (0.5 mL min^{-1}). The injection volume was 10 mL. Following collection, the samples were centrifuged at a speed of 10,000 rpm for 10 min. The resulting supernatant was then filtered through a membrane filter with a pore size of 0.45 μm, and then it was examined by HPLC. Simultaneously, the culture medium without PNP was used as a blank control. For total organic carbon (TOC) detection, TOC-vcph total organic carbon was used, the test culture solution (2 mL) was taken, centrifuged at 4000 r min^{-1} for 10 min, and then the supernatant was taken, diluted 10 times, and the TOC content was directly measured by injection.

The biomass was determined by the dry weight of cells per unit volume: Briefly, first, the uniformly mixed culture (20 mL) was centrifuged at 5000 rpm for cell sedimentation. Next, the supernatant was discarded, the pellet was rinsed with deionized water, and this process was repeated three times. The pellet was then subjected to infrared drying at 105 °C for 30 min, and subsequently, the dry weight was measured [24,25]. The physicochemical characterization of the bacteria was carried out according to the Manual of Microbiological Identification [26]. The PNP degradation products were assessed by gas chromatography/mass spectroscopy (GC/MS, TRACE GC UltraTM/Polaris Q system, Thermo Fisher Scientific Co., Ltd. (Waltham, MA, USA)) combined with an electron ionization (EI) source. The chromatographic separation was achieved using an HP-5 MS column (50 m × 0.25 mm × 0.25 μm). For the gas phase section: injector temperature was 260 °C; column temperature—initial temperature 70 °C, held for 1 min, increased to 140 °C at a rate of 5 °C·min^{-1}, then to 220 °C at a rate of 15 °C·min^{-1}, finally to 240 °C at a rate of 5 °C·min^{-1}, and held for 2 min. High-purity helium gas with a column flow rate of 2.0 mL·min^{-1} was used as the carrier gas; no splitting occurred during injection. Mass spectrometry section: transfer line temperature was 250 °C, electron ionization (EI) source temperature was 250 °C, ionization energy was 70 eV, and scanning range was 45 to 500 amu.

2.3. Screening and Enrichment of PNP-Degrading Bacterial Strain

Isolation process: Sludge (5 g) was taken from the wastewater sedimentation pond generated due to marine food processing in Zhejiang Province, China, and then the sludge was added to an EM and cultured at 32 °C and 150 rpm for 5 days. A small amount of bacteria-enriched culture was taken and transferred to the MM with PNP as the sole carbon source to acclimate the enriched PNP-degrading bacteria. Initially, the concentration of PNP was set to 5 mg L^{-1} and the bacteria were domesticated by a uniform increase to 100 mg L^{-1} to improve the tolerance of the bacteria. The last generation of EM-containing bacteria was gradually diluted and coated on PSM containing 50 mg L^{-1} PNP, which was cultured at a constant temperature of 32 °C. After the growth of a single colony, it was continuously streaked and purified to obtain pure PNP-degrading bacteria, which were stored in a refrigerator at 4 °C for future use.

2.4. Identification of Bacteria

2.4.1. Identification of Bacterial Morphs

Transmission electron microscopy (TEM) was used to observe the cell morphology. The transmission electron microscope adopted the electron microscope at the Institute of Biotechnology, Zhejiang University (JEOL JEM-1230, Akishima, Japan), for morphological characterization. Further, the physiological characteristics of the bacteria were tested according to standard methods (Biolog Micro Work station (GN), Hayward Company, Hayward, CA, USA).

2.4.2. 16S rDNA Gene Identification

Polymerase chain reaction (PCR) amplification was carried out using the obtained cell lysis fragment as a template, which was ultimately used for the amplification of 16S rDNA sequences [27]. PCR primers are a pair of universal primers, Pf 5′-agagtttgatcctggctcag-3′, and reverse primer Pf: 5′-cggctaccttgttacacttc -3′. PCR products were prepared and sequenced by Shanghai Yes Service Biotech, Inc (Shanghai, China).

2.4.3. Effects of Different Factors on the Degradation of PNP by *Bordetella* sp.

The obtained pure bacteria were put into MM and cultured in a 500 mL flask for testing. The concentration of PNP in the solution was increased from 50 to 100 mg L^{-1} every 3 days to make *Bordetella* adapt to the degradation of PNP.

The best degradation parameters such as temperature, pH, salinity, and ventilation were determined by a single-factor method. The flask was filled with MM (100 mL) and inoculated with culture solution of the *Bordetella* strain (15 mL), and the hydraulic retention time (HRT) was 72 h. The basic conditions for the degradation experiment were set as follows: temperature 30 °C, ventilation volume 0.3 $m^3 \cdot (m^3 \cdot min)^{-1}$, pH 7.5, salinity 3%. Then, the effects of various parameters on the degradation of PNP were studied through a single-factor variable. For a comprehensive evaluation of the effect of different initial pH values on the degradation of PNP, a pH range of 5.5–9.0 was set, with the unit difference of two contiguous pH values being 0.5, and the pH of the culture medium was adjusted with NaOH or HCl (2 mol L^{-1}). For systematic exploration of the effects of salinity, NaCl concentration was set in the range from 0 to 8% and the unit difference of two contiguous NaCl values was 1%. To study the effects of different temperatures on the degradation of PNP, the set temperature range was between 18 and 42 °C and the difference between the temperature values in the experiment was 4 °C. All tests were conducted under sterile conditions with a residence time of 72 h, and 2-milliliter samples were analyzed by HPLC.

We chose to use sodium acetate, glucose, sodium succinate, yeast extract, ammonium chloride, pantothenic acid, and other compounds for the co-metabolism experiment. On this basis, we focused on the promotion of pantothenic acid and on the degradation of PNP by *Bordetella*. The intermediate products in the process of degradation were measured and analyzed, and the possible metabolic pathways were speculated.

2.5. Scale-Up of PNP Degradation

A self-made reaction system was used to study the biodegradation of PNP at different initial concentrations (Figure S1). The reaction system was equipped with a 12-liter bioreactor, an adjustable air pump, a ceramic ventilator, a feed pump, and a buffer tank. The bioreactor temperature was maintained at 30 °C and the air inflow was 0.3 m^3 $(m^3{\cdot}min)^{-1}$. Simulated PNP wastewater consisted of a culture solution containing PNP culture solution A. The salinity of the simulated wastewater was controlled at 3%, which was adjusted by adding NaCl. Four different PNP concentrations of 120, 240, 360, and 480 mg L^{-1} were used for the degradation tests. Other than PNP, NaCl, and MM, there were no other nutrients in the simulated PNP wastewater. The pH of the simulated wastewater was in the range of 5.5–9.0 and 0~8% (*w*/*v*) of NaCl was used. PNP wastewater flowed from bottom to top in the bioreactor, fed at the bottom, discharged from the top, and the HRT was kept at 72 h. After each discharge of degraded wastewater, the remaining 15% solution was used as the next seed. The daily change in the concentrations of PNP at the outlet of the effluent wastewater was measured in the experiment to observe the degradation of wastewater in the tank.

The degradation rate, denoted as $R_{a,w}$, can be calculated by using the following formula:

$$R_{a,w} = \frac{Q_{e,air}}{V_{w,w}} \tag{1}$$

where $Q_{e,air}$ (m^3 min^{-1}) is the air flow rate and $V_{w,w}$ (m^3) denotes the volume of bacterial degradation wastewater.

The degradation rate of PNP was calculated as follows:

$$\text{Degradation rate of PNP }(\%) = \frac{C_0 - C_t}{C_0} \times 100\% \tag{2}$$

where C_t (mg L^{-1}) and C_0 (mg L^{-1}) denote the terminal and original concentrations of PNP post-treatment and pre-biodegradation, respectively.

$$\text{Mineralization rate }(\%) = \frac{TOC_0 - TOC_t}{TOC_0} \times 100\% \tag{3}$$

where TOC_0 (mg L^{-1}) and TOC_t (mg L^{-1}) represent the initial concentrations of TOC in the treatment and the residual concentrations of TOC during the test, respectively.

2.6. Biodegradation of PNP and Kinetics Model

To study the trend of the bacterial degradation of PNP, the first-order kinetic reaction model was used for analysis:

$$C = C_0 \times e^{-k_1 t} \tag{4}$$

$$t_{1/2} = \frac{\ln 2}{k_1} \tag{5}$$

where C (mg L^{-1}) refers to the concentration of PNP at degradation time t; C_0 (mg L^{-1}) is the initial PNP concentration; time parameter t is associated with the microbial degradation process; k_1 defines the reaction rate constant; and $t_{1/2}$ is the time required for the concentration of PNP in the reaction system to decrease by half.

3. Results and Discussion

3.1. Identification of Physiological Characteristics

The isolated bacterium could use PNP as the sole carbon source for growth after the obtained strain was grown on PSM for 48 h. The isolated colony was characterized as transparent, semi-convex, convex, infiltrated and with irregular edges, and rod-like with a length of about 0.5 × 1.5 μm. The *Bordetella* strain was found to be Gram-negative, demonstrating positive outcomes in the citric acid test, glucose fermentation test, methyl red test, urea test, and Voges-Proskauer test. These results are generally in line with prior research reports [28,29]. Furthermore, the 16S rDNA sequence of the strain was deposited in GenBank with the accession number MN524132. The obtained strain exhibited high homology with *Bordetella* sp. and the bacterium was confirmed to be the phylogenetic tree derived from the 16S rDNA gene sequence of the *Bordetella* strain after comparative analysis (Figure 1). The strain domesticated here was designated as *Bordetella* S113 by the laboratory. The electron microscopy image of *Bordetella* S113 is provided in Figure S2 of Supplementary Materials. Under electron microscopy, the strain appeared ovoid in shape, with no flagella observed, and black granular contents were visible internally.

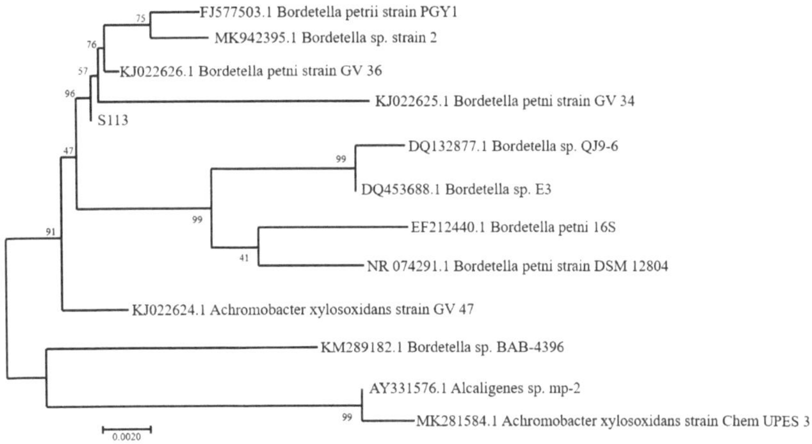

Figure 1. Phylogenetic tree derived from 16S rDNA gene sequence of *Bordetella* S113 strain.

3.2. *Effects of Different Factors on PNP Degradation*

The experiment adhered to the principle of single-variable control, with PNP degradation being conducted solely by *Bordetella* S113. Figure 2 depicts the changes in degradation rate and dry cell weight as influenced by various operational parameters, including temperature, pH, salinity, and aeration. Figure 2A illustrates that *Bordetella* S113 could grow in the range of 18–42 °C and that it grew best at 30 °C. When the temperature was low, the bacteria could grow slowly at a low temperature, such as the PNP degradation rate of 41% observed at 18 °C. However, at 42 °C, the degradation rate of *Bordetella* S113 reached 62.3%. The experiment showed that the temperature exhibited a significant impact on bacterial degradation, and the degradation ability was good in the range of 22–42 °C, which is consistent with results in the literature [21].

The effects of different pH values on the degradation of PNP by *Bordetella* S113 are shown in Figure 2B. When the pH was 5.5, the degradation rate of PNP was only 43%. With the increase in pH, the degradation rate increased continuously. When the pH was 8.0, the degradation rate of PNP reached 98%, and the dry weight of bacteria in the solution was 2.2 g L^{-1}, reaching the best degradation effect on PNP; however, when the pH exceeded 8.0, the degradation rate of PNP decreased with the increase in pH, and when the pH was 9.0, the degradation rate of PNP was only 78%. Figure 2B illustrates that when the solution was alkaline, the dry weight of bacteria in the solution was larger, indicating that *Bordetella* S113 exhibited a good growth and proliferation state and a high degradation efficiency of PNP. The trend of changes in dry cell weight was found to be the same as the degradation rate (Figure 2C). When the solution was acidic, the weight of the stem cells in the solution was lower, only reaching 1.2 g L^{-1} at pH 5.5.

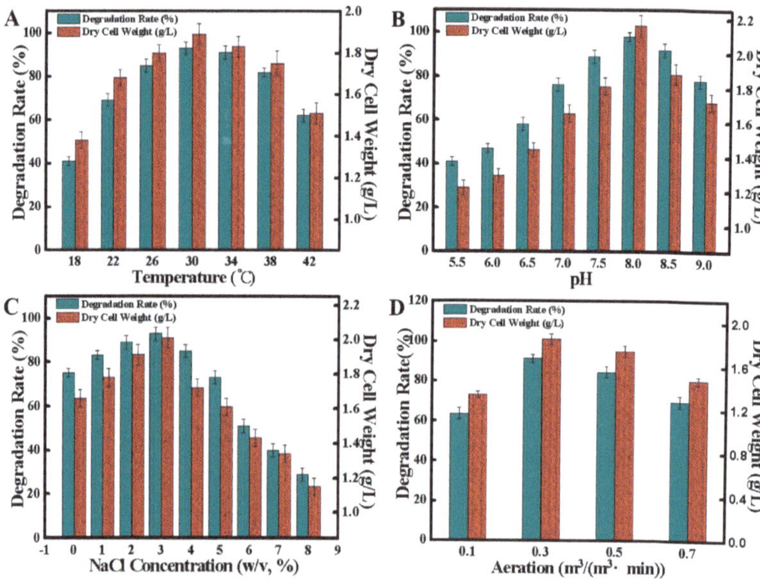

Figure 2. Effects of temperature (**A**), pH (**B**), salinity (**C**), and aeration (**D**) on the metabolic activity and PNP degradation of *Bordetella* S113 strains.

In view of the fact that *Bordetella* S113 was obtained from the wastewater pool from marine food processing, in this study, the effect of salinity on the degradation of PNP by *Bordetella* S113 was tested and the results indicate a high salt tolerance of the strain. With the increase in NaCl content, the number of bacterial cells in the reaction system gradually increased, and the degradation rate of PNP also increased. When the concentration of NaCl was 3.0% (w/v), the dry weight of bacteria in the solution was 2.1 g L^{-1}, and the maximum degradation rate of bacteria was 93%.

The aeration condition showed a profound impact on the rate of pollutant degradation and proliferation of bacteria in wastewater treatment systems, serving as a key factor influencing treatment efficacy. The effects of different aeration rates are shown in Figure 2D. The results indicate that the bacterium was at the stage of reproduction adaptation at the beginning, and after 36 h, the degradation rate of PNP accelerated. The abovementioned test showed that the optimal aeration rate for *Bordetella* to degrade PNP was 0.3 m^3 $(m^3{\cdot}min)^{-1}$, and the degradation rate of PNP was maximum at 92%. However, when the aeration rate was too large, the degradation rate decreased. Aeration plays two roles: first, aeration is beneficial to the transfer and distribution of nutrients in waste liquid, and second, it provides oxygen to waste liquid but excessive ventilation may produce growth inhibition.

3.3. *Effects of Co-Substrate Supplementation on PNP Degradation*

Figure 3A illustrates the effects of adding 20 mg L^{-1} sodium acetate, glucose, sodium succinate, yeast extract, ammonium chloride, and pantothenic acid on the degradation of PNP by *Bordetella* S113. Various co-substrate supplementations promoted the degradation effect of PNP and the following results were obtained after the addition of the aforementioned six compounds: sodium acetate enhanced the degradation of PNP by 3.6%; glucose led to 2.4% enhancement; sodium succinate increased the degradation rate by 6.0%; yeast extract provided a 12.1% increase; pantothenic acid enhanced the rate by 19.3%; and ammonium chloride increased the degradation of PNP by 1.2%. These additives were beneficial to the degradation of PNP by *Bordetella* S113 to varying degrees, among which the effect of pantothenic acid was the most obvious. Based on the abovementioned experiments, pantothenic acid with different concentrations of 10, 20, 30, and 40 mg L^{-1} were added to evaluate their effects on promoting the degradation of PNP. The experimental setup involved an inorganic salt medium with varying concentrations of PNP, specifically 50, 150, 250, 350, and 450 mg L^{-1}. The bacterial strains were grown under optimized conditions, i.e., a temperature maintained at 30 °C, a pH adjusted to 8.0, an aeration rate set at 0.3 $m^3 \cdot (m^3 \cdot min)^{-1}$, and a salinity level of 3% (NaCl, w/v). Figure 3B shows the relationship between the variation in the degradation rate of PNP and the increase in PNP concentration after adding pantothenic acid to wastewater. In the waste solution with low concentrations of PNP, the degradation rate of *Bordetella* S113 was relatively high and the relative increase rate was not obvious. However, when the PNP concentration in the waste solution was 50 and 150 mg L^{-1}, the addition of 30 mg L^{-1} pantothenic acid maximized the degradation rate, which assisted almost complete degradation by *Bordetella* S113. When the PNP concentration in the waste solution was 450 mg L^{-1}, the original degradation rate increased from 29% to 53%, corresponding to the most obvious improvement of 82.8%. Although the degradation rate of PNP decreased with the increase in the initial concentration of PNP, the experimental results indicated that the degradation rate of PNP by *Bordetella* S113 was effectively promoted by the addition of pantothenic acid, and the relative degradation rate of PNP by *Bordetella* S113 increased significantly under the condition of high concentrations of PNP.

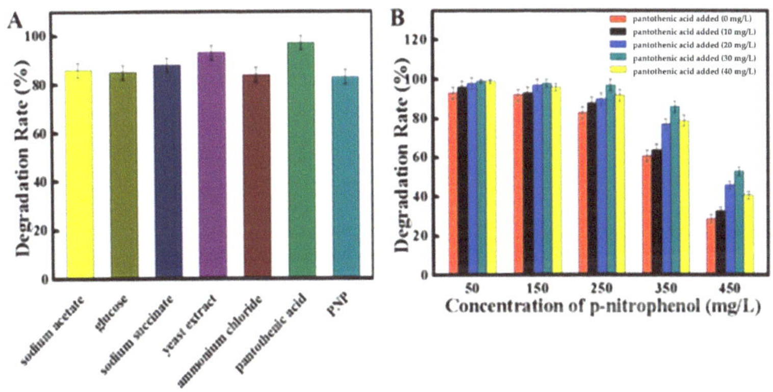

Figure 3. (**A**) Degradation rate after adding co-substrate supplementation. (**B**) Effect of the concentration of pantothenic acid on PNP degradation.

Furthermore, the effect mechanism of pantothenic acid was analyzed. Figure 4 demonstrates that pantothenic acid is a precursor of the microbe coenzyme A (CoA). CoA can be acylated to acetyl CoA, and both CoA and acetyl CoA are important enzymes in the process of PNP degradation [30]. In the process of cracking PNP, β-ketone adipic is generated, which binds with CoA and is then converted to β-ketohexanediate monoacyl CoA; this is further decomposed into succinic acid and acetyl-CoA, and then enters the Krebs cycle for oxidation. This degradation process plays an important role in the transfer of acetyl groups and acyl groups. CoA is also involved in the multiple enzyme center between the acyl transfer and participates in the metabolism of sugar, fat, and the synthesis of protein. Pantothenic acid is a precursor of the acyl-carrier protein (ACP) in microorganisms [31]. ACP is related to a variety of protein complex enzymes and is also related to fatty acid metabolism. As an acyl carrier, ACP carries the acyl chain to complete the enzymatic reactions such as condensation, reduction, and dehydrogenation. It is the cofactor of the acyl-ACP desaturation reaction of different acyl-chain-length fatty acids and the plastids acyltransferase action. Lipid molecules can be closely related to the phospholipid bilayer through the cell membrane, which affects the cell's tolerance to toxic and harmful substances. Therefore, pantothenic acid affects the production of acetyl CoA and ACPs and then impacts the electron transfer of the enzyme system in microorganisms, the generation and transfer of multiple intermediate components, and the metabolism of sugar, fat, and protein. Importantly, the complexity of pantothenic acid in the degradation process of organisms can be observed [32,33].

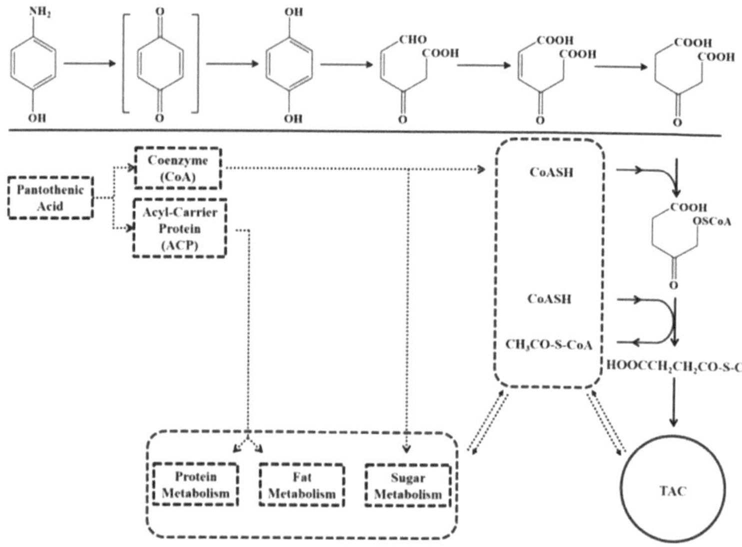

Figure 4. The possible biodegradation pathways of PNP by *Bordetella* S113 and the functioning process of pantothenic acid.

In the degradation process, CoA and acetyl CoA can be considered to coexist in an enzyme library in bacteria. Moreover, they, together with ACPs, participate in the regulation of synthesis and the metabolism of various substances such as sugar, fat, and protein, and shuttle in multiple related enzyme systems [34]. These experimental results can deepen the understanding of the role of pantothenic acid in the metabolic pathway and help improve the efficiency of microbial degradation. Currently, although the cost of pantothenic acid is relatively high, with improvements in the field of biotechnology, the price is gradually decreasing, and its use-value will be gradually reflected. These results can expand the selection range of co-metabolites and provide some references for the use of vitamin compounds as co-metabolites.

3.4. The Possible Pathway of Degradation of PNP by Bordetella S113

It is generally believed that the degradation process of PNP involves two metabolic pathways. The first PNP degradation pathway is affected by flavinprotein monooxygenase that removes the nitro group to obtain hydroquinone, which is then decomposed into adipic acid and ketoxime by catechol 2,3-dioxygenase, etc. The diacid finally enters the tricarboxylic acid cycle. In general, the primary degradation pathway for PNP in Gram-negative bacteria is the hydroquinone cleavage pathway [35,36], as seen in strains *Maraxella* sp. and *Burkholderia cepacia* PKJ 200. In the second route [37], PNP is first decomposed into 4-Nitrocatecho and finally enters the tricarboxylic acid cycle.

After 18 h of biodegradation of PNP by *Bordetella* S113, the degradation solution was extracted with dichloromethane and concentrated for GC/MS analysis. Figure 3 shows the presence of intermediates with retention times of 8.07 and 9.08 min. The retention time of 8.07 min was found to be consistent with that of the standard hydroquinone. Moreover, the peak retention time was 9.88 min, which is consistent with that of the standard PNP. The corresponding organic molecules in the GC/MS spectrum are PNP and hydroquinone, and the hydroquinone produced is the key intermediate to judge the PNP degradation process as the cracking pathway of hydroquinone [38]. Moreover, GC/MS results of the intermediate products obtained during the degradation of PNP are provided in Figure S3 of Supplementary Materials.

3.5. Kinetics of Biodegradation of PNP

As evident from Figure 5, the initial concentration of PNP exerts a substantial influence on the degradation process. In this study, the first-order kinetic reaction model was utilized to study the biodegradation of PNP and the corresponding results are presented in Table S2. Through fitting calculations, the values of K_1 and $T_{1/2}$ were determined for initial PNP concentrations of 120, 240, 360, and 480 mg L^{-1}. Specifically, K_1 values were 1.70, 1.55, 1.26, and 0.67, respectively, and the corresponding $T_{1/2}$ values were 0.41, 0.45, 0.55, and 1.04 days. Notably, when the PNP concentration was below 210 mg L^{-1}, the correlation coefficients surpassed 0.99, indicating that the biodegradation of PNP could be accurately characterized by first-order reaction kinetics. Table S2 evidently presents that an increase in the initial PNP concentration led to a significant decrease in the bacterial degradation reaction rate constant K, suggesting that high concentrations of PNP inhibit bacterial growth and degradation activity. The mineralization rates of wastewater with different concentrations are shown in Figure 5. The mineralization rates were 91.6%, 88.2%, 73.7%, and 39.5% when the PNP concentration was 120, 240, 360, and 480 mg L^{-1}, respectively. The TOC removal rate and degradation performance of PNP with different initial concentrations are provided in Table S1 of Supplementary Materials.

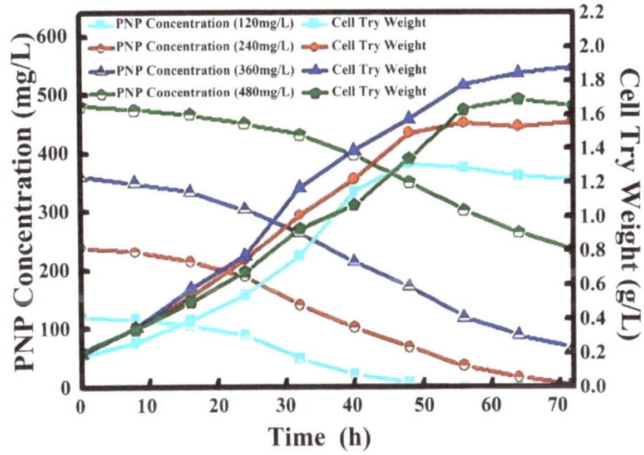

Figure 5. Effects of initial concentrations on the degradation process of PNP.

3.6. Comparison of Salt Tolerance with Other Reported PNP-Degrading Strains

Noteworthy, PNP possesses toxicity toward microorganisms, thus limiting the ability of species such as *Nocardioides* sp. ZS2 [23], *Pseudomonas* sp. [22,39], and activated sludge [40] to degrade PNP in wastewater at instances where its concentration is below 100 mg L^{-1} (Table 1). When the PNP concentration was increased to 300 mg L^{-1}, the removal rate significantly decreased by 47% within 96 h. By contrast, *Bordetella* S113—the bacterium used in this study—tolerated high concentrations of PNP and could remove 85.9% of PNP (350 mg L^{-1}) within 72 h with an average removal rate of 4.18 mg $(L \cdot h)^{-1}$, outperforming the results of most of the studies in the literature. Compared to other strains, *Bordetella* S113 exhibits superior salt tolerance, being capable of withstanding salinities ranging from 0 to 8% NaCl (w/v). Moreover, it achieves the highest PNP degradation rate of over 80% at a salinity level of 3% NaCl (w/v). Furthermore, coupling of the immobilized material coconut coir with the PNP-degrading bacterial consortium consisting of *Paenarthrobacter* spp. and *Zoogloea* spp. significantly enhanced the PNP removal rate to 54.29 mg $(L \cdot h)^{-1}$ [41]. This improvement is attributed to the increased degradation efficiency of the bacterial consortium toward PNP facilitated by coconut coir, as well as the adsorption capacity of the coir for PNP. The future integration of PNP-degrading bacteria with immobilized materials to further augment the degradation rate of PNP holds promising potential.

Table 1. Diversity in PNP degradability and halotolerance among microorganisms.

Microorganism	Carrier	Salinity (mg L^{-1})	Initial Concentration (mg L^{-1})	Efficiency	PNP Removal Rate (mg·(L·h)$^{-1}$)	Ref.
Nocardioides sp. ZS2	Hydrophilic sponge carriers	-	2	97.9%/36 h	0.054	[23]
Activated sludge	Fe carrier	-	14	100%/8 h	1.75	[40]
Bacillus pantothenticus	3% alginate beads	-	25	100%/36 h	0.69	[42]
Autotrophic denitrification sludge	-	-	50	85%/4.4 h	9.65	[43]
Pseudomonas alloputida	-	-	70	100%/24 h	2.92	[22]
Pseudomonas putida DLL-E4	-	-	76	100%/16 h	4.75	[39]
Polypore Trametes versicolor	Luffa aegyptiaca sponge	-	139	97%/72 h	1.87	[44]
Paenarthrobacter spp., *Zoogloea* spp.	Coconut coir	-	250	99%/4.65 h	54.29	[41]
Arthrobacter ureafaciens DSM 20126	-	-	400	100%/168 h	2.38	[45]
Enterococcus gallinarum JT-02	-	TDS 205.3	300	47%/96 h	1.47	[21]
Methylobacterium sp. C1	-	NaCl 1000	200	100%/120 h	1.67	[36]
Bordetella S113	-	NaCl 30,000	350	85.9%/72 h	4.18	This study

4. Conclusions

In this study, a new genus was separated from the precipitate of wastewater discharged from an aquaculture company, which exhibited high efficiency toward the degradation of PNP and was named *Bordetella* sp. PNP with a concentration of 350 mg L^{-1} could be degraded up to 85.9% by *Bordetella* sp. within 72 h under optimized experimental conditions and 3% high salinity (NaCl, w/v) without any pretreatment. Moreover, it was also found that, due to the co-metabolism mechanism, when the PNP concentration was as high as 450 mg L^{-1}, by adding co-metabolites such as pantothenic acid, the degradation rate of PNP by *Bordetella* sp. increased by 82.5%. The structure and residence time of the product during the degradation process were characterized, which proved that the degradation of PNP by *Bordetella* sp. mainly occurred due to the lysis of hydroquinone. The results show that this PNP-degrading bacteria isolated from wastewater sediments exhibit good prospects in the treatment of high-salinity wastewater containing high concentrations of organic pollutants, which provides novel ideas and research bases for efficient large-scale biodegradation on high-salinity wastewater.

Supplementary Materials: The following supporting information can be downloaded at https://www.mdpi.com/article/10.3390/w16233360/s1, Table S1: TOC removal rate and degradation performance of PNP with different initial concentrations; Table S2: Kinetics parameters of PNP biodegradation by *Bordetella* S113 strain; Figure S1: Schematic diagram of experimental bioreactor; Figure S2: Photograph of electron microscopy of *Bordetella* S113; Figure S3: GC/MS spectrometry of intermediate products in degradation of PNP.

Author Contributions: H.L., P.T., and X.Y., Experiments. Y.T. and Z.F., Analysis of data and visualization. L.Q., Writing—original draft and polishing. T.L., Supervision. J.T., Formal analysis. Y.W. and L.J., Investigation. All authors have read and agreed to the published version of the manuscript.

Funding: This work was supported by the Zhejiang Provincial Natural Science Foundation of China (Grant Nos. LTGS24D060001 and LTGS23B070001), the Zhejiang Province San-Nong-Jiu-Fang Science and Technology Cooperation Project (Grant No. 2024SNJF052), Zhejiang Province Key Research and

Development Programs (Grant Nos. 2024C03234 and 2024C03282(SD2)), the National Natural Science Foundation of China (Grant No. 22278373), and Zhejiang Provincial Postdoctoral Science Foundation. A patent application related to this work has been filed.

Data Availability Statement: The data presented in this study are available in the result sections in this article; further inquiries can be directed to the corresponding authors.

Conflicts of Interest: Authors Jia Tan and Yiwei Wang was employed by the company Zhejiang Fenghe Detection Technology Co., Ltd. The remaining authors declare that the research was conducted in the absence of any commercial or financial relationships that could be construed as a potential conflict of interest.

References

1. Grace Pavithra, K.; Sundar Rajan, P.; Arun, J.; Brindhadevi, K.; Hoang Le, Q.; Pugazhendhi, A. A review on recent advancements in extraction, removal and recovery of phenols from phenolic wastewater: Challenges and future outlook. *Environ. Res.* **2023**, *237*, 117005. [CrossRef] [PubMed]
2. Qian, M.; Zhang, Y.; Bian, Y.; Feng, X.; Zhang, Z. Nitrophenols in the environment: An update on pretreatment and analysis techniques since 2017. *Ecotoxicol. Environ. Saf.* **2024**, *281*, 116611. [CrossRef] [PubMed]
3. Guo, Z.; Hu, X.; Sun, W.; Peng, X.; Fu, Y.; Liu, K.; Liu, F.; Meng, H.; Zhu, Y.; Zhang, G.; et al. Mixing state and influence factors controlling diurnal variation of particulate nitrophenol compounds at a suburban area in northern China. *Environ. Pollut.* **2024**, *344*, 123368. [CrossRef] [PubMed]
4. Wei, J.; Lu, T.; Dong, F.; Zhang, C.; Zhang, Y. Gene expression profiles of two testicular somatic cell lines respond differently to 4-nitrophenol mediating vary reproductive toxicity. *Toxicology* **2021**, *463*, 152991. [CrossRef]
5. Wang, J.; Yin, J.; Peng, D.; Zhang, X.; Shi, Z.; Li, W.; Shi, Y.; Sun, M.; Jiang, N.; Cheng, B.; et al. 4-Nitrophenol at environmentally relevant concentrations mediates reproductive toxicity in Caenorhabditis elegans via metabolic disorders-induced estrogen signaling pathway. *J. Environ. Sci.* **2025**, *147*, 244–258. [CrossRef]
6. Sagban, F.O.T. Impacts of wastewater sludge amendments in restoring nitrogen cycle in p-nitrophenol contaminated soil. *J. Environ. Sci.* **2011**, *23*, 616–623. [CrossRef] [PubMed]
7. El Messaoudi, N.; Miyah, Y.; Benjelloun, M.; Georgin, J.; Franco, D.S.P.; Şenol, Z.M.; Ciğeroğlu, Z.; El Hajam, M.; Knani, S.; Nguyen-Tri, P. A comprehensive review on designing nanocomposite adsorbents for efficient removal of 4-nitrophenol from water. *Nano-Struct. Nano-Objects* **2024**, *40*, 101326. [CrossRef]
8. Tran, H.N. Adsorption Technology for Water and Wastewater Treatments. *Water* **2023**, *15*, 2857. [CrossRef]
9. Hu, X.; Zhang, X.; Ngo, H.H.; Guo, W.; Wen, H.; Li, C.; Zhang, Y.; Ma, C. Comparison study on the ammonium adsorption of the biochars derived from different kinds of fruit peel. *Sci. Total Environ.* **2020**, *707*, 135544. [CrossRef]
10. Bampi, J.; Da Silva, T.C.; Da Luz, C.; Pasquali, G.D.L.; Dervanoski, A.; Tochetto, G. Study of the competitive effect on the adsorption of phenol and 4-nitrophenol in a batch reactor and fixed-bed column using coconut shell activated carbon. *J. Water Process Eng.* **2024**, *65*, 105825. [CrossRef]
11. Hossain, M.A.; Lee, G.; Jhung, S.H. Covalent-organic framework derived nitrogen-enriched carbon: A remarkable adsorbent to remove 4-nitrophenol from water. *Sep. Purif. Technol.* **2024**, *328*, 125068. [CrossRef]
12. Chen, C.-Y.; Kuo, E.-W.; Nagarajan, D.; Dong, C.-D.; Lee, D.-J.; Varjani, S.; Lam, S.S.; Chang, J.-S. Semi-batch cultivation of Chlorella sorokiniana AK-1 with dual carriers for the effective treatment of full strength piggery wastewater treatment. *Bioresour. Technol.* **2021**, *326*, 124773. [CrossRef] [PubMed]
13. Palanivel, D.; Venugopal, A.; Ranganathan, S.; Cingaram, R.; Natesan Sundramurthy, K. Biogenic synthesis and characterization of zinc oxide nanoparticles for effective and rapid catalytic reduction of 4-nitrophenol. *J. Indian Chem. Soc.* **2024**, *101*, 101337. [CrossRef]
14. Safira, A.R.; Alluhayb, A.H.; Aadil, M.; Alkaseem, M.; Fattah-alhosseini, A.; Kaseem, M. Enhanced photocatalytic reduction of p-nitrophenol by polyvinylpyrrolidone-modified MOF/porous MgO composite heterostructures. *Compos. Part B Eng.* **2024**, *284*, 111710. [CrossRef]
15. Bao, T.; Wang, Q.; Jiang, Y.; Zhao, X.; Cao, Y.; Cao, J.; Li, Q.; Ji, X.; Si, W. Multicomponent nanoparticles decorating lignin-derived biochar composite for 4-nitrophenol sensing. *J. Environ. Chem. Eng.* **2024**, *12*, 113596. [CrossRef]
16. Han, L.; Ma, J.; Lin, H.; Chen, C.; Teng, J.; Li, B.; Zhao, D.; Xu, Y.; Yu, W.; Shen, L. A novel flower-like nickel-metal-organic framework (Ni-MOF) membrane for efficient multi-component pollutants removal by gravity. *Chem. Eng. J.* **2023**, *470*, 144311. [CrossRef]
17. Wu, K.; Atasoy, M.; Zweers, H.; Rijnaarts, H.; Langenhoff, A.; Fernandes, T.V. Impact of wastewater characteristics on the removal of organic micropollutants by Chlorella sorokiniana. *J. Hazard. Mater.* **2023**, *453*, 131451. [CrossRef]
18. Arora, P.K.; Srivastava, A.; Singh, V.P. Bacterial degradation of nitrophenols and their derivatives. *J. Hazard. Mater.* **2014**, *266*, 42–59. [CrossRef]
19. Rusănescu, C.O.; Rusănescu, M.; Constantin, G.A. Wastewater Management in Agriculture. *Water* **2022**, *14*, 3351. [CrossRef]
20. Han, L.; Shen, L.; Lin, H.; Cheng, T.; Wen, J.; Zeng, Q.; Xu, Y.; Li, R.; Zhang, M.; Hong, H.; et al. Three dimension-printed membrane for ultrafast oil/water separation as driven by gravitation. *Nano Energy* **2023**, *111*, 108351. [CrossRef]

21. Tempestti, J.C.M.; Mohan, H.; Muthukumar Sathya, P.; Lee, S.-W.; Venkatachalam, J.; Oh, B.-T.; Seralathan, K.-K. Detoxification of p-nitrophenol (PNP) using Enterococcus gallinarum JT-02 isolated from animal farm waste sludge. *Environ. Res.* **2023**, *231*, 116289. [CrossRef] [PubMed]
22. Arora, P.K.; Saroj, R.S.; Mishra, R.; Omar, R.A.; Kumari, P.; Srivastava, A.; Garg, S.K.; Singh, V.P. Draft genome sequence data of a 4-nitrophenol-degrading bacterium, Pseudomonas alloputida strain PNP. *Data Brief* **2021**, *38*, 107390. [CrossRef]
23. Wang, Y.; Bai, Y.; Su, J.; Xu, L.; Ren, Y.; Ren, M.; Hou, C.; Cao, M. Enhanced denitrification and p-nitrophenol removal performance via hydrophilic sponge carriers fixed with dual-bacterial: Optimization, performance, and enhancement mechanism. *J. Hazard. Mater.* **2024**, *475*, 134922. [CrossRef]
24. Li, R.; Wu, Y.; Lou, X.; Li, H.; Cheng, J.; Shen, B.; Qin, L. Porous Biochar Materials for Sustainable Water Treatment: Synthesis, Modification, and Application. *Water* **2023**, *15*, 395. [CrossRef]
25. Qin, L.; Gao, M.; Zhang, M.; Feng, L.; Liu, Q.; Zhang, G. Application of encapsulated algae into MBR for high-ammonia nitrogen wastewater treatment and biofouling control. *Water Res.* **2020**, *187*, 116430. [CrossRef] [PubMed]
26. Xiang, Z.; Chen, X.; Bai, J.; Rong, H.; Li, H.; Zhao, Y.; Huang, X. A novel anaerobic/aerobic-moving bed-dynamic membrane combined biofilm reactor (A/O-MB-DMBR) treating mariculture wastewater with chitosan (CTS): Performance, control and microbial community. *Environ. Technol. Innov.* **2023**, *29*, 103009. [CrossRef]
27. Hoogenkamp, M.A.; Brandt, B.W.; Laheij, A.M.G.A.; De Soet, J.J.; Crielaard, W. 16S rDNA sequencing and metadata of Dutch dental unit water. *Data Brief* **2021**, *37*, 107221. [CrossRef] [PubMed]
28. Ashfaq, M.Y.; Da'na, D.A.; Al-Ghouti, M.A. Application of MALDI-TOF MS for identification of environmental bacteria: A review. *J. Environ. Manag.* **2022**, *305*, 114359. [CrossRef]
29. Fan, Z.; Qin, L.; Zheng, W.; Meng, Q.; Shen, C.; Zhang, G. Oscillating membrane photoreactor combined with salt-tolerated Chlorella pyrenoidosa for landfill leachates treatment. *Bioresour. Technol.* **2018**, *269*, 134–142. [CrossRef]
30. Song, W.; Li, Z.; Ding, Y.; Liu, F.; You, H.; Qi, P.; Wang, F.; Li, Y.; Jin, C. Performance of a novel hybrid membrane bioreactor for treating saline wastewater from mariculture: Assessment of pollutants removal and membrane filtration performance. *Chem. Eng. J.* **2018**, *331*, 695–703. [CrossRef]
31. Wang, Y.; Su, J.; Ali, A.; Chang, Q.; Bai, Y.; Gao, Z. Enhanced nitrate, manganese, and phenol removal by polyvinyl alcohol/sodium alginate with biochar gel beads immobilized bioreactor: Performance, mechanism, and bacterial diversity. *Bioresour. Technol.* **2022**, *348*, 126818. [CrossRef] [PubMed]
32. Wang, Y.; Zhou, J.; Zhang, Z.; Huang, L.; Zhang, B.; Liu, Z.; Zheng, Y. Efficient carbon flux allocation towards D-pantothenic acid production via growth-decoupled strategy in Escherichia coli. *Bioresour. Technol.* **2024**, *411*, 131325. [CrossRef]
33. Xia, Y.; Zhou, X.-Q.; Wu, P.; Jiang, W.-D.; Liu, Y.; Tang, J.-Y.; Zhang, R.-N.; Zhang, L.; Mi, H.-F.; Feng, L. The role of pantothenic acid in alleviating hypoxia-induced liver injury of sub-adult grass carp (Ctenopharyngodon idellus): Possible mechanisms and implication. *Aquaculture* **2024**, *590*, 741083. [CrossRef]
34. Qin, L.; Liu, Q.; Meng, Q.; Fan, Z.; He, J.; Liu, T.; Shen, C.; Zhang, G. Anoxic oscillating MBR for photosynthetic bacteria harvesting and high salinity wastewater treatment. *Bioresour. Technol.* **2017**, *224*, 69–77. [CrossRef]
35. Sun, Y.; Zhou, J.; Cai, W.; Zhao, R.; Yuan, J. Hierarchically porous NiAl-LDH nanoparticles as highly efficient adsorbent for p-nitrophenol from water. *Appl. Surf. Sci.* **2015**, *349*, 897–903. [CrossRef]
36. Yue, W.; Chen, M.; Cheng, Z.; Xie, L.; Li, M. Bioaugmentation of strain Methylobacterium sp. C1 towards p-nitrophenol removal with broad spectrum coaggregating bacteria in sequencing batch biofilm reactors. *J. Hazard. Mater.* **2018**, *344*, 431–440. [CrossRef]
37. Zheng, Y.; Liu, D.; Liu, S.; Xu, S.; Yuan, Y.; Xiong, L. Kinetics and mechanisms of p-nitrophenol biodegradation by Pseudomonas aeruginosa HS-D38. *J. Environ. Sci.* **2009**, *21*, 1194–1199. [CrossRef]
38. Zhou, M.; Lei, L. The role of activated carbon on the removal of p-nitrophenol in an integrated three-phase electrochemical reactor. *Chemosphere* **2006**, *65*, 1197–1203. [CrossRef]
39. Huang, Y.; Tu, H.; Zheng, W.; Duan, Y.; Li, Z.; Cui, Z. PnpB involvement in the regulation of temperature-sensitive para-nitrophenol degradation in Pseudomonas putida MT54 via PnpA. *Biochem. Biophys. Res. Commun.* **2018**, *503*, 1575–1580. [CrossRef]
40. Cao, L.; Zhu, G.; Tao, J.; Zhang, Y. Iron carriers promote biofilm formation and p-nitrophenol degradation. *Chemosphere* **2022**, *293*, 133601. [CrossRef]
41. Le, B.-N.T.; Nguyen, V.-A.T.; Nguyen, N.-P.; Nguyen, H.-N.; Phan, T.-T.H.; Nguyen, T.-H.; Pham, T.-P.T.; Nguyen, H.-D.P. Enhanced degradation performance toward para-nitrophenol of adapted immobilized microbial community on coconut coir. *Int. Biodeterior. Biodegrad.* **2025**, *196*, 105923. [CrossRef]
42. Sreenivasulu, C.; Megharaj, M.; Venkateswarlu, K.; Naidu, R. Degradation of p-nitrophenol by immobilized cells of *Bacillus* spp. isolated from soil. *Int. Biodeterior. Biodegrad.* **2012**, *68*, 24–27. [CrossRef]
43. Yenilmez, A.E.; Ertul, S.; Yilmaz, T.; Ucar, D.; Di Capua, F.; Sahinkaya, E. Co-removal of P-nitrophenol and nitrate in sulfur-based autotrophic and methanol-fed heterotrophic denitrification bioreactors. *J. Environ. Chem. Eng.* **2023**, *11*, 111325. [CrossRef]

44. Levin, L.; Carabajal, M.; Hofrichter, M.; Ullrich, R. Degradation of 4-nitrophenol by the white-rot polypore Trametes versicolor. *Int. Biodeterior. Biodegrad.* **2016**, *107*, 174–179. [CrossRef]
45. Qiu, X.; Wu, P.; Zhang, H.; Li, M.; Yan, Z. Isolation and characterization of Arthrobacter sp. HY2 capable of degrading a high concentration of p-nitrophenol. *Bioresour. Technol.* **2009**, *100*, 5243–5248. [CrossRef]

water

Article

Analysis on Operation and Water Quality Characteristics of Centralized Wastewater Treatment Plants of Industrial Parks in Yellow River Basin, China

Yanjun Wang [1,2,†], Yue Yuan [1,†], Hao Xue [1], Yin Yu [1], Yang Shi [3], Huina Wen [4,*] and Min Xu [1,*]

1 Chinese Research Academy of Environmental Sciences, Beijing 100012, China; 15601147849@163.com (Y.W.); yuan.yue@craes.org.cn (Y.Y.); xh715810629@163.com (H.X.); yuyinphoebe@163.com (Y.Y.)
2 Jiangsu Changzhou Environment Monitoring Center, Changzhou 213000, China
3 Yellow River Basin Ecological and Environmental Administration Ministry of Ecology and Environment, Zhengzhou 450004, China; shi.yang@yreea.mee.gov.cn
4 Ecological and Environmental Monitoring and Scientific Research Center, Yellow River Basin Ecological and Environmental Administration Ministry of Ecology and Environment, Zhengzhou 450004, China
* Correspondence: wen.huina@yreea.mee.gov.cn (H.W.); renyumeiwen1987@163.com (M.X.); Tel.: +86-371-660-23-429 (H.W.); +86-10-849-10-433 (M.X.)
† These authors contributed equally to this work.

Abstract: The Yellow River basin serves as an important economic belt and industrial base in China, featuring numerous industrial parks. However, alongside its economic significance, the basin struggles with significant water environmental challenges. This study analyzed the operational status, influent water quality, and energy consumption of 63 centralized wastewater treatment plants (WWTPs) from 54 major industrial parks in the Yellow River basin. The scale of these WWTPs was primarily within the range of 1×10^4~5×10^4 m^3/d, with an average hydraulic loading rate of 53.8%. Aerobic treatment processes are predominant. The influent concentrations of chemical oxygen demand (COD), biochemical oxygen demand (BOD), ammonia nitrogen (NH_3-N), total nitrogen (TN), and total phosphorus (TP) in the WWTPs exhibited a right-skewed distribution. The BOD/COD ratio of the WWTPs fluctuated between 0.1 and 1.6, and 75% of the WWTPs showed a COD/TN ratio lower than eight. The average BOD_5/TN was 2.7, and the probability of influent $BOD_5/TP > 20$ was 84.6%. A significant linear correlation exists between the influent BOD and COD concentrations, while moderate linear relationships are also observed among NH_3-N, TN and TP, emphasizing the importance of maintaining appropriate nitrogen and phosphorus levels for efficient pollutant removal. The average electricity consumption of WWTPs in the Yellow River basin in 2023 was 1.1 kWh/m^3. It is important to upgrade these WWTPs and reduce their energy consumption. Further strengthening the construction of industrial wastewater collection and treatment facilities based on regional characteristics is recommended to promote the high-quality development of industrial wastewater treatment in the Yellow River basin.

Keywords: industrial wastewater treatment plant; Yellow River basin; construction and operational status; influent water quality; energy consumption

Citation: Wang, Y.; Yuan, Y.; Xue, H.; Yu, Y.; Shi, Y.; Wen, H.; Xu, M. Analysis on Operation and Water Quality Characteristics of Centralized Wastewater Treatment Plants of Industrial Parks in Yellow River Basin, China. *Water* **2024**, *16*, 806. https://doi.org/10.3390/w16060806

Academic Editor: Daniel Mamais

Received: 20 February 2024
Revised: 3 March 2024
Accepted: 6 March 2024
Published: 8 March 2024

1. Introduction

Water is a vital natural and strategic resource, directly related to the sustainable development of both the economy and society. Globally, water scarcity poses a pressing challenge, projected to affect two-thirds of the world's population by 2025 [1]. The Yellow River ranks as the fifth largest river in the world and serves as an important water source in northern China [2]. Approximately 15% of China's agriculturally irrigated land and 12% of its population rely on the Yellow River for water supply [3]. Furthermore, it facilitates long-distance water supply to key regions such as Beijing, Tianjin and Hebei [4]. Ecological preservation and green development in the Yellow River basin are two of China's major national strategies. The Chinese government issued the "Outline of Ecological Protection and High-Quality Development Plan for the Yellow River Basin" in 2021, emphasizing environmental improvement and sustainable development [5].

The Yellow River basin is not only an important ecological barrier and economic region in China but also an important energy, chemical industry, raw material and basic industrial base. It accommodates a large number of high-energy-consuming and high-pollution industries and has become an important industrial infrastructure cluster [6,7]. Factor endowments and development circumstances determine that traditional industries in the Yellow River basin primarily thrive on resource mining and processing, shaping a resource- and energy-reliant industrial framework based on coal, oil, natural gas, metal mining and processing [8–10].

Industry is the main driver of economic and social development in the Yellow River basin, concurrently serving as a key contributor to water resource consumption and water environmental pollution. The accelerated pace of economic expansion has exacerbated the risks of water resource depletion and water pollution in this region. The per capita water resource of the Yellow River basin is 905 m^3, approximately one-third of the national per capita water, and below the standard for water-scarce areas (1000 m^3) [11,12]. Over the past two decades, the total water consumption in the basin has increased from 112.74 billion m^3 in 2000 to 125.87 billion m^3 in 2020, accompanied by a rise in wastewater discharge from 9.06 to 21.3 billion tons [13]. Industrial wastewater discharge accounts for nearly one-third of the total wastewater discharge in this area. In recent years, escalating tensions stemming from carrying capacity shortages, water shortages, and water environment issues in the Yellow River basin have become prominent [3,14]. According to the 2022 China Ecological Environment Status Bulletin, 2.3% of the 263 water quality sections across the Yellow River were categorized as below Class V, significantly surpassing the national average of 0.7% [15]. The contradiction between water supply and demand, coupled with water pollution problems, has gradually become an important factor impeding the sustainable economic and social development of the Yellow River basin.

Industrial parks represent a ubiquitous aspect of global industrial development [16]. With China's rapid urbanization and industrialization, the development of industrial parks has been accelerated in recent decades. China currently has 2543 industrial parks, with approximately a quarter situated within the Yellow River basin [17]. The establishment of shared infrastructure, such as centralized wastewater treatment plants (WWTPs) in industrial parks, is a global practice and constitutes a fundamental strategy to promote industrial symbiosis [18,19]. Centralized industrial WWTPs are centrally managed facilities or independently operated units that specialize in treating industrial wastewater originating from industrial parks as well as neighboring enterprises or industrial facilities. Given the concentration of industrial activities in diverse industrial parks, the construction of centralized industrial WWTPs is growing rapidly. The number of centralized industrial WWTPs and their wastewater treatment capacities in 2017 were 3.6 times and 3.2 times those of 2007, respectively [20]. Centralized industrial WWTPs play an important role in controlling industrial water pollution and maintaining the safety of the water environment in the Yellow River basin, which serves as an indispensable safeguard for promoting the green and high-quality development of industrial parks.

Current research on centralized industrial WWTPs mainly focuses on the process design or improvement of individual or multiple plants, as well as suggestions for supervision and management policies [21–23]. However, comprehensive analyses and research on centralized WWTPs in industrial parks in the Yellow River basin have been absent. This study took 63 centralized WWTPs in 54 national and provincial industrial parks in the Yellow River basin as the research objects and investigated aspects including the scale and operation of the WWTPs, the characteristics and correlations of the main pollutants in the influent, and their energy consumption and operational costs. The objective is to identify challenges in the construction and operation of centralized WWTPs in industrial parks in the Yellow River basin and to provide insights for the high-quality development of industrial wastewater treatment in the basin.

2. Methods

2.1. Data Collection

The Yellow River basin includes nine provincial districts, namely Qinghai, Sichuan, Gansu, Ningxia, Inner Mongolia, Shaanxi, Shanxi, Henan and Shandong. In this study, 54 industrial parks in the Yellow River basin were selected as the research objects due to their importance among all parks. Specifically, all selected parks are classified as national or provincial and have centralized industrial WWTPs instead of sharing with municipal WWTPs or treating water separately. Figure 1 depicts the geographical locations of 54 industrial parks in the Yellow River basin. The leading industries in these industrial parks included the petrochemical industry, the coal chemical industry, the fine chemical industry, the smelting industry, the pharmaceutical sector and the pesticide industry. The research objects comprised a total of 63 centralized WWTPs located within these industrial parks.

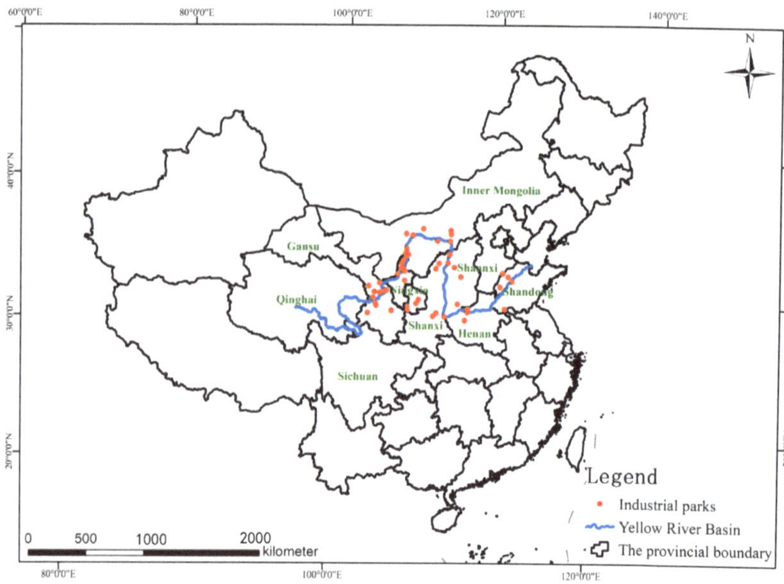

Figure 1. Map of geographical locations of 54 industrial parks in the Yellow River basin.

Considering the provincial administrative divisions and the geographical division of the upper, middle and lower reaches of the Yellow River basin, in this study, 43 centralized industrial WWTPs were identified in the upper reaches of the Yellow River basin, encompassing Gansu, Ningxia and Inner Mongolia, while 11 centralized industrial WWTPs were situated in the middle reaches, spanning Shaanxi and Shanxi, with an additional 9 centralized industrial WWTPs located in the lower reaches, including Henan and Shandong.

Overall, the operation and water quality data of 63 centralized industrial WWTPs in 2023 were obtained from surveys. However, due to the diverse operating realities, the water quality index varied between parks, and some of the data were missing or abnormal. Thus, all the data were pretreated and screened to eliminate outliers, and a random sampling survey was carried out to ensure the reliability of the data.

2.2. Data Analysis

In this study, the construction and operational status were analyzed considering regional distribution, design processing capacity, actual wastewater treatment volume, and wastewater treatment processes. The influent quality parameters of 63 centralized industrial WWTPs, including the chemical oxygen demand (COD), biochemical oxygen demand (BOD), ammoniacal–nitrogen (NH_3-N), total nitrogen (TN) and total phosphorus (TP), were analyzed, as their concentrations and ratios are vital for wastewater treatment process selection and design. Influent water quality characteristics were measured, and a correlation analysis was conducted. Statistical analyses on the wastewater influent quality at industrial WWTPs in the Yellow River basin were performed by OriginPro 8.5 (developed by OriginLab Corporation, Northampton, MA, USA).

3. Results and Discussion

3.1. Operating Status

3.1.1. Treatment Capacity of Centralized Industrial WWTPs

The quality of industrial wastewater was characterized by its complexity and significant variations in concentration. Centralized industrial WWTPs, as specialized units dedicated to handling complex industrial wastewater, play a crucial role in industrial wastewater treatment. In 2023, the 63 centralized industrial WWTPs of the major cities in the Yellow River basin had a total designed processing capacity of 123.06×10^4 m^3/d, accounting for approximately 4.34% and 5.29%, respectively, in terms of both the number and processing capacity compared to China's overall statistics on centralized industrial WWTPs. Centralized industrial WWTPs could be divided into five grades according to their processing capacity: $\geq 10 \times 10^4$, 5×10^4~10×10^4, 1×10^4~5×10^4, 0.5×10^4~1×10^4 and $<0.5 \times 10^4$ m^3/d. The distribution of centralized industrial WWTPs in the Yellow River basin is shown in Table 1. The centralized industrial WWTPs in the Yellow River basin were primarily designed with a capacity ranging from 1×10^4 to 5×10^4 m^3/d. These facilities constitute 68.25% of the total number of centralized industrial WWTPs and contribute approximately 71.03% to the overall design processing capacity in the Yellow River basin. Only a few of the centralized industrial WWTPs had a capacity greater than 5×10^4 m^3/d. These two categories accounted for only 6.35% of the total number in the Yellow River basin but represented a significant portion (24.79%) of the total designed processing capacity. The proportion of centralized industrial WWTPs with a capacity less than 1×10^4 m^3/d accounted for 25.40% of the total quantity. However, their designed processing capacity only represented 4.19% of the Yellow River basin's total.

Table 1. The number and treatment capacity of centralized industrial WWTPs in the Yellow River Basin.

Water Treatment Capacity (10^4 m^3/d)	Number						
	Gansu	Ningxia	Inner Mongolia	Shaanxi	Shanxi	Henan	Shandong
<0.5	6	0	4	0	2	0	0
0.5~1	2	1	0	0	1	0	0
1~5	9	12	6	5	2	4	5
5~10	0	1	1	1	0	0	0
≥10	0	0	1	0	0	0	0

In different regions, the construction of centralized industrial WWTPs in the upper reaches of the Yellow River basin was relatively well developed, with a total of 43 plants and a combined processing capacity of 81.55×10^4 m^3/d. In the middle reaches of the Yellow River basin, there were a total of 11 centralized industrial WWTPs with a combined processing capacity of 18.01×10^4 m^3/d. In the lower reaches, there were nine centralized industrial WWTPs with a total processing capacity of 23.5×10^4 m^3/d.

The hydraulic loading rate refers to the proportion between the actual volume of wastewater treated and the designed treatment capacity during a specific operational period. It is a fundamental metric that reflects the stable functioning of WWTPs. Figure 2 illustrates the wastewater treatment conditions of the centralized industrial WWTPs in the Yellow River basin. The average hydraulic loading rate of the centralized industrial

WWTPs in the Yellow River basin in 2023 was 53.8%. There were significant variations in the hydraulic loading rates among different cities, ranging from 10% to 112.5%. Notably, Jiyuan, Liaocheng and Heze exhibited the highest hydraulic loading rates at 112.5%, 100% and 100%, respectively. Conversely, Wuwei, Bayannur and Lanzhou demonstrated the lowest hydraulic loading rates at 10%, 12.01% and 17.45%, respectively. In terms of geographical distribution, the average hydraulic loading rates of major cities located in the upper, middle and lower reaches of the Yellow River basin were recorded as 43.94%, 48.09% and 81.23%, respectively.

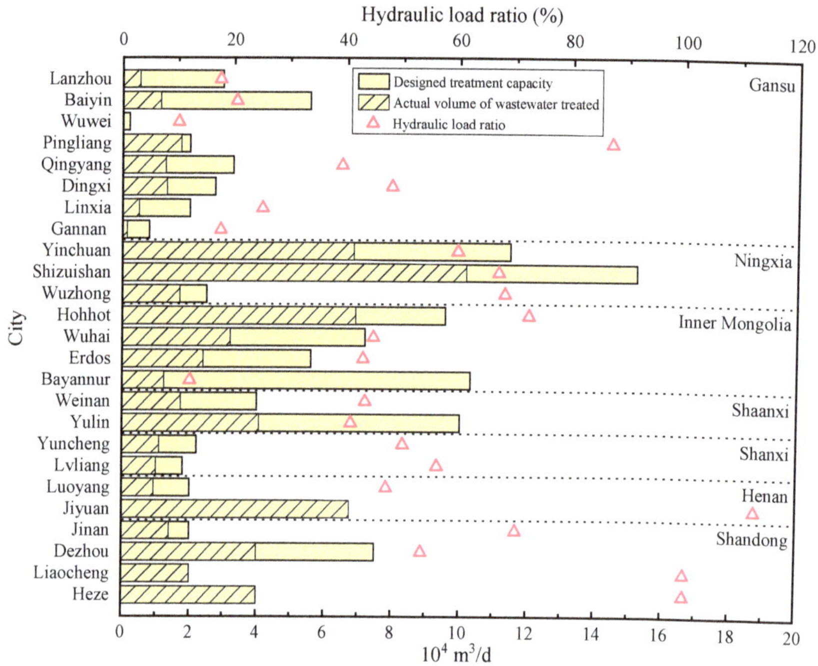

Figure 2. Wastewater treatment status of centralized industrial WWTPs in the Yellow River basin.

In summary, most centralized industrial WWTPs in the Yellow River basin were small to medium-sized facilities with a processing capacity of less than 5×10^4 m^3/d. However, their wastewater collection rates exhibited a relatively low level. Notably, around 95.45% of the centralized industrial WWTPs had a wastewater treatment rate below 90%. The actual water volume treated by the centralized industrial WWTPs fell significantly below its designed capacity, indicating a substantial potential for enhancing wastewater treatment utilization. The load rate of the centralized industrial WWTPs in the lower reaches of the Yellow River basin significantly exceeded that in the middle and upper reaches. The operational capacity of Jiyuan, Liaocheng and Heze's industrial wastewater treatment plants was either at or over their limits, indicating a pressing need for the further expansion of these facilities.

3.1.2. Wastewater Treatment Process

The predominant treatment processes used in the centralized industrial WWTPs were aerobic and anaerobic biological treatments. In the Yellow River basin, 91.84% of these plants combined two or more of these processes. The continuous advancement of water quality improvement objectives has led to increasingly stringent emission limits for water pollutants, resulting in the growing adoption of advanced treatment processes. Consequently, 36.73% of the centralized industrial WWTPs had implemented advanced treatment processes. The main emphasis of this article is on the analysis of centralized industrial WWTPs that utilize biological treatment systems. In the Yellow River basin, aerobic biological treatment processes were predominantly utilized in a significant proportion (55.1%) of the centralized industrial WWTPs. The primary processes used were the anaerobic–anoxic–oxic (AAO) process, the anaerobic–oxic (AO) process, an oxidation ditch, and the membrane bioreactor (MBR) process.

3.2. Characteristics of Influent Quality

3.2.1. Influent Quality of Industrial, Centralized WWTPs

The concentrations of five basic water quality parameters of the 63 centralized industrial WWTPs in the Yellow River basin are presented in Figure 3a. The average influent COD, BOD, NH_3-N, TN and TP concentrations were 323.3, 145.9, 30.2, 53.9 and 3.4 mg/L, respectively, while the median influent COD, BOD, NH_3-N, TN and TP concentrations were 252.0, 116.4, 28.1, 44.7 and 2.9 mg/L, respectively. For all the parameters, the average concentration is higher than the median concentration, exhibiting a right-skewed distribution for the influent quality. The 5~95% ranges of the influent COD, BOD, NH_3-N, TN and TP concentrations were 70.0~1050.7, 22.9~435, 5.7~66.1, 7.4~87.0 and 0.2~8.1 mg/L, respectively. This considerable fluctuation in water quality fully reflects the complex composition, wide variation, and unpredictable characteristics of industrial wastewater. The 25~75% ranges of the influent COD, BOD, NH_3-N, TN and TP concentrations of the centralized industrial WWTPs were 167.4~355.6, 60.0~155.0, 15.9~43.6, 25.7~56.7 and 1.5~5.0 mg/L, respectively. These data indicate that the influent COD and BOD concentrations were relatively low, which might lead to an insufficient carbon source for the biological processes of the WWTPs and consequently inhibit the removal of nitrogen and phosphorus [24]. A possible explanation for the relatively low COD and BOD concentrations is that the studied industrial parks are generally large in scale and have complex industrial structures. The composition of the wastewater produced by enterprises in different industries varies, and after being mixed into the centralized WWTPs, COD and BOD concentrations are likely to decrease due to mutual dilution. Therefore, it is necessary to carry out classified collection through the industry- and quality-based treatment of wastewater in large industrial parks.

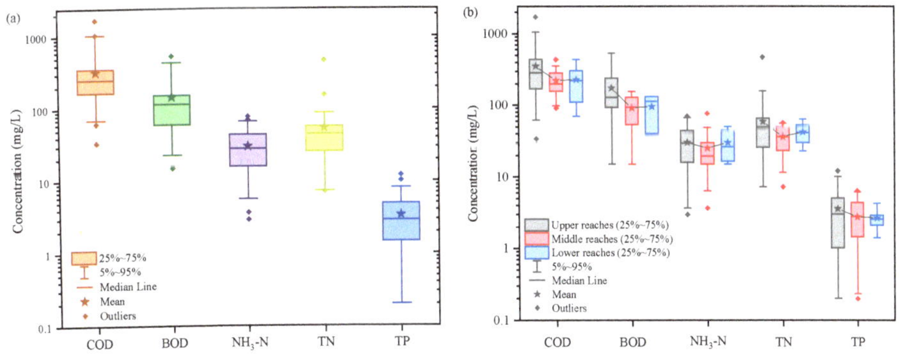

Figure 3. Wastewater influent quality of centralized industrial WWTPs: (**a**) in the Yellow River basin and (**b**) in the upper, middle and lower reaches.

The influent water quality of the centralized industrial WWTPs showed regional variations in the Yellow River basin, as illustrated in Figure 3b. Generally, the mean and median concentrations of parameters in the upper reaches (including Gansu, Ningxia and Inner Mongolia) were marginally higher than those in the middle reaches (including Shaanxi and Shanxi) and the lower reaches (including Henan and Shandong), likely attributable to the heightened industrial activity in the region. This increased industrialization also correlates with the greater volume of data for the upper reaches, resulting in a wider range of variation in the pollutant concentrations.

3.2.2. Ratios of Water Quality Parameters

The ratios of different water quality parameters of the centralized industrial WWTPs in the Yellow River basin were calculated, and the analyzed results are shown in Figure 4.

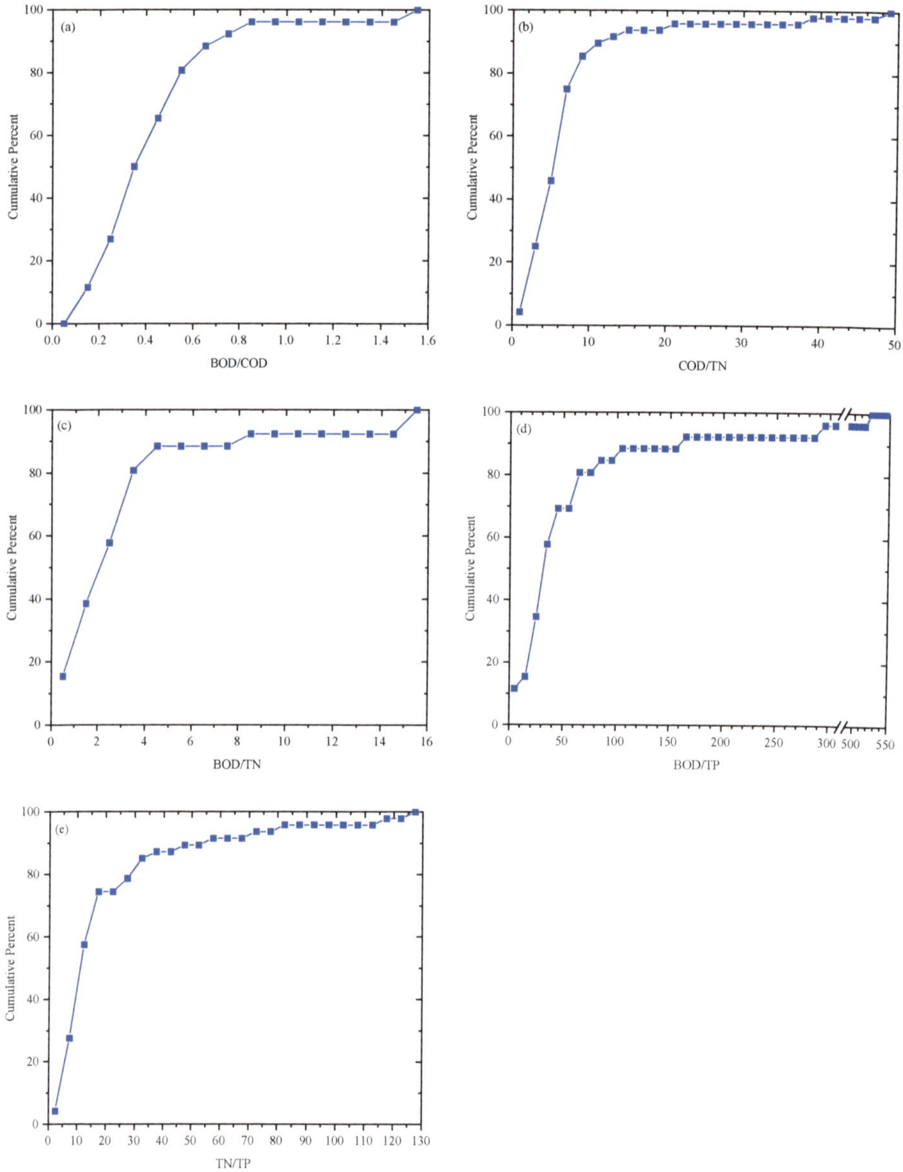

Figure 4. Ratios of water quality parameters of centralized industrial WWTPs in the Yellow River basin. (**a**) BOD/COD; (**b**) COD/TN; (**c**) BOD/TN; (**d**) BOD/TP; and (**e**) TN/TP.

The BOD/COD ratio can reflect the biodegradability of sewage. When the BOD/COD ratio is between 0.4 and 0.6, the biodegradability of the wastewater is considered good, and lower values indicate that the wastewater is poorly biodegradable [25]. The BOD/COD ratio of the WWTPs fluctuates between 0.1 and 1.6, with an average value of 0.5 and a medium value of 0.4, which means that 50% of the influent of the WWTPs was not favorable for biological treatment (Figure 4a). The overall biodegradability of the influent was poor as the studied industrial parks are mainly focused on chemical, metallurgical, pharmaceutical and other industries, which usually produce non-degradable or refractory organic pollutants.

COD represents a constraining factor in the process of denitrification. Effective denitrification can be achieved when the influent COD/TN ratio falls between 8 and 12, and if the COD/TN ratio is too low, the carbon source is considered insufficient and additional supplementation is required [26]. As shown in Figure 4b, the COD/TN ratio ranges from 0.6 to 49.4; however, 75% of the WWTPs showed values lower than 8, revealing that insufficient carbon sources are a vital problem for centralized industrial wastewater treatment. In fact, 37 of the 63 centralized industrial WWTPs surveyed in this study have been using additional carbon sources to effectively remove nitrogen and phosphorus from wastewater, resulting in a total additional carbon source consumption of 17,372.5 ton per year.

The BOD/TN ratio can also reflect whether there is enough organic matter in the wastewater influent for the efficient removal of TN. The carbon source is considered adequate when BOD/TN > 4 [27]. Figure 4c shows that the range of the BOD/TN ratio is 0.8~15.8, the average value is 2.7, and the median value is 3.7. Similar to COD, 80.8% of the BOD/TN ratios for the WWTPs were below four, once again proving the lack of carbon sources in the wastewater from the industrial parks in the Yellow River basin.

The BOD/TP ratio can be used to assess the feasibility of biological phosphorus removal. Generally, a BOD/TP ratio greater than 20 ensures good phosphorus removal efficiency [28]. As presented in Figure 4d, the BOD/TP ratio varies significantly from 8.9 to 529.4, with an average value of 63.9 and a medium value of 32.0. Only 15.4% of WWTPs exhibited a BOD/TP ratio less than 20, indicating that phosphorus could be effectively removed during biotreatment.

From Figure 4e, the centralized industrial WWTPs had a TN/TP ratio between 3.0 and 126.6. The balance of C, N and P in the influent is the key to effective biological treatment processes. It is generally believed that when the BOD/TN/TP ratio is between 100:5:1 and 100:10:1, an aerobic process is the most efficient, and when the BOD/TN/TP ratio is 250:5:1, an anaerobic process is the most efficient [26]. In this study, the TN/TP ratio of 72.3% of the WWTPs exceeded 10, which did not meet the optimum conditions for microbial growth.

Overall, the centralized industrial WWTPs in the Yellow River basin exhibit challenges, including low concentrations of COD and BOD, limited biodegradability and imbalanced compositions of C, N and P in their wastewater influent, all of which hinder the effect of the biological treatment process. Pretreatment techniques to improve biodegradability and additional carbon sources are essential, but substantial carbon source consumption not only diverges from the objectives outlined in the Global Sustainable Development Goals but also amplifies the financial strain of operational costs.

3.2.3. Correlation of Influent Water Quality Parameters

Based on the influent levels of BOD, COD, NH_3-N, TN and TP in the centralized industrial wastewater treatment plants (WWTPs) within the Yellow River basin, a linear regression analysis was conducted using the least squares method. The relationships among these water quality parameters, along with the corresponding regression equations and correlation coefficients (R^2), are presented in Figure 5 and Table 2. Figure 5a demonstrates a significant linear correlation between BOD and COD (R^2 = 0.73118), while the associations between NH_3-N (R^2 = 0.12457) and TN (R^2 = 0.24157) are moderate. In contrast, no linear correlation is observed between BOD and TP (R^2 = 0.05991). The correlation analysis of COD and BOD reveals a proportional relationship, where the COD concentration increases with rising BOD levels. Figure 5b illustrates weak linear relationships between COD and NH_3-N (R^2 = 0.1309), TN (R^2 = 0.10026), and TP (R^2 = 0.11159), respectively. Additionally, Figure 5c demonstrates a clear correlation between NH_3-N and TN (R^2 = 0.426), underscoring the necessity of controlling influent NH_3-N and enhancing nitrification and denitrification efficiency to achieve effective total nitrogen removal in WWTPs. A general linear relationship is found between NH_3-N and TP (R^2 = 0.26939) as well as TN and TP (R^2 = 0.22071) from Figure 5c,d. Thus, maintaining appropriate nitrogen and phosphorus levels is crucial for the efficient removal of pollutants.

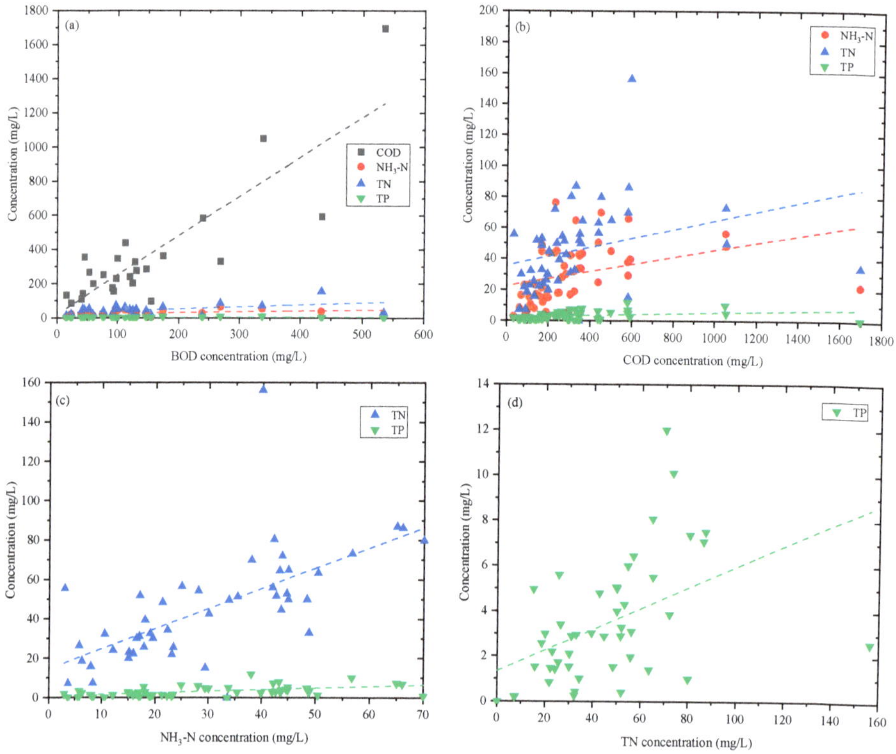

Figure 5. Relevance of influent water quality parameters: (a) BOD; (b) COD; (c) NH_3-N; and (d) TN.

Table 2. Correlation and regression analysis of influent water quality parameters.

Index	COD	NH_3-N	TN	TP
BOD	Y = 2.3079x + 20.544 R^2 = 0.73118	y = 0.0435x + 24.028 R^2 = 0.12457	y = 0.11587x + 30.322 R^2 = 0.24157	y = 0.00472x + 2.7249 R^2 = 0.05991
COD		y = 0.02323x + 22.626 R^2 = 0.1309	y = 0.02913x + 35.744 R^2 = 0.10026	y = 0.00306x + 2.3627 R^2 = 0.11159
NH3-N			y = 1.0262x + 14.548 R^2 = 0.426	y = 0.07964x + 1.0196 R^2 = 0.26939
TN				y = 0.04608x + 1.3470 R^2 = 0.22071

3.3. Energy Consumption and Operational Cost

The statistical results for energy consumption, operational costs and their correlation in the centralized industrial WWTPs in the Yellow River basin are presented in Figure 6. The energy consumption of the WWTPs primarily comprises electricity consumption and fuel energy consumption, with electricity consumption constituting the largest share. The carbon emissions resulting from the wastewater treatment process represent approximately 1~2% of China's total carbon emissions [29]. A narrow distribution of electricity consumption is observed in Figure 6; the 25~75% range of energy consumption is 0.7~1.1 kWh/m^3, while the average and median values are 1.1 and 2.5 kWh/m^3, respectively. In 2020, the national median electricity consumption for municipal WWTPs was 0.36 kWh/m^3, while the median electricity consumption for municipal WWTPs in major cities within the Yellow River basin varied from 0.26 to 1.11 kWh/m^3, with approximately 66.8% of the WWTPs in the Yellow River basin exceeding the national median value [30]. The intricate composition of pollutants and the limited biodegradability of industrial wastewater necessitate the utilization of additional treatment units, particularly certain physicochemical technologies that consume more energy compared to biotreatment methods, in centralized industrial WWTPs. Consequently, it is understandable that their energy consumption surpasses that of municipal WWTPs. Moreover, the relatively low influent COD concentration also contributes to the high energy consumption to a certain extent. Nevertheless, the general operational energy consumption of industrial WWTPs in the Yellow River basin remains on the high side, indicating potential for optimization and enhancement. To reduce energy consumption, measures such as adopting energy-efficient technologies, optimizing treatment processes and investing in renewable energy sources can be implemented.

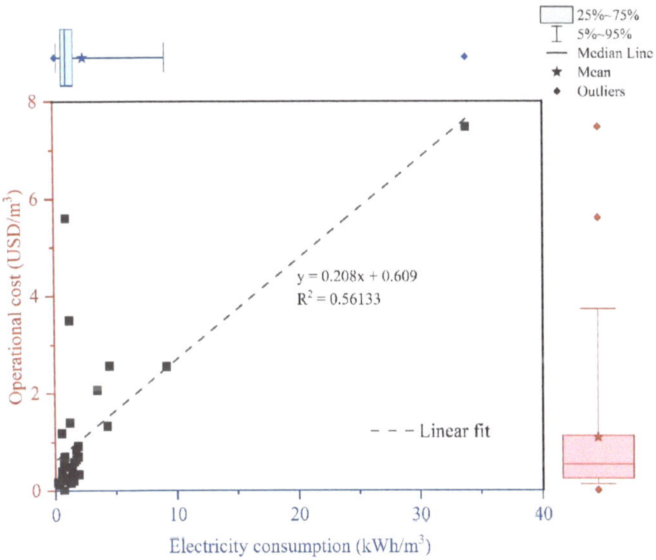

Figure 6. Energy consumption, operational cost and their correlation in centralized industrial WWTPs in the Yellow River basin.

Electricity consumption not only stands out as a significant contributor to carbon emissions in WWTPs but also substantially contributes to the operational costs of these facilities. As shown in Figure 6, the 25~75% range of operational costs is 0.2~1.1 USD/m^3, while the average and median values are 1.1 and 0.5 USD/m^3, respectively. An apparent linear correlation between the operational cost and the electricity consumption was obtained, with R^2 = 0.56133, further confirming that electricity consumption contributes significantly to operational costs. Thus, it is imperative to take effective measures such as improving the influent water quality and optimizing the treatment process to save energy and reduce the cost of the centralized industrial WWTPs in the Yellow River basin.

4. Conclusions

The scales of the centralized WWTPs of industrial parks in the Yellow River basin were generally small, mainly ranging from 1×10^4 to 5×10^4 m^3/d, and the average hydraulic loading rate of the 63 WWTPs was 53.81%, surpassing the national average level. The hydraulic loading rate demonstrated a certain scale effect. However, there was a noticeable regional disparity in the development of wastewater treatment facilities, with slower progress in the construction of facilities in the upper and middle reaches. Both the treatment capacity and collection rate of wastewater require further enhancement. Aerobic biological treatment processes dominated in the Yellow River basin, with their usage accounting for 55.1% of the all plants. The predominant processes include the AAO, AO, oxidation ditch and MBR processes.

The influent pollutant concentrations of the centralized industrial WWTPs in the Yellow River basin displayed a positively skewed distribution overall. The substantial fluctuation in water quality indices, coupled with challenges such as poor biodegradability, insufficient carbon sources and imbalanced nitrogen-to-phosphorus ratios, warrants particular attention. Consequently, emphasis should be placed on controlling the influent water quality during forthcoming upgrades and renovations. Furthermore, the energy consumption of the centralized industrial WWTPs in the Yellow River basin exceeded the national median value. Optimizing the operation of these plants is important to enhance their operational stability and efficiency.

This study only conducted an overall analysis of the water quality characteristics of centralized WWTPs operating in major industrial parks in the Yellow River basin. Given the limitations in data acquisition, it is recommended that subsequent studies continue to enhance the analysis of industrial parks with different water treatment processes and various effluent discharge standard limits, thereby helping to improve the stability of effluent water quality from centralized industrial wastewater treatment facilities.

Author Contributions: Conceptualization, Y.W.; data curation, M.X.; formal analysis, Y.Y. (Yue Yuan) and M.X.; funding acquisition, H.W.; investigation, H.W. and Y.S.; methodology, H.X.; project administration, H.W.; resources, H.W.; software, Y.Y. (Yin Yu); supervision, H.X.; validation, M.X.; visualization, Y.Y. (Yue Yuan) and M.X.; writing—original draft, Y.W.; writing—review and editing, M.X. and Y.Y. (Yin Yu). All authors will be informed about each step of manuscript processing, including submission, revision, revision reminder, etc., via emails from our system or the assigned Assistant Editor. All authors have read and agreed to the published version of the manuscript.

Funding: This research was funded by the Joint Research Program for Ecological Conservation and High-Quality Development of the Yellow River Basin, China (No. 2022-YRUC-01-0405).

Data Availability Statement: The data are not publicly available due to confidentiality agreement.

Conflicts of Interest: The authors declare no conflicts of interest.

References

1. Eliasson, J. The rising pressure of global water shortages. *Nature* **2015**, *517*, 6. [CrossRef]
2. Zhang, Q.; Xu, C.Y.; Yang, T. Variability of water resource in the Yellow River basin of past 50 years, China. *Water Resour. Manag.* **2009**, *23*, 1157–1170. [CrossRef]
3. Chen, Y.; Fu, B.; Zhao, Y.; Wang, K.; Zhao, M.M.; Ma, J.; Wu, J.; Xu, C.; Liu, W.; Wang, H. Sustainable development in the Yellow River Basin: Issues and strategies. *J. Clean. Prod.* **2020**, *263*, 121223. [CrossRef]
4. Liu, J.; Zang, C.; Tian, S.; Liu, J.; Yang, H.; Jia, S.; You, L.; Liu, B.; Zhang, M. Water conservancy projects in China: Achievements, challenges and way forward. *Glob. Environ. Chang.* **2013**, *23*, 633–643. [CrossRef]
5. Central Committee of the Communist Party of China and the State Council. Outline of Ecological Protection and High-Quality Development Plan for the Yellow River Basin (China State Council Publication No. ADM 30-1749). Available online: https://www.gov.cn/gongbao/content/2021/content_5647346.htm (accessed on 20 February 2024).
6. Lu, D.; Sun, D. Development and management tasks of the Yellow River Basin: A preliminary understanding and suggestion. *Acta Geogr. Sin.* **2019**, *74*, 2431–2436.
7. Hu, J.; Liu, Y.; Fang, J.; Jing, Y.; Liu, Y.; Liu, Y. Characterizing pollution-intensive industry transfers in China from 2007 to 2016 using land use data. *J. Clean. Prod.* **2019**, *223*, 424–435. [CrossRef]
8. Yang, X.; Feng, Z.; Chen, Y. Evaluation and obstacle analysis of high-quality development in Yellow River Basin and Yangtze River Economic Belt, China. *Humanit. Soc. Sci. Commun.* **2023**, *10*, 757. [CrossRef]
9. Chen, Y.; Zhu, M.; Lu, J.; Zhou, Q.; Ma, W. Evaluation of ecological city and analysis of obstacle factors under the background of high-quality development: Taking cities in the Yellow River Basin as examples. *Ecol. Indic.* **2020**, *118*, 106771. [CrossRef]
10. Feng, Z.; Chen, Y.; Yang, X. Measurement of Spatio-Temporal Differences and Analysis of the Obstacles to High-Quality Development in the Yellow River Basin, China. *Sustainability* **2022**, *14*, 14179. [CrossRef]

11. Wang, H.; Ma, T. Optimal water resource allocation considering virtual water trade in the Yellow River Basin. *Sci. Rep.* **2024**, *14*, 79. [CrossRef]
12. Engelman, R.; LeRoy, P. *Sustaining Water: Population and Future of Renewable Water Supplies*; Population Action International: Washington, DC, USA, 1993; Volume 23, p. 296.
13. Ministry of Water Resources of the People's Republic of China. China Water Resource Bulletin. 2020. Available online: http://dsgl.mwr.gov.cn/dszs/202212/P020221228567515805249.pdf (accessed on 20 February 2024).
14. Sun, J.; Wang, X.; Shahid, S.; Yin, Y.; Li, E. Spatiotemporal changes in water consumption structure of the Yellow River Basin, China. *Phys. Chem. Earth* **2022**, *126*, 103112. [CrossRef]
15. Ministry of Ecology and Environment of the People's Republic of China. China Ecological Environment Status Bulletin. 2022. Available online: https://www.mee.gov.cn/hjzl/sthjzk/zghjzkgb/202305/P020230529570623593284.pdf (accessed on 20 February 2024).
16. Sakr, D.; Baas, L.; El-Haggar, S.; Huisingh, D. Critical success and limiting factors for eco-industrial parks: Global trends and Egyptian context. *J. Clean. Prod.* **2011**, *19*, 1158–1169. [CrossRef]
17. National Development and Reform Commission, China. Audit Notice Catalogue of Chinese Development Zones. 2018. Available online: https://www.ndrc.gov.cn/fggz/lywzjw/zcfg/201803/t20180302_1047056.html (accessed on 20 February 2024).
18. Chertow, M.R. Industrial symbiosis: Literature and taxonomy. *Annu. Rev. Energy Environ.* **2000**, *25*, 313–337. [CrossRef]
19. Tian, J.; Liu, W.; Lai, B.; Li, X.; Chen, L. Study of the performance of eco-industrial park development in China. *J. Clean. Prod.* **2014**, *64*, 486–494. [CrossRef]
20. Liu, J.; Liu, T.; Su, Y.; Wang, J.; Chen, M. Analysis on operation and water quality characteristics of industrial wastewater treatment plants in China. *Water Wastewater Eng.* **2021**, *47*, 92–96, 103.
21. Kuok, K.K.; Chiu, P.C.; Rahman, M.R.; Bakri, M.K.B.; Chin, M.Y. Effectiveness of centralized wastewater treatment plant in removing emerging contaminants: A case study at Kuching, Malaysia. *J. Water Resour. Prot.* **2022**, *14*, 650–663. [CrossRef]
22. Wang, H.; Qi, T.; Qiao, X.; Li, X.; Ding, S.; Liu, Y. Method for ensuring the safety and effectiveness of wastewater treatment under centralized treatment mode by using a petrochemical park as case study. *J. Water Process Eng.* **2023**, *56*, 104421. [CrossRef]
23. Liu, H.; Wang, H.; Zhou, X.; Fan, J.; Liu, Y.; Yang, Y. A comprehensive index for evaluating and enhancing effective wastewater treatment in two industrial parks in China. *J. Clean. Prod.* **2019**, *230*, 854–861. [CrossRef]
24. Zhao, W.; Wang, M.; Li, J.; Huang, Y.; Li, B.; Pan, C.; Li, X.; Peng, Y. Optimization of denitrifying phosphorus removal in a predenitrification anaerobic/anoxic/post-aeration + nitrification sequence batch reactor (pre-A2NSBR) system: Nitrate recycling, carbon/nitrogen ratio and carbon source type. *Front. Environ. Sci. Eng.* **2018**, *12*, 8. [CrossRef]
25. Zhang, B.; Ning, D.; Yang, Y.; Van Nostrand, J.D.; Zhou, J.; Wen, X. Biodegradability of wastewater determines microbial assembly mechanisms in full-scale wastewater treatment plants. *Water Res.* **2020**, *169*, 115276. [CrossRef]
26. Zhou, J.; Wang, Y.N.; Zhang, W.; Shi, B. Nutrient balance in aerobic biological treatment of tannery wastewater. *J. Am. Leather Chem. Assoc.* **2014**, *109*, 154–160.
27. Ding, S.Z.; Bao, P.; Bo, W.; Zhang, Q.; Peng, Y.Z. Long-term stable simultaneous partial nitrification, anammox and denitrification (SNAD) process treating real domestic sewage using suspended activated sludge. *Chem. Eng. J.* **2018**, *339*, 180–188. [CrossRef]
28. Zou, L.; Li, H.; Wang, S.; Zheng, K.; Wang, Y.; Du, G.; Li, J. Characteristic and correlation analysis of influent and energy consumption of wastewater treatment plants in Taihu Basin. *Front. Environ. Sci. Eng.* **2019**, *13*, 83. [CrossRef]
29. Yan, P.; Shi, H.X.; Chen, Y.P.; Gao, X.; Fang, F.; Guo, J.S. Optimization of recovery and utilization pathway of chemical energy from wastewater pollutants by a net-zero energy wastewater treatment model. *Renew. Sustain. Energy Rev.* **2020**, *133*, 110160. [CrossRef]
30. Xu, A.; Wu, Y.; Chen, Z.; Hao, S.; Hu, H. Analysis of the current situation of construction and operation of municipal wastewater treatment plants in the Yellow River basin. *Water Wastewater Eng.* **2022**, *48*, 27–36.

water

Article

Rapid Removal of Cr(VI) from Wastewater by Surface Ionized Iron-Based MOF: Ion Branching and Domain-Limiting Effects

Chen Wang, Jiakun Chen and Qi Yang *

School of Water Resources and Environment, China University of Geosciences, Beijing 100083, China; 17603363661@163.com (C.W.); cjk2245986367@outlook.com (J.C.)
* Correspondence: yq@cugb.edu.cn; Tel.: +86-13691491158; Fax: +86-10-82321081

Abstract: Exploring the ratio of metal centers to organic ligands and the amount of DMF are important to improve the stability and adsorption efficiency of MOF materials as adsorbents. In this work, MIL101(Fe)-Na_2CO_3 was successfully obtained by modification with formic acid, sodium carbonate, carbon nanotubes, and moieties. The adsorption efficiency of MIL-101(Fe) on Cr(VI) was greatly improved, and the removal efficiency was able to reach 100% in 20 min with a maximum adsorption capacity of 20 mg/g. The inhibition order of the competing anions for the removal of hexavalent chromium was as follows: $Cl^- < NO_3^- < SO_4^{2-}$. The analysis of the adsorption thermodynamic model found that the adsorption of MIL101(Fe)-Na_2CO_3 for Cr(VI) showed spontaneous heat-absorbing and entropy-increasing chemisorption behavior. When using NaOH as the eluent and HCl as the regeneration stabilizer, MIL-101(Fe)-Na_2CO_3 had good adsorption capacity in multiple cycles.

Keywords: MOF materials; Cr(VI); adsorption kinetics; adsorption thermodynamic

check for updates

Citation: Wang, C.; Chen, J.; Yang, Q. Rapid Removal of Cr(VI) from Wastewater by Surface Ionized Iron-Based MOF: Ion Branching and Domain-Limiting Effects. *Water* **2024**, *16*, 25. https://doi.org/10.3390/w16010025

Academic Editors: Tao Zhang, Jing Yuan, Huu-Tuan Tran and Muhammad-Jamal Alhnidi

Received: 14 November 2023
Revised: 14 December 2023
Accepted: 18 December 2023
Published: 20 December 2023

1. Introduction

Cr(VI) is highly migratory and easily soluble in water, and is considered a carcinogenic and mutagenic agent that is harmful to the ecological environment [1,2]. The toxic effects of Cr(VI) on the human body are mainly in terms of skin irritation, induction of lung cancer, and damage to the kidney, liver, stomach, etc.; the negative effects on the environment are mainly in terms of making certain plants sprout and grow less, causing algae to die, etc. [3–5]. The U.S. Environmental Protection Agency (USEPA) has identified Cr(VI) as one of the most toxic pollutants in the water system, and has published a maximum limit of 0.05 mg/L for Cr(VI) in water [6].

With the deep development of industry, the pollution of Cr(VI) has spread all over the regions and even further affected the life of local people. The diffusivity and toxicological relevance of Cr(VI) itself make water bodies produce a serious and permanent pollution state. Therefore, it is urgent to solve Cr(VI) pollution, and it is important to find a cost-effective Cr(VI) pollution removal technology. Currently, researchers at home and abroad have proposed a variety of methods to remove Cr(VI) from wastewater, mainly including chemical reduction precipitation, adsorption, ion exchange resin method, solvent extraction, membrane separation, etc. [7–11].

Metal-organic frameworks (MOFs) are coordination polymers making up a class of highly ordered porous crystalline hybrid materials consisting of metal clusters and polyfunctional organic linkers. Compared with conventional adsorbents, MOFs present captivating merits due to their varying compositions and structures, such as higher surface area and pore volume, massive porosity, adjustable pore size, and favorable dispersion of metal ions in the framework [12]. Their characteristics include adjustable topology, high porosity, low density, and high thermal and chemical stability [13]. In recent years, MOFs have been used in a wide range of applications such as gas separation/storage, water purification, chemical sensors, optics, drug transport, bioreactors, and adsorption [14–21].

MOFs are effective adsorbents for the removal of contaminants from aqueous solutions because of their extraordinary structural and surface properties [22]. For the adsorption of organic pollutants, various interactions between the host and the guest occur, such as electrostatic, acid-base, H-bonding, hydrophobicity, and coordination with open metal sites, all of which have important effects on adsorption [23,24]. In a previous study, a variety of MOFs were used to adsorb two drugs, furosemide and salazosulfapyridine, in the aqueous phase, then the adsorption efficiency and stability were compared. It was found that MIL-101(Cr) was the most stable and had the highest adsorption capacity, and there was a synergistic effect between the drug and MIL-101(Cr) [25]. Regarding the removal of inorganic pollutants, different MOFs have different adsorption capacities for different metal ions, and numerous studies have been previously performed. AMOF-1 material has been synthesized as an effective adsorbent for Cd(II) with a maximum adsorption capacity of 41 mg/g, while Fe_3O_4@MIL-100(Fe) material has been synthesized for Cr(VI) removal with a maximum adsorption capacity of 18 mg/g [26,27]. Therefore, MOFs may be a promising material for adsorption of hexavalent chromium. However, few studies have reported the removal of Cr(VI) using MOFs.

In this paper, we successfully modified MIL-101(Fe)-Na_2CO_3 by controlling and deploying the ratio of metal centers to organic ligands and the DMF dosage of MIL-101 in order to realize the rapid degradation of Cr(VI) through ionic branching and domain-limiting effects. In addition, this paper clarifies the deep-seated mechanism (electrostatic attraction, adsorption on monomolecular layer) for the removal of Cr(VI), which reveals the potential role of modified MIL-101 in practical applications.

MIL-101(Fe) MOFs were selected as the main object of study due to the pollution status of Cr(VI), and were applied to the adsorption of Cr(VI) in water through appropriate modification in order to achieve a better adsorption effect. First, we sought to investigate the effects of the ratio of metal center to organic ligand, DMF dosage, and reaction time on the stability and adsorption efficiency of the synthesized adsorbent and to select the best conditions for the synthesis of MIL-101(Fe). Second, MIL-101(Fe)-Na_2CO_3 adsorbent material was successfully obtained by modification with formic acid, sodium carbonate, carbon nanotubes, and groups, and its morphology was characterized. Third, the adsorption efficiency of the adsorbent for Cr(VI) was investigated by kinetic and thermodynamic adsorption models.

2. Materials and Preparation

2.1. Chemicals

Sodium hydroxide (NaOH, ≥98.0%), hydrochloric acid (HCl, ≥98.0%), anhydrous ethanol (EtOH), potassium dichromate (Potassium dichromate standard solution, 0.0999 mol/L), acetone (C_3H_6O, ≥99.5%), diphenylcarbonyl dihydrazide ($C_{13}H_{14}N_4O$, AR), sulfuric acid (H_2SO_4, 95.0~98.0%), phosphoric acid (H_3PO_4, ≥85.0%), ferric chloride ($FeCl_3$, 98.0%), terephthalic acid (H_2BDC, ≥99.0%), amino-terephthalic acid ($C_8H_7NO_4$, ≥98.0%), formic acid (GCS, ≥99.5%), N,N-dimethylformamide (DMF, 99.8%), acetic acid (Electronic grade G2), and zinc nitrate hexahydrate (Sigma-228737, 98%) were from Sinopharm (Shanghai, China).

2.2. Adsorbent Preparation

The materials were prepared using a hydrothermal autoclave; the preparation flowchart is shown in Figure 1, and the preparation method was as follows.

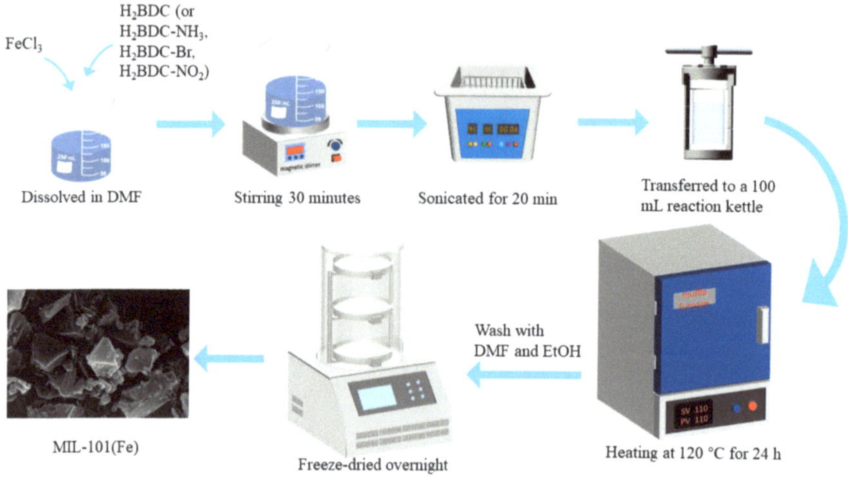

Figure 1. Hydrothermal autoclave preparation process.

2.2.1. MIL-101 Preparation

A certain amount of anhydrous ferric chloride (4.9 mM) was added to 15 mL of N,N-dimethylformamide (DMF) solvent and a certain amount of terephthalic acid (2.45 mM) was added to 15 mL of DMF; the above two mixed solutions were sonicated for 20 min until completely dissolved. Then, the two solutions were mixed and sonicated for 20 min until they were completely mixed, after which the mixed solutions were transferred to a 100 mL reaction kettle and the temperature of the kettle was maintained at 120 °C for 24 h. After the reaction, the solutions were removed and cooled to room temperature in a ventilated area, a certain speed was set for centrifugal separation, then they were washed three times with hot ethanol and DMF, placed in a vacuum dryer, and freeze-dried to produce MIL-101(Fe) [28].

2.2.2. Group Modification

The adsorbent introduction group was prepared mainly by adding a certain amount of amino (nitro, bromine atom)-terephthalic acid (2.45 mM) to 15 mL of DMF solvent and anhydrous ferric chloride (5 mM) to 15 mL of DMF, sonicating the above two solutions for 20 min, then mixing the sonication for 20 min until completely mixed. The mixed solution was then transferred to a 100 mL reactor, the temperature of the reactor was maintained at 120 °C for 24 h, and the reaction was taken out and cooled to room temperature in a ventilated place, centrifuged at a certain speed, washed three times with DMF and hot ethanol, and placed in a vacuum dryer to freeze and dry. This resulted in three MIL-101(Fe) moieties [29].

2.2.3. CNT@MIL-101 Preparation

The loaded CNTs were prepared by dispersing different mass fractions of carbon nanotubes (CNTs) into 9 mL of ethanol solvent and sonicating for 10 min until the CNTs were completely dispersed, followed by adding the above-mentioned CNTs with different mass fractions into a DMF mixture of terephthalic acid and ferric chloride and sonicating for 5 min until they were completely mixed. The above mixed solution was transferred to a reactor and maintained at 120 °C for 24 h. After standing and cooling, the solution was centrifuged, washed three times with hot ethanol and DMF, and placed in a vacuum dryer to freeze and dry, i.e., CNT@MIL-101(Fe) loaded with different mass fractions was produced. CNT synthetic adsorbents of 2, 5, 10, 15, and 20 wt% were taken and named as follows:

2CNT@MIL-101(Fe), 5CNT@MIL-101(Fe), 10CNT@MIL-101(Fe), 15CNT@MIL-101(Fe), and 20CNT@MIL-101(Fe).

2.3. Reaction Conditions and Analytical Methods

The quantitative Cr(VI) stock solution was placed in the reaction flask. Experiments were carried out by controlling single variables such as the pH, reaction temperature, and initial contaminant concentration The reaction solutions were filtered and sieved for different time periods using a 0.22 μm filter membrane; the pH was adjusted by 0.1 mol/L HCl and 0.1 mol/L NaOH by controlling the thermostatic oscillator temperature to maintain different reaction temperatures. The speed of the thermostatic oscillator was set to 180 rpm, then the concentration of Cr(VI) in the solution was detected by the color development reaction and the removal rate of Cr(VI) and the adsorption amount of the adsorbent were calculated by Equations (1) and (2):

$$\text{Removal efficiency}(\%) = \frac{C_0 - C_e}{C_0} \times 100\% \tag{1}$$

$$q_e = \frac{C_0 V_0 - C_e V_e}{m} \tag{2}$$

where q_e(mg/g) is the equilibrium adsorption capacity, C_0 and *Ce* (mg/L) are the initial concentration and final concentration of Cr(VI), respectively, m (g) is the mass of dry hydrogel, and V_0 and V_e (L) are the initial volume and equilibrium volume during the adsorption process, respectively [30].

In this paper, experiments were conducted by controlling a single variable and all experiments were conducted three times.

3. Results and Discussion

3.1. Characterization

Figure 2 presents the SEM image of the adsorbent. Figure 2a shows MIL-101(Fe); it can be seen that it has a typical octahedral structure, which is in agreement with previous studies that successfully synthesized MIL-101(Fe) [31]. Figure 2b–d show three different groups; it can be seen that the MIL-101(Fe)-NH2 and MIL-101(Fe)-NO2 particles have good crystallinity that is similar to MIL-101(Fe) and have similar octahedral structures, while MIL-101(Fe)-Br generates a crystal-like layer of MIL-101(Fe) has poorer crystallinity. After the incorporation of carbon nanotubes into the MIL-101(Fe), the morphology of CNT@MIL-101(Fe) shows hybrid characteristics, with CNT entangled in the MIL-101(Fe) crystals (Figure 2e; due to the large size, it could not be clearly photographed at 200 nm; thus, 1 μm was used instead). In addition, the size of CNT@MIL-101(Fe) is increased compared to MIL-101(Fe) due to the attachment and reinforcement of CNT [32].

As can be seen in Figure 3a, MIL-101(Fe) has a better crystalline phase structure and the main characteristic diffraction peaks appear at 2θ of 9.44°, 12.61°, 16.62°, and 18.86°, which is similar to previous results from the literature [33]. The positions of the characteristic diffraction peaks of the MOFs modified by -NH_2, -NO_2, and carbon nanotubes are almost the same as the positions of the main peaks of MIL-101(Fe), indicating that the addition of CNTs, etc., did not prevent the generation of MIL-101 crystals. A few diffraction peaks in MIL-101(Fe)-Br disappeared, and there were only weak characteristic diffraction peaks at 9.58° and 16.76°, which suggests that the addition of bromine atoms caused a change in the crystal structure of the MIL-101(Fe) system.

Figure 2. SEM images of adsorbents: (**a**) MIL-101(Fe), (**b**) MIL-101(Fe)-NH_2, (**c**) MIL-101(Fe)-NO_2, (**d**) MIL-101(Fe)-Br, (**e**) 5CNT@MIL-101(Fe), (**f**) MIL-101-Na_2CO_3.

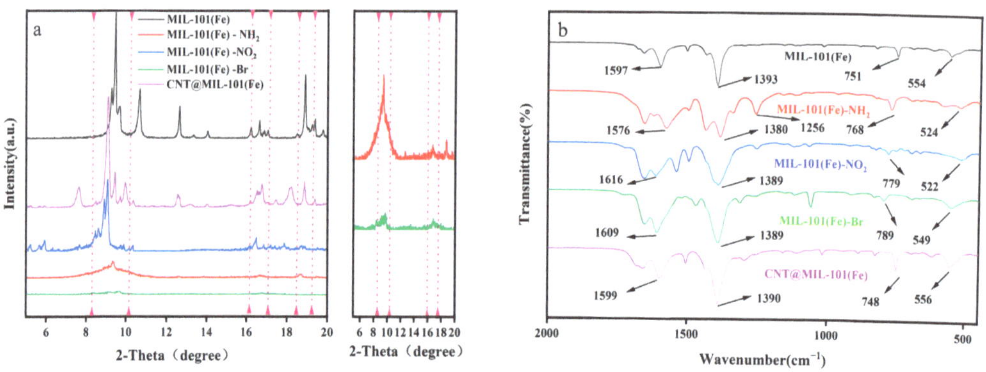

Figure 3. (**a**) XRD diffractogram of MOFs (the red dotted line is the MIL-101(Fe) eigenvalue) and (**b**) FT-IR of MOFs.

The FT-IR plots of MIL-101(Fe), MIL-101(Fe)-NH2, MIL-101(Fe)-NO2, MIL-101(Fe)-Br, and 5CNT@MIL-101(Fe) are shown in Figure 3b. MIL-101(Fe) shows a distinct FT-IR pattern, with peaks at 554, 751, 1393, and 1597 cm^{-1}. The peaks at 751 cm^{-1} and 554 cm^{-1} indicate the C-H bond [34] and Fe-O bond [35] on the benzene ring, respectively. The bands at 1597 and 1393 cm^{-1} represent the asymmetric and symmetric stretching of O–C–O [36].

These characteristic peaks are present in the remaining four materials as well. In addition to the typical peaks of MIL-101(Fe), the FT-IR spectra may vary due to changes in the ligand functional groups, where in MIL-101(Fe)-NH2 there is an MIL-101(Fe)-deficient peak at the 1256 cm^{-1} band which represents the stretching of aromatic C-N. Upon addition of CNT, no significant shift of the characteristic peak was observed, similar to the characteristic peak of MIL-101(Fe) [37].

Figure 4 shows that the N_2 adsorption–desorption isotherms of MIL-101(Fe), MIL-101(Fe)-NH_2, MIL-101(Fe)-NO_2, MIL-101(Fe)-Br, and 5CNT@MIL-101(Fe) are type I isotherms with H_4 hysteresis loops. Five isotherms indicate the presence of meso- and micropores in these samples. Table S1 lists the BET surface areas of the MIL-101(Fe), MIL-101(Fe)-NH_2, MIL-101(Fe)-NO_2, MIL-101(Fe)-Br, and 5CNT@MIL-101(Fe) samples, which were 1730.85, 1724.85, 1652.52, 2318.37, and 2443.35 m^2g^{-1}, respectively. The total pore volumes of the samples were 0.601, 0.511, 1.021, 1.578, and 1.127 cm^3g^{-1}, respectively.

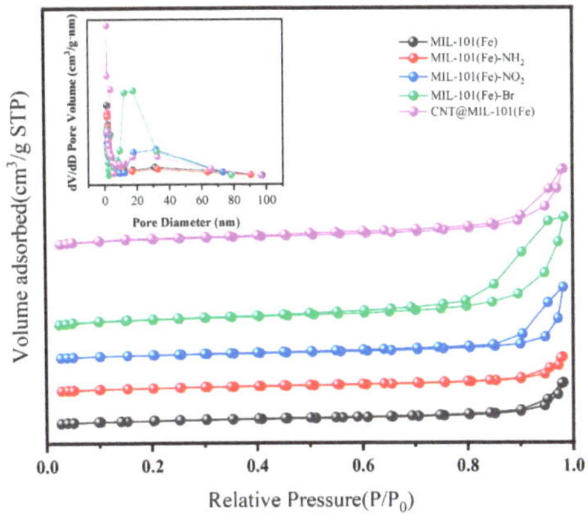

Figure 4. N_2 adsorption–desorption isotherms of MOFs.

3.2. *MIL-101 Preparation, Modulation, and Modification*

As can be seen from Figure 5a and Table S2, the adsorption effect of the synthesized adsorbent on Cr(VI) varied with the variation of the n(H_2BDC):n($FeCl_3$) ratio; the adsorption removal rate of Cr(VI) was the largest when the ratio of n(H_2BDC):n($FeCl_3$) was 1:2, reaching 50%. MOFs of the MIL series are structures composed of metal nodes (clusters) and organic connections; thus, the amounts of different ratios of metal to organic ligand are important for the formation of the MOF structure. The reason for the maximum removal rate of Cr(VI) adsorption when the ratio of n(H2BDC):n($FeCl_3$) is 1:2 may be because the crystallinity of the MIL-101(Fe) adsorbent under this ratio is the highest; with the continuous increase of $FeCl_3$, the excess $FeCl_3$ cannot be fully ligated with the H2BDC connection, resulting in the synthetic MIL-101(Fe) adsorbent. The crystallinity of the synthetic MIL-101(Fe) adsorbent decreases, the purity of the adsorbent decreases, and the adsorption effect then decreases.

According to the structure of MIL-101(Fe) and the chemical stoichiometry characteristics of Fe_3OCl-$(DMF)_2$$(H2BDC)_3$, it is known that the amount of DMF has an important influence in the synthesis of MIL-101(Fe) [27]. As described above, the best input ratio of n(H_2BDC):n($FeCl_3$) is 1:2; thus, while keeping this ratio constant during the synthesis of MIL-101(Fe), 10 mL, 20 mL, 30 mL, 40 mL, and 50 mL of DMF were added while selecting the same Cr(VI) concentration of 10 mg/L in the adsorption reaction. Under the condition of ensuring the minimum dissolved amount, the amount of DMF was increased and the adsorption effect was observed. It can be seen from Figure 5b and Table S3 that the adsorption and removal rate of Cr(VI) changed significantly with the increasing amount of added DMF; the adsorption rate increased by about 10% when the amount of DMF added was

20 mL, and reached 71% when the amount of DMF added was 30 mL, while the adsorption rate was maintained at about 72% when the amount of DMF added was 40 mL and 50 mL.

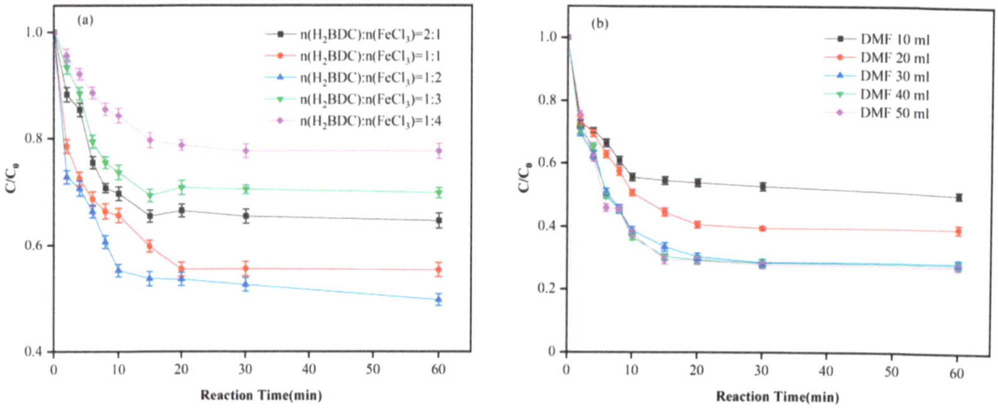

Figure 5. (**a**) Effect of the ratio of metal to organic ligand synthesis on the removal rate of Cr(VI); (**b**) effect of different DMF dosages on the removal rate of Cr(VI) (pH = 5.77, T = 30 °C, dosage of adsorbent = 1 g/L, concentration of Cr(VI) = 10 mg/L).

When 10 mL and 20 mL of DMF were added, it was difficult to dissolve the organic ligand completely, affecting the coordination of the organic ligand and metal and failing to achieve sufficient connection coordination, thereby affecting the crystallinity of the adsorbent. With the addition of 30 mL of DMF, the organic ligand and metal center were fully dissolved and coordinated, the best ratio of n(H_2BDC):n($FeCl_3$) could be maintained, and the best adsorption effect was obtained. It is worth noting that excess DMF dilutes the concentration of both the metal and the organic ligand, which has a certain hindering effect on their coordination. When the dosage of DMF continues increasing to excess, this further dilutes the concentration of metals and organic ligands; the resulting hindering effect on the coordination of metals and organic ligands means that the adsorption effect grows insignificantly.

Introducing functional groups on the adsorbent may increase certain active adsorption sites and strengthen the bonding energy between the adsorption bonds. However, it was found in our adsorption experiments (Figure 6a) that the adsorption and removal rate of pollutants decreased relative to the initial MIL-101(Fe) after the introduction of different functional groups. This was probably because the addition of -NH_2, -NO_2, and -Br functional groups to the MIL-101(Fe) adsorbent meant that the surface area of the adsorbent was occupied by the functional groups due to their large specific gravity, which reduced the space available for self-use of the adsorbent, in turn resulting in a decrease in the specific surface area and porosity of the adsorbent. Notably, the specific gravity of the -NH_2 functional group was smaller than that of the -NO_2 and -Br functional groups [38]. Therefore, while the adsorbent with the introduction of -NH2 functional group was slightly better than the adsorbents with the introduction of -NO_2 and -Br functional groups, none were as effective as the original MIL-101(Fe).

The adsorption effect on Cr(VI) was observed by adding 10 mL, 20 mL, and 30 mL of formic acid to it while maintaining the optimal preconditions (Figure 6b). The adsorption removal rate increased from 71% to 79% when 10 mL of formic acid was added. The adsorption removal rate of Cr(VI) was reduced to within 10% when the amount of formic acid was increased to 20 mL and 30 mL. The reason for this may be that the structure of the MIL-101(Fe) adsorbent was basically destroyed after the addition of 20 mL and 30 mL of formic acid, and the structural voids collapsed and the porosity was greatly reduced, leading to deterioration of the adsorption effect. The addition of 10 mL of formic acid may have increased the crystallinity of MIL-101(Fe) while increasing the specific surface area of the adsorbent, which in turn increased the adsorption removal effect of Cr(VI) by less than 10% over the adsorption removal effect of the original MIL-101(Fe) [39].

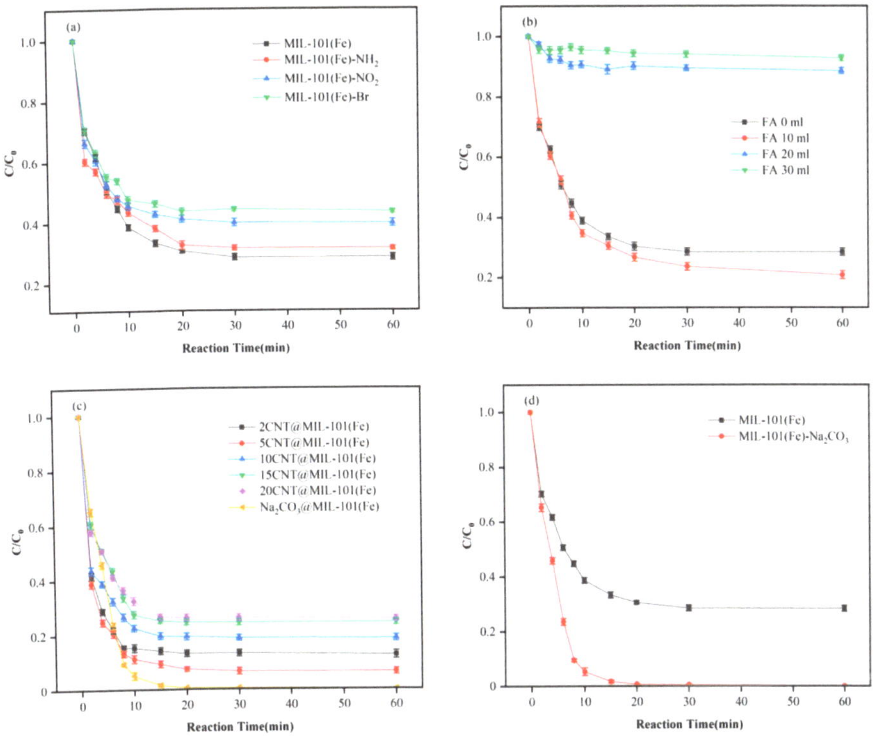

Figure 6. (**a**) MIL-101 modification by different groups, (**b**) different FA dosages, (**c**) different CNT dosages, and (**d**) Cr(VI) removal by adsorption with Na_2CO_3 modification.

Figure 6c shows that the adsorbents with carbon nanotube loadings of 2%, 5%, 10%, 15%, and 20% had adsorption and removal rates of 86%, 92%, 81%, 74%, and 73%, respectively, for the pollutant Cr(VI). The trend of the adsorption and removal rates shows that the best effect was achieved when the loading rate of carbon nanotubes was 5%. The adsorption of carbon nanotubes on the pollutant Cr(VI) is mainly carried out through the functional groups on the surface, and belongs to a potential surface adsorbent, which itself has a certain adsorption effect on Cr(VI). By loading a certain amount of carbon nanotubes on MOFs, it is intended to expand the specific surface area of the adsorbent, thereby increasing the adsorption sites and promoting the adsorption and removal of the pollutant Cr(VI). When the loading rate of carbon nanotubes was 2% and 5%, the adsorption and removal rate of Cr(VI) increased, while when the loading rate was 10%, 15%, and 20% the adsorption and removal rate of Cr(VI) decreased continuously. This may be due to the excessive loading of carbon nanotubes to agglomerate on the surface of MOFs, leading to a poor dispersion effect and covering the adsorption sites on the surface of MOFs, thereby blocking the adsorption pore channel of MOFs and causing the adsorption and removal rate of the adsorbent MOFs for Cr(VI) to decrease [40].

A certain amount of Na_2CO_3 was added to the adsorbent MIL-101(Fe) as a mineralizing agent. Compared with other mineralizing agents, such as HF and TMAOH substances, Na_2CO_3 itself is not toxic, and the introduction of Na_2CO_3 can both improve the crystallinity of the adsorbent and increase the synthetic yield of the adsorbent [41]. It can be seen from Figure 6d that the adsorption effect of the adsorbent after the addition of Na_2CO_3 is significantly better than that of the original adsorbent; the adsorption efficiency reaches 100% under specific conditions, and the adsorption effect has been significantly improved.

It can be seen from Figure 7 that the removal rates of Cr(VI) by adsorption after modification of MIL-101 by different groups take the following order over a time: MIL-101-Na_2CO_3 (100%) > 5CNT@MIL-101(Fe) (92%) > MIL-101-FA (79%) > MIL-101 (71%) > MIL-101-NH_2 (68%). Therefore, in the modification experiment of the MIL-101(Fe) adsorbent, the best adsorption effect on Cr(VI) was the adsorbent modified by Na_2CO_3. In order to further investigate its adsorption characteristics, a single-factor control experiment was carried out with this adsorbent.

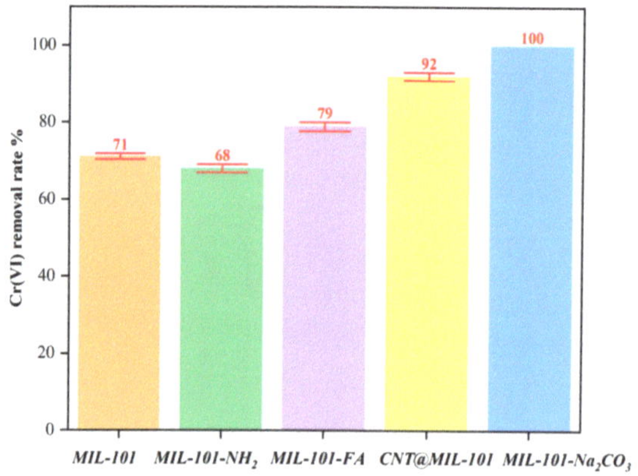

Figure 7. Removal rates of Cr(VI) adsorption by MIL-101 modified with different groups.

3.3. Adsorption Performance Study

The adsorption experiments were carried out at pH 2, 5, 7, and 9,with an unadjusted solution pH = 5.77 (Figure 8a). The adsorption and removal rates of Cr(VI) pollutants at different pH conditions were 87%, 81%, 80%, 76%, and 74%, respectively, with the highest adsorption and removal rate at pH = 2. The adsorption and removal rates showed a trend of gradual decrease with increasing pH. Chromium mainly exists in aqueous solution in the form of anions such as $HCrO_4^-$, CrO_4^{2-}, $Cr_2O_7^{2-}$, etc. [42]. When pH = 2, the content of H^+ in water is greater, meaning that a protonation reaction takes place on the surface of the adsorbent and chromium ions are adsorbed to its surface through the effect of electrostatic gravity, which can promote the adsorption and removal of Cr(VI) pollutants. With continuous increase of the pH, more OH^- is generated, which competes with CrO_4^{2-} and other anions for adsorption, causing the adsorption and removal rate of Cr(VI) to decrease [43].

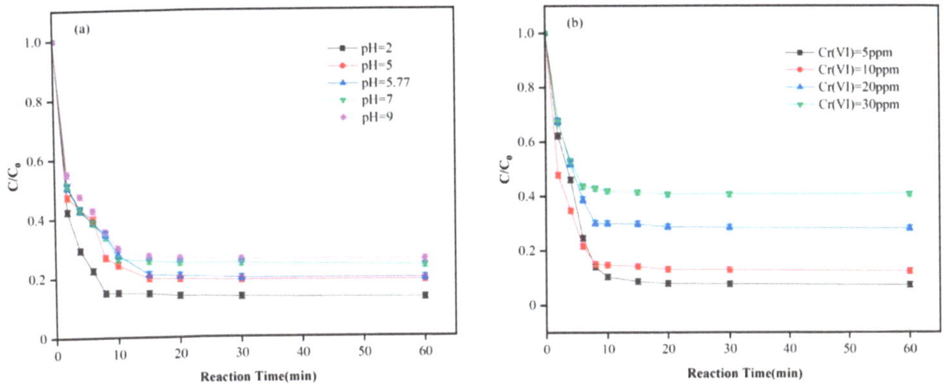

Figure 8. (**a**) Effect of different pH levels on the Cr(VI) removal rate and (**b**) effect of different Cr(VI) levels.

In this study, experiments were conducted using Cr(VI) concentrations of 5 mg/L, 10 mg/L, 20 mg/L, and 30 mg/L. The adsorption removal rates shown in Figure 8b indicate that the removal rate decreased from 92% to 59% while the adsorption capacity increased from 4.63 mg/g to 17.42 mg/g with the increasing concentration of pollutants at the same adsorbent dosing.

The adsorption mechanism was explored by adding three anions of different intensities to the adsorption system. It can be seen from Figure 9 that SO_4^{2-} has the greatest effect on the adsorption system, followed by NO_3^-, while Cl^- has the least effect. In terms of ionic strength, as the ionic strength of the same ion increases, its inhibitory effect on the adsorption system becomes more obvious, probably because increasing the ionic strength reduces the chance of pollutants coming into contact with the adsorbent, which makes the adsorption effect decrease. Furthermore, because electrostatic force is the key to the adsorption process, the coexisting anions compete for adsorption with the pollutant ions, meaning that less pollutant ions are adsorbed at the active potential on the surface of the adsorbent, in turn leading to a decrease in the pollutant adsorption and removal rates [44].

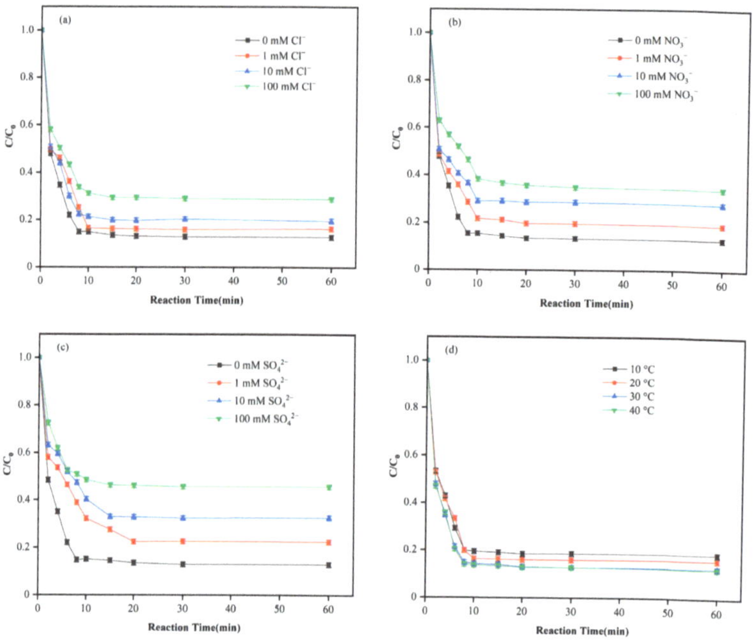

Figure 9. (**a**–**c**) Effect of different anion concentrations on Cr(VI) removal and (**d**) effect of temperature on Cr(VI) removal.

From Figure 9d, it can be concluded that the degree of influence of temperature on the adsorption efficiency of the adsorbent is not very large. The adsorption capacity of the adsorbent on Cr(VI) increases slightly with increasing temperature, probably because high temperature increases the frequency of collisions between Cr(VI) and the adsorbent. As the temperature increases, the adsorption removal rate for Cr(VI) increases continuously; from 10 to 30 °C, the adsorption removal rate increases by about 10%, while from 30 to 40 °C the adsorption removal rate essentially does not change. Thus from the economic point of view, a reaction temperature of 30 °C should be selected.

3.4. Adsorption Kinetics and Isotherms

Adsorption kinetics is an important factor in the investigation of Cr(VI) adsorption, as it determines the mass transfer rate of adsorption. As shown in Figure 10, based on the adsorption kinetic model (pseudo-second-order and pseudo-first-order), the results of the adsorption experiments were calculated and fitted in order to determine the adsorption mechanism and the control of the adsorption process [45].

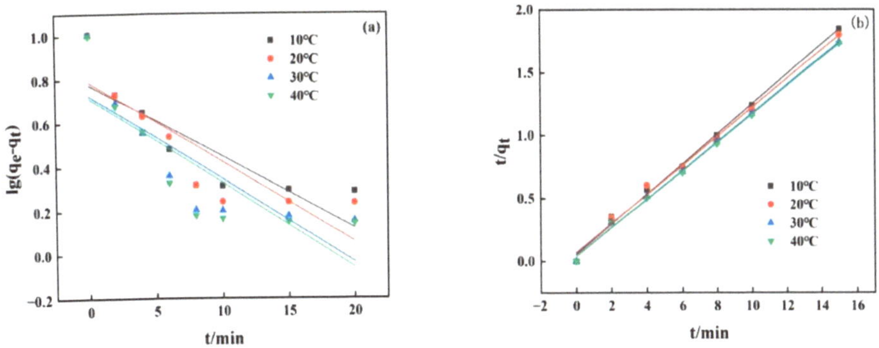

Figure 10. (**a**) Pseudo-first-order dynamics model and (**b**) pseudo-second-order dynamics model.

It can be seen from Table 1 that the adsorption process of MIL-101(Fe)-Na_2CO_3 on Cr(VI) is more consistent with the proposed secondary kinetic model, with an R^2 (0.998) significantly higher than that of the proposed primary kinetic model (0.745). A proposed secondary kinetic model was established at the adsorption rate limiting step, which is mainly used to describe the reactive chemisorption process. This chemisorption process represents the existence of electron sharing or electron exchange between the adsorbent and the pollutant; therefore, it is known from the fitted kinetic model that the adsorption of this adsorbent for Cr(VI) mainly conforms to the proposed secondary kinetics. Analysis of the whole adsorption process of Cr(VI) by the adsorbent can be divided into three stages. The adsorption rate is faster in the beginning stage, slows down gradually as time advances, and finally levels out. The reason for this may be that the concentration of Cr(VI) in the solution is high at first, and the presence of a large number of adsorption sites on the adsorbent surface helps the adsorbent to activate quickly when it comes into contact with the pollutant Cr(VI). This makes the adsorption rate relatively fast; with the gradual extension of the adsorption time, the concentration of Cr(VI) gradually decreases with the combination of adsorption sites, and the adsorption rate of Cr(VI) gradually decreases. The adsorption curve then flattens out, indicating that the adsorption sites on the adsorbent surface are close to saturation.

Table 1. MIL-101–Na_2CO_3 kinetic model fitting parameters.

T (°C)	Pseudo-First-Order			Pseudo-Second-Order		
	K_1	q_e	R^2	K_2	q_2	R^2
10 °C	0.075	5.97	0.697	0.0245	8.42	0.996
20 °C	0.084	6.15	0.745	0.0186	8.71	0.993
30 °C	0.088	5.34	0.694	0.0309	8.84	0.998
40 °C	0.089	5.19	0.677	0.0325	8.89	0.998

Adsorption isotherm models for describing the process of adsorption can be classified into Freundlich, Langmuir, Temkin, and Dubinin–Radushkevich (D-R) models (Figure 11). The Freundlich model assumes that the molecules of the adsorbed substance are adsorbed on the surface of the heterogeneous adsorbent as single or multilayer molecules, and that there is a process of interaction between the adsorbed substance molecules. The Langmuir model assumes that the adsorbent surface is homogeneous, that only one adsorbed substance molecule is adsorbed by the adsorbent on a reaction center with the same energy, and that there is no process of interaction between the molecules of the adsorbed substance. The Temkin model assumes that there is an indirect interaction between the adsorbed substance molecules and the adsorbent. Finally, the D-R model assumes that the adsorbed substance molecules and the adsorbent initially bind by adsorption at the most favorable sites, then subsequently undergo a multilayer adsorption process [46].

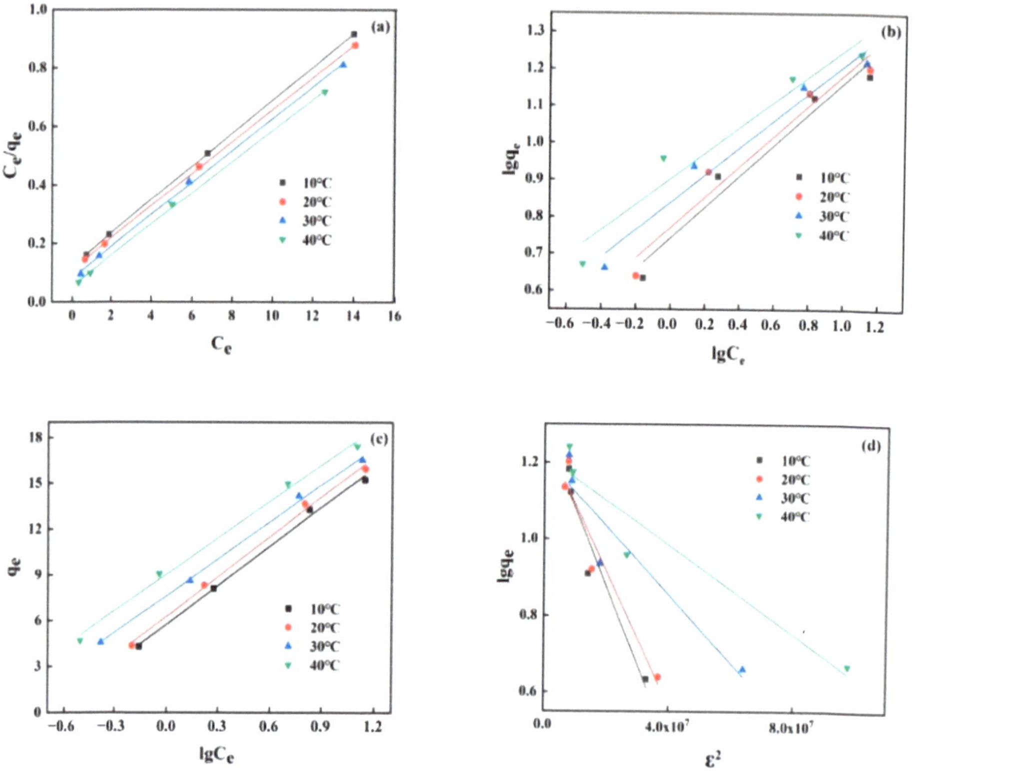

Figure 11. (**a**) Freundlich (**b**) Langmuir (**c**) Temkin (**d**) Dubinin-Radushkevich(D-R). Fitting of different isotherm models to the adsorption process.

According to the relevant data, the R^2 values obtained from the four models can be fitted separately; it can be seen that the R^2 values of the four models range from 0.894 to 0.999, among which the R^2 value of the Langmuir adsorption isotherm model is the highest and reaches 0.999, which is the best fit, indicating that the adsorbent conforms to this adsorption model in the process of Cr(VI) adsorption. The q_m (theoretical maximum adsorption amount) of 18.76 mg/g in the Langmuir adsorption isotherm model is close to the adsorption amount of 20 mg/g obtained in the actual experimental process. The simulation results of the remaining three models are slightly worse, with the D-R model the worst, indicating that the adsorbent does not fit well with the other three models during its adsorption of Cr(VI).

The adsorption process of the Na_2CO_3–MIL-101(Fe) adsorbent on Cr(VI) is more consistent with the Langmuir adsorption model, indicating that the adsorption mechanism of this adsorbent can be considered as a monolayer adsorption with a uniform surface. Its theoretical maximum adsorption amount is about 18 mg/g, which is similar to the adsorption amount actually sought for this adsorbent. From the data analysis of the Langmuir adsorption separation constant R_L, when $R_L > 1$, it is unfavorable for adsorption; when $R_L = 1$, it shows linear adsorption; when $R_L < 1$, it is favorable for adsorption; and when $R_L = 0$, it is irreversible reaction. Therefore, it can be concluded from the above data that $R_L < 1$ favors adsorption.

3.5. Thermodynamics and Recycling

The relevant calculations of the thermodynamic parameters are important for the study of spontaneous reactions and thermodynamic changes in the adsorption process. The thermodynamic changes in the adsorption process were understood by calculating the relevant adsorption thermodynamic parameters. When fitting the thermodynamic equations to the adsorption process at four temperatures (10 °C, 20 °C, 30 °C, and 40 °C), the R^2 values of the fitted curves reached 0.995 (Figure 12).

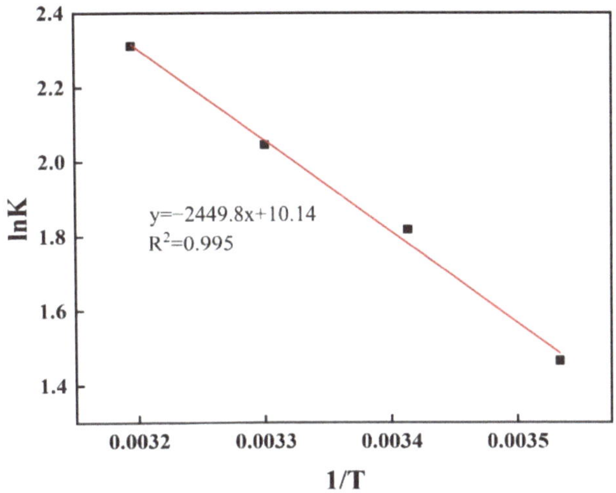

Figure 12. Van' Hoff plots of Na_2CO_3–MIL-101(Fe) adsorptive Cr(VI).

From the thermodynamic data in Table S4, it can be concluded that the free energy $\Delta G < 0$ indicates that the adsorption process of Cr(VI) adsorption by this adsorbent is feasible and can proceed spontaneously. It can be seen that ΔG decreases with the increase in temperature, indicating that there is a positive effect on the adsorption process at higher temperatures and that the spontaneity of the reaction is increasing, which is the same as the conclusion obtained from the earlier one-way temperature experiment. Enthalpy $\Delta H > 0$ indicates that the adsorption process of this adsorbent is a heat absorption reaction; entropy $\Delta S > 0$ indicates that the orderliness between the solid–liquid interface present in the adsorption process decreases.

NaOH (0.01 M) solution, HCl (0.01 M) solution, and deionized water were selected as the detergents of Cr(VI); 0.1 g of the adsorbent was added to 50 mL of the above three solutions, then the three mixed solutions were placed in the shaker and shaken for 4 h at room temperature to remove the Cr(VI) pollutants adsorbed on the surface of the adsorbent. The regenerated adsorbent was collected and drained through lyophilizer after 4 h, then the adsorption experiments were performed again. The adsorption removal rate after washing by three kinds of detergent demonstrates that the adsorption effect of regenerated adsorbent on Cr(VI) is NaOH (91.5%) > H_2O (16%) > HCl (2%). These results show that the adsorbent is more stable under acidic and neutral conditions, while desorption occurs best under alkaline conditions (Figure 13).

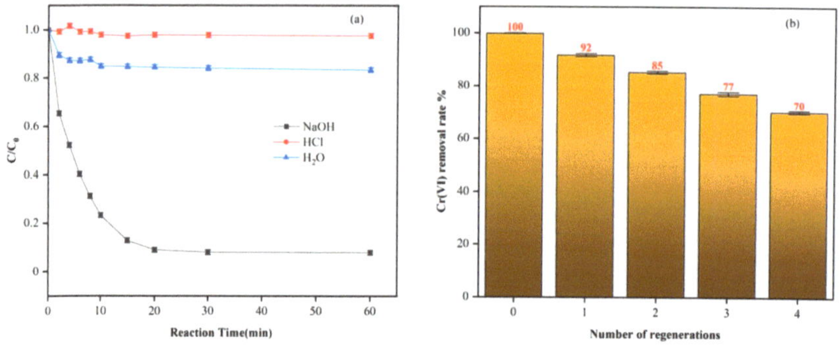

Figure 13. (**a**) Selection of eluent and (**b**) cyclic regeneration experiments.

The regenerative availability of MIL-101(Fe)–Na_2CO_3 was investigated by four repeated adsorption experiments, with 0.01 M NaOH solution selected as the eluent and 0.01 M HCl chosen as the regenerative stabilizer of the adsorbent. After a certain amount of adsorbent had been treated with detergent, the adsorbent was washed three times with HCl to better regenerate the adsorbent, followed by adsorption experiments, and so on, repeatedly. In this way, we conducted four adsorption experiments to observe the adsorption effect of the regenerated adsorbent on Cr(VI). The results showed that the adsorption removal rate of the adsorbent decreased from 100% to 92%, 85%, 77%, and 70% in order, indicating that MIL-101(Fe)–Na_2CO_3 had better regenerative adsorption capacity for Cr(VI).

Table 2 shows a comparison of the adsorption capacity of several adsorbents for Cr. The data indicate that MIL-101(Fe)-Na_2CO_3 has relatively good adsorption capacity compared to other adsorbents.

Table 2. Comparison of adsorption capacities of MIL-101(Fe)–Na_2CO_3 for Cr(VI) removal with other recently reported adsorbents.

Adsorbent	Experimental Conditions			Adsorption Capacity (mg/g)	References
	C0 (mg Cr/L)	Dose (g/L)	Time (h)		
MIL-101(Fe)-Na_2CO_3	10	0.5	0.3	18.76	This study
α-Fe_2O_3/γ-Al_2O_3	5	1.0	1	3.83	[47]
Nano-γ-Al_2O_3 adsorbent	20	4.0	4	13.3	[48]
γ-Al_2O_3	90	0.8	6	6.70	[49]
Activated alumina	10	10.0	-	7.44	[50]
Sphere-like γ-Al_2O_3	30	1.6	4	5.70	[51]
Fe-modified T. natans	20	1.5	8	11.83	[52]

In order to judge the selectivity of MIL-101-Na_2CO_3, we carried out adsorption experiments on other heavy metals. From Figure 14, it can be observed that when adsorbing 10 mg/L of Cr(VI), Pb(II), Cd(II), and V(V), only Cr(VI) was completely adsorbed within 60 min, being adsorbed within 20 min. The adsorption of other heavy metals was the best as well, with Pb(II) reaching 55% within 60 min. For Cd(II) and V(V), the adsorption effect was only the best with Pb(II), reaching 55% within 60 min. Only Pb(II) had the best adsorption effect, reaching 55% within 60 min, while the adsorption effect of Cd(II) and V(V) was only 35% and 28%. The above results indicate that the MIL-101–Na_2CO_3 adsorbent has good selectivity for Cr(VI).

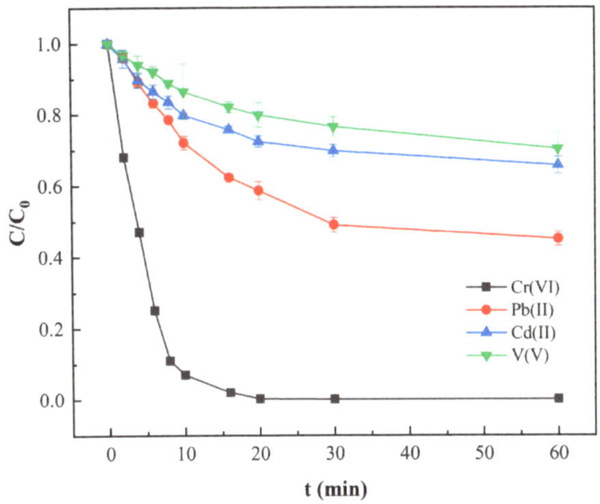

Figure 14. Adsorption capacity of MIL-101–$NaCO_3$ on different heavy metals.

3.6. Mechanistic Analysis

In order to better verify the MIL-101–Na_2CO_3 mechanism, the XPS spectra before and after the adsorption reaction using MIL-101–Na_2CO_3 were obtained, as shown in Figure 15. The full-size XPS spectra of MIL-101–Na_2CO_3 before the reaction and after adsorption are shown in Figure 15a, where the combination of each element is corrected by the c 1 s peak (284.8 eV). As can be seen in the figure, the general trend of unused and used MIL-101–Na_2CO_3 in the different systems is consistent, with only minor differences. The two main peaks occupying 711.73 and 724.01 eV are consistent with Fe $2p_{3/2}$ and Fe $2p_{1/2}$, as seen in the high-resolution XPS spectrum of Fe 2p, where the satellite peaks of the Fe cation oscillate at 718 eV (Figure 15b). Most of the characteristic peaks shift after the reaction, confirming the involvement of metallic substances in the adsorption reaction.

According to the Na 1 s spectra before and after the catalytic reaction presented in Figure 15c, the presence of the Na 1s peaks verifies the successful participation of Na_2CO_3 in the MIL-101 system. The changes of the Na 1 s peaks and the shifts of the characteristic peaks before and after the reaction verify that elemental Na plays an important role in the MIL-101-Na_2CO_3 adsorption system. The O 1 s spectra obtained before and after the reaction show three characteristic peaks corresponding to chemisorbed oxygen, oxygen atom vacancies, and lattice oxygen, respectively. The incorporation of O during the reaction is verified by Figure 15d.

As can be seen in Figure 2f above, the addition of sodium carbonate to MIL-101(Fe) resulted in the growth of a similar crystal-like layer of MIL-101(Fe) from the DMF solution, which grew smaller crystals in the size range of about 200–400 nm.

From Figure S1, it can be seen that the diffraction peaks 2θ = 5.11°, 8.03°, 8.99°, 16.25° for the adsorbent are consistent with the diffraction data described in the literature [29]. The diffraction peaks of the adsorbent before and after adsorption appear at basically the same position, with the peaks broadening slightly after the adsorption reaction, indicating structural defects on the surface of the adsorbent. This might be due to breakage of coordination bonds between the central metal ion and the organic ligand, resulting in reduced surface area of the adsorbent.

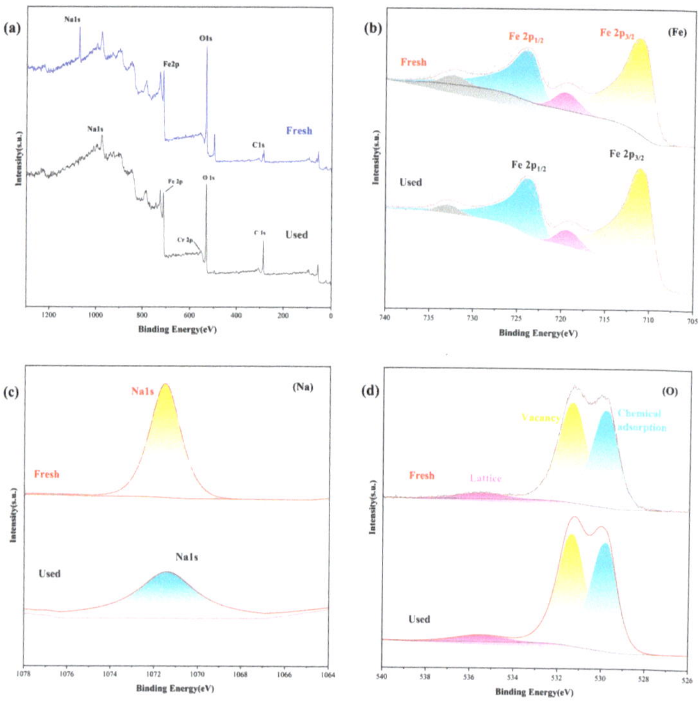

Figure 15. High-resolution XPS spectra of (**a**) element spectrum, (**b**) Fe, (**c**) Na, and (**d**) O.

4. Conclusions

In this paper, MIL-101(Fe) adsorbent was synthesized for the adsorption of hexavalent chromium pollutants by a hydrothermal method. A series of modifications were carried out in order to observe the induced adsorption effects: MIL-101-Na_2CO_3 > 5CNT@MIL-101(Fe) > MIL-101-FA > MIL-101 > MIL-101-NH_2. In the adsorption process of MIL-101–Na_2CO_3, the adsorption equilibrium time for hexavalent chromium was 20 min, the optimum pH for the adsorption experiment was 2, and the adsorption process was mainly electrostatic adsorption. The optimum temperature of the adsorption reaction was 30 °C, and the adsorption removal rate showed an increasing trend during the temperature change from 10 °C to 40 °C. These results indicate that the adsorption experiment involved a heat absorption reaction. From the fitting results of the kinetic model and isotherm model, the adsorption process is consistent with the Langmuir model of monomolecular layer adsorption and the proposed secondary kinetic model of chemical bonding force. Based on the characterization of the adsorbent material, the adsorbent showed structural residues and a decrease in specific surface area after the reaction, which is consistent with a decrease in adsorption capacity.

The successfully modified MIL-101(Fe)–Na_2CO_3 material has potential for development in the treatment of Cr(VI) wastewater. Although preliminary progress has been made in the treatment of Cr(V) in wastewater in this paper, there remain many issues to be further explored, including the following:

1. While the modified MIL-101(Fe)–Na_2CO_3 material has good prospects, its stability needs to be improved, and the preparation method needs to be improved in future work to find a simpler and more economical experimental preparation method.
2. Future work could explore different alloy materials to replace the single metal, reducing the cost of preparing the f materials while providing improved adsorption performance.

3. Other carrier materials could be explored in order to find the most effective carrier and further optimize the removal of Cr(VI).
4. The significant advantages of metal–organic frameworks can be utilized to broaden the modification methods and improve the adsorption capacity for a variety of heavy metals.

Supplementary Materials: The following supporting information can be downloaded at: https://www.mdpi.com/article/10.3390/w16010025/s1, Figure S1: XRD pattern of MIL-101-Na_2CO_3; Table S1: BET results of MIL-101(Fe), MIL-101(Fe)-NH_2, MIL-101(Fe)-NO_2, MIL-101(Fe)-Br, CNT@MIL-101(Fe); Table S2: Effect of the ratio of metal to organic ligand synthesis on the removal rate of Cr(VI); Table S3: Effect of different DMF dosage on the removal rate of Cr(VI); Table S4: MIL-101-Na_2CO_3 thermodynamic model fitting parameters.

Author Contributions: Conceptualization, C.W. and J.C.; validation, C.W., J.C. and Q.Y.; formal analysis, C.W.; data curation, C.W. and J.C.; writing—original draft preparation, C.W.; writing—review and editing, Q.Y. All authors have read and agreed to the published version of the manuscript.

Funding: This research received no external funding.

Data Availability Statement: The data presented in this study are available on request from the corresponding author.

Conflicts of Interest: The authors declare that they have no known competing financial interest or personal relationships that could have appeared to influence the work reported in this paper.

References

1. John, B.C.; Viswambaram, V.A.; Raj, S.S.; Mankunipoyil, S.A. Adsorptive removal of Cr (VI) using mesoporous iron-aluminum oxyhydroxide-polyvinyl alcohol self-supporting film: Kinetics, optimization studies and mechanism. *Mater. Today Commun.* **2023**, *34*, 105315. [CrossRef]
2. Sheth, Y.; Dharaskar, S.; Khalid, M.; Walvekar, R. Investigating chromium Cr(VI) removal using imidazolium based ionic liquid-chitosan composite adsorptive film. *J. Mol. Liq.* **2022**, *347*, 118317. [CrossRef]
3. Lawal Usman, U.; Kumar Allam, B.; Bahadur Singh, N.; Banerjee, S. Adsorptive removal of Cr(VI) from wastewater by hexagonal boron nitride-magnetite nanocomposites: Kinetics, mechanism and LCA analysis. *J. Mol. Liq.* **2022**, *354*, 118833. [CrossRef]
4. Sun, Y.; Gui, Q.; Zhang, A.; Shi, S.; Chen, X. Polyvinylamine-grafted polypropylene membranes for adsorptive removal of Cr(VI) from water. *React. Funct. Polym.* **2022**, *170*, 105108. [CrossRef]
5. Sanchayanukun, P.; Muncharoen, S. Chitosan coated magnetite nanoparticle as a working electrode for determination of Cr(VI) using square wave adsorptive cathodic stripping voltammetry. *Talanta* **2020**, *217*, 121027. [CrossRef] [PubMed]
6. Acharya, R.; Lenka, A.; Parida, K. Magnetite modified amino group based polymer nanocomposites towards efficient adsorptive detoxification of aqueous Cr (VI): A review. *J. Mol. Liq.* **2021**, *337*, 116487. [CrossRef]
7. El Kaim Billah, R.; Shekhawat, A.; Mansouri, S.; Majdoubi, H.; Agunaou, M.; Soufiane, A.; Jugade, R. Adsorptive removal of Cr(VI) by Chitosan-SiO_2-TiO_2 nanocomposite. *Environ. Nanotechnol. Monit. Manag.* **2022**, *18*, 100695. [CrossRef]
8. Korde, S.; Tandekar, S.; Jeyaseelan, C.; Saravanan, D.; Jugade, R. Mesoporous magnetic Chitosan-Zirconia-Iron oxide nanocomposite for adsorptive removal of Cr(VI) ions. *Mater. Lett.* **2022**, *311*, 131513. [CrossRef]
9. Omer, A.M.; Abd El-Monaem, E.M.; Abd El-Latif, M.M.; El-Subruiti, G.M.; Eltaweil, A.S. Facile fabrication of novel magnetic ZIF-67 MOF@aminated chitosan composite beads for the adsorptive removal of Cr(VI) from aqueous solutions. *Carbohydr. Polym.* **2021**, *265*, 118084. [CrossRef]
10. Doagoo, F.; Peyravi, M.; Khalili, S. Photo-degradation and discoloration of nitrogen substituted ZnO via multilayer adsorptive membrane for Cr(VI) removal. *J. Environ. Chem. Eng.* **2021**, *9*, 105153. [CrossRef]
11. Ma, J.; Liu, C.; Chen, K. Removal of Cr(VI) species from water with a newly-designed adsorptive treatment train. *Sep. Purif. Technol.* **2020**, *234*, 116041. [CrossRef]
12. Zhong, Y.P.; Liu, S.Y.; Huang, C.; Li, X.X.; Chen, L.; Li, L.; Zhu, J. Effect of amylose/amylopectin ratio of esterified starch-based films on inhibition of plasticizer migration during microwave heating. *Food Control* **2017**, *82*, 283–290. [CrossRef]
13. Joseph, L.; Jun, B.-M.; Jang, M.; Park, C.M.; Muñoz-Senmache, J.C.; Hernández-Maldonado, A.J.; Heyden, A.; Yu, M.; Yoon, Y. Removal of contaminants of emerging concern by metal-organic framework nanoadsorbents: A review. *Chem. Eng. J.* **2019**, *369*, 928–946. [CrossRef]
14. Amenaghawon, A.N.; Anyalewechi, C.L.; Osazuwa, O.U.; Elimian, E.A.; Eshiemogie, S.O.; Oyefolu, P.K.; Kusuma, H.S. A comprehensive review of recent advances in the synthesis and application of metal-organic frameworks (MOFs) for the adsorptive sequestration of pollutants from wastewater. *Sep. Purif. Technol.* **2023**, *311*, 123246. [CrossRef]
15. Chen, C.; Kosari, M.; Jing, M.; He, C. Microwave-assisted synthesis of bimetallic NiCo-MOF-74 with enhanced open metal site for efficient CO_2 capture. *Environ. Funct. Mater.* **2023**, *1*, 253–266. [CrossRef]

16. Yu, Y.-P.; Pan, M.-M.; Jiang, M.; Yu, X.; Xu, L. Facile synthesis of self-assembled three-dimensional flower-like Cu-MOF and its pyrolytic derivative Cu-N-C450 for diverse applications. *J. Environ. Chem. Eng.* **2023**, *11*, 109400. [CrossRef]
17. Mohan, B.; Kamboj, A.; Virender; Singh, K.; Priyanka; Singh, G.; Pombeiro, A.J.L.; Ren, P. Metal-organic frameworks (MOFs) materials for pesticides, heavy metals, and drugs removal: Environmental safety. *Sep. Purif. Technol.* **2023**, *310*, 123175. [CrossRef]
18. Zhang, Y.; Wang, W.; Liu, C.; Shen, P.; Liu, Z.; Hu, J.; Shi, F. Facile synthesis of HPW@MOF-199 embedded in SBA-15 functionalized with -COOH groups as a steady catalyst for the esterification reaction. *Fuel* **2023**, *340*, 127563. [CrossRef]
19. Nagajyothi, P.C.; Ramaraghavulu, R.; Pavani, K.; Shim, J. Catalytic reduction of methylene blue and rhodamine B using Ce-MOF-derived CeO_2 catalyst. *Mater. Lett.* **2023**, *336*, 133837. [CrossRef]
20. Hou, W.; Xing, Y.; Li, C.; Li, T.; Liu, D.; Ernawati, L.; Sunarso, J.; Meng, X. Achieving high-efficiency and broad bandwidth with low filler loading for hierarchical Fe_3O_4/Co-MOF absorbers. *Mater. Res. Bull.* **2023**, *161*, 112171. [CrossRef]
21. Gorgani, L.; Mohammadi, M.; Darzi, G.N.; Raoof, J.B. Electrochemical aptasensor based on bimetallic CuZr-MOF for ultrasensitive detection of miR-21. *Sens. Actuators B Chem.* **2023**, *378*, 133194. [CrossRef]
22. Prakash Tripathy, S.; Acharya, R.; Das, M.; Acharya, R.; Parida, K. Adsorptive remediation of Cr (VI) from aqueous solution using cobalt ferrite: Kinetics and isotherm studies. *Mater. Today Proc.* **2020**, *30*, 289–293. [CrossRef]
23. Salehi, S.; Ehsani, M.H.; Aghazadeh, M. Direct electrosynthesis of Ni-, Co-, and Ni,Co-MOF onto porous support for high-performance supercapacitors. *J. Alloys Compd.* **2023**, *940*, 168885. [CrossRef]
24. Lu, J.; Wang, S.; Zhao, Y.; Ge, K.; Wang, J.; Cui, H.; Yang, Y.; Yang, Y. Photocatalytic reduction of CO2 by two-dimensional Zn-MOF-NH2/Cu heterojunctions. *Catal. Commun.* **2023**, *175*, 106613. [CrossRef]
25. Cychosz, K.A.; Matzger, A.J. Water Stability of Microporous Coordination Polymers and the Adsorption of Pharmaceuticals from Water. *Langmuir* **2010**, *26*, 17198–17202. [CrossRef]
26. Chakraborty, A.; Achari, A.; Eswaramoorthy, M.; Maji, T.K. MOF-aminoclay composites for superior CO_2 capture, separation and enhanced catalytic activity in chemical fixation of CO_2. *Chem. Commun.* **2016**, *52*, 11378–11381. [CrossRef]
27. Yang, Q.H.; Xu, Q.; Jiang, H.L. Metal-organic frameworks meet metal nanoparticles: Synergistic effect for enhanced catalysis. *Chem. Soc. Rev.* **2017**, *46*, 4774–4808. [CrossRef]
28. Monforte, F.; Falsaperna, M.; Pellegrino, A.L.; Bongiorno, C.; Motta, A.; Mannino, G.; Condorelli, G.G. Direct Growth on Si(100) of Isolated Octahedral Mil-101(Fe) Crystals for the Separation of Aromatic Vapors. *J. Phys. Chem. C* **2019**, *123*, 28836–28845. [CrossRef]
29. Tang, J.; Yang, M.; Yang, M.; Wang, J.J.; Dong, W.J.; Wang, G. Heterogeneous Fe-MIL-101 catalysts for efficient one-pot four-component coupling synthesis of highly substituted pyrroles. *New J. Chem.* **2015**, *39*, 4919–4923. [CrossRef]
30. Li, Q.; You, Y.; Hu, X.; Lu, D.; Wen, Q.; Yu, G.; Wang, W.; Xu, T. Preparation of amino-modified carbon quantum dots-ZnO/cellulose nanofiber multifunctional hydrogel: Enhanced adsorption synergistic photoreduction and reversible fluorescence response visual recognition of Cr(VI). *Int. J. Biol. Macromol.* **2023**, *254*, 128068. [CrossRef]
31. He, L.; Dong, Y.; Zheng, Y.; Jia, Q.; Shan, S.; Zhang, Y. A novel magnetic MIL-101(Fe)/TiO_2 composite for photo degradation of tetracycline under solar light. *J. Hazard. Mater.* **2019**, *361*, 85–94. [CrossRef] [PubMed]
32. Yan, D.Y.; Hu, H.; Gao, N.Y.; Ye, J.S.; Ou, H.S. Fabrication of carbon nanotube functionalized MIL-101(Fe) for enhanced visible-light photocatalysis of ciprofloxacin in aqueous solution. *Appl. Surf. Sci.* **2019**, *498*, 9. [CrossRef]
33. Ma, Y.W.; Lu, Y.F.; Hai, G.T.; Dong, W.J.; Li, R.J.; Liu, J.H.; Wang, G. Bidentate carboxylate linked TiO_2 with NH2-MIL-101(Fe) photocatalyst: A conjugation effect platform for high photocatalytic activity under visible light irradiation. *Sci. Bull.* **2020**, *65*, 658–669. [CrossRef] [PubMed]
34. Hu, H.; Zhang, H.X.; Chen, Y.; Chen, Y.J.; Zhuang, L.; Ou, H.S. Enhanced photocatalysis degradation of organophosphorus flame retardant using MIL-101(Fe)/persulfate: Effect of irradiation wavelength and real water matrixes. *Chem. Eng. J.* **2019**, *368*, 273–284. [CrossRef]
35. Zhao, F.P.; Liu, Y.P.; Ben Hammouda, S.; Doshi, B.; Guijarro, N.; Min, X.B.; Tang, C.J.; Sillanpaa, M.; Sivula, K.; Wang, S.B. MIL-101(Fe)/g-C3N4 for enhanced visible-light-driven photocatalysis toward simultaneous reduction of Cr(VI) and oxidation of bisphenol A in aqueous media. *Appl. Catal. B Environ.* **2020**, *272*, 14. [CrossRef]
36. Li, J.; Wang, L.J.; Liu, Y.Q.; Zeng, P.; Wang, Y.; Zhang, Y.Z. Removal of Berberine from Wastewater by MIL-101(Fe): Performance and Mechanism. *ACS Omega* **2020**, *5*, 27962–27971. [CrossRef]
37. Zhang, T.; Li, P.; Fang, C.; Jiang, R. Phosphate recovery from animal manure wastewater by struvite crystallization and CO_2 degasification reactor. *Ecol. Chem. Eng. S-Chem. I Inz. Ekol. S* **2014**, *21*, 89–99. [CrossRef]
38. Boontongto, T.; Burakham, R. Evaluation of metal-organic framework NH2-MIL-101(Fe) as an efficient sorbent for dispersive micro-solid phase extraction of phenolic pollutants in environmental water samples. *Heliyon* **2019**, *5*, e02848. [CrossRef]
39. Wang, Y.L.; Zhang, N.; Chen, D.N.; Ma, D.; Liu, G.G.; Zou, X.G.; Chen, Y.P.; Shu, R.J.; Song, Q.Y.; Lv, W.Y. Facile synthesis of acid-modified UiO-66 to enhance the removal of Cr (VI) from aqueous solutions. *Sci. Total Environ.* **2019**, *682*, 118–127. [CrossRef]
40. Maksimchuk, N.V.; Zalomaeva, O.V.; Skobelev, I.Y.; Kovalenko, K.A.; Fedin, V.P.; Kholdeeva, O.A. Metal-organic frameworks of the MIL-101 family as heterogeneous single-site catalysts. *Proc. R. Soc. A-Math. Phys. Eng. Sci.* **2012**, *468*, 2017–2034. [CrossRef]
41. Li, Z.; Liu, X.; Jin, W.; Hu, Q.; Zhao, Y. Adsorption behavior of arsenicals on MIL-101(Fe): The role of arsenic chemical structures. *J. Colloid Interface Sci.* **2019**, *554*, 692–704. [CrossRef] [PubMed]
42. Guo, J.; Li, J.-J.; Wang, C.-C. Adsorptive removal of Cr(VI) from simulated wastewater in MOF BUC-17 ultrafine powder. *J. Environ. Chem. Eng.* **2019**, *7*, 102909. [CrossRef]

43. Wang, C.C.; Ren, X.; Wang, P.; Chang, C. The state of the art review on photocatalytic Cr(VI) reduction over MOFs-based photocatalysts: From batch experiment to continuous operation. *Chemosphere* **2022**, *303*, 134949. [CrossRef] [PubMed]
44. Karbassiyazdi, E.; Kasula, M.; Modak, S.; Pala, J.; Kalantari, M.; Altaee, A.; Esfahani, M.R.; Razmjou, A. A juxtaposed review on adsorptive removal of PFAS by metal-organic frameworks (MOFs) with carbon-based materials, ion exchange resins, and polymer adsorbents. *Chemosphere* **2023**, *311*, 136933. [CrossRef] [PubMed]
45. Qasem, K.M.A.; Khan, S.; Chinnam, S.; Saleh, H.A.M.; Mantasha, I.; Zeeshan, M.; Manea, Y.K.; Shahid, M. Sustainable fabrication of Co-MOF@CNT nano-composite for efficient adsorption and removal of organic dyes and selective sensing of Cr(VI) in aqueous phase. *Mater. Chem. Phys.* **2022**, *291*, 126748. [CrossRef]
46. He, Y.; Sun, H.; Liu, W.; Yang, W.; Lin, A. Study on removal effect of Cr(VI) and surface reaction mechanisms by bimetallic system in aqueous solution. *Environ. Technol.* **2020**, *41*, 1867–1876. [CrossRef] [PubMed]
47. Mikhaylov, V.I.; Maslennikova, T.P.; Ugolkov, V.L.; Krivoshapkin, P.V. Hydrothermal synthesis, characterization and sorption properties of Al/Fe oxide-oxyhydroxide composite powders. *Adv. Powder Technol.* **2016**, *27*, 756–764. [CrossRef]
48. Shokati Poursani, A.; Nilchi, A.; Hassani, A.H.; Shariat, M.; Nouri, J. A novel method for synthesis of nano-γ-Al_2O_3: Study of adsorption behavior of chromium, nickel, cadmium and lead ions. *Int. J. Environ. Sci. Technol.* **2015**, *12*, 2003–2014. [CrossRef]
49. Cai, W.; Yu, J.; Jaroniec, M. Template-free synthesis of hierarchical spindle-like γ-Al_2O_3 materials and their adsorption affinity towards organic and inorganic pollutants in water. *J. Mater. Chem.* **2010**, *20*, 4587–4594. [CrossRef]
50. Yi, H.; Jin-Song, Z.; Yan, Z.; Ting-Lin, H.; Qiu-Xun, T. Adsorption of Cr(VI) from aqueous solution by activated alumina and activated carbon. *J. Xi'an Univ. Archit. Technol. Nat. Sci. Ed.* **2011**, *43*, 864–868.
51. Ge, J.; Deng, K.; Cai, W.; Yu, J.; Liu, X.; Zhou, J. Effect of structure-directing agents on facile hydrothermal preparation of hierarchical γ-Al_2O_3 and their adsorption performance toward Cr(VI) and CO_2. *J. Colloid Interface Sci.* **2013**, *401*, 34–39. [CrossRef] [PubMed]
52. Liu, W.; Zhang, J.; Zhang, C.; Wang, Y.; Li, Y. Adsorptive removal of Cr (VI) by Fe-modified activated carbon prepared from Trapa natans husk. *Chem. Eng. J.* **2010**, *162*, 677–684. [CrossRef]

water

MDPI

Article

Analysis of Hotel Water-Use Behavior Based on the MLP-SEM Model

Rong Cai [1,2], Xue Bai [1,2,*], Jialin Liu [1,2] and Mengting Hu [1,2]

1 China National Institute of Standardization, Beijing 100191, China
2 Key Laboratory of Energy Efficiency, Water Efficiency and Greenization for State Market Regulation, Beijing 102200, China
* Correspondence: bai_xue2022@163.com

Abstract: As a representative service industry, the hotel industry has a complex water-use structure and high water consumption. It is of great significance to investigate the mechanisms determining hotel water-use behavior for demand analysis, as this would make it possible to enhance water-use efficiency and enact targeted water-saving measures. Using Spearman's hierarchical correlation coefficient, the multi-layer perceptron (MLP) neural network model, and the structural equation model (SEM), in this study, we explored the mechanism determining hotel consumers' water-use behavior from different dimensions and constructed a typical water-use behavior model based on the MLP-SEM model. In terms of individual water-use behavior, the results showed that individual characteristics, water-conservation awareness, and consumption behavior possessed significant differences regarding their influence on and correlation with various water-use behaviors. The most relevant factors influencing each behavior, namely washing up, hand washing, and drinking, were daily stay in the hotel, education, and income. Gender had the greatest impact on bathing and toilet-flushing water-use behaviors. The importance of daily stay in the hotel was 0.181, which meant that this was the most significant factor influencing the direct water-use behavior of hotel guests. The following factors were identified: hotel type, income, age, and gender. Typical individual characteristics had a significant impact on main water-use behaviors, whereas typical consumption behaviors had no effect. These results can provide a foundation for relevant research in other industries and serve as a basis for a prediction model of water consumption in hotels based on water-use behavior. Furthermore, they provide a basis for the delicate management of water-use behavior in hotels, making it possible to effectively guide the public to consciously adopt water-saving habits, thus improving water efficiency, which could alleviate the shortage of water resources in the long-term.

Keywords: correlation; hotel; multi-layer perceptron neural network model; structural equation model; water-use behavior

Citation: Cai, R.; Bai, X.; Liu, J.; Hu, M. Analysis of Hotel Water-Use Behavior Based on the MLP-SEM Model. *Water* **2023**, *15*, 1534. https://doi.org/10.3390/w15081534

Academic Editor: Stefano Alvisi

Received: 10 February 2023
Revised: 4 April 2023
Accepted: 4 April 2023
Published: 14 April 2023

1. Introduction

Water is an indispensable natural resource for human survival as well as social and economic development. With the rapid urbanization of China and continuous improvement in living standards, the proportion of the national economy represented by the service industry has been increasing. In 2015, China's service industry accounted for more than 50% of its gross domestic product (GDP). Pressure for water demand from the rapid development of the service industry has also gradually emerged. In addition, with the improvement in people's living standards, consumer demand for the service industry is increasing, which further increases the pressure on water-resource management. Therefore, strengthening the sustainable use of water resources for the future development of the service industry has become an inevitable trend.

To strengthen water-use management in the service industry, scholars in China and elsewhere have conducted research on water-intake monitoring, management, and demand forecasting [1,2]. As the micro-basis of demand-side management of water resources [3], the study of water-use behavior is of great significance for promoting changes in consumption behavior and water-use modes, improving water-use efficiency, and alleviating water shortages [4]. In recent years, scholars have investigated the factors influencing water-use behavior through large-scale behavioral research and have analyzed the drivers and influencing paths based on statistical models, such as the structural equation model and the LPA analysis method [5,6]. Studies have found that individual characteristics, including gender, age, educational level, income, awareness of water conservation, etc. [7–11], have a great impact on household water-use behavior. Water habits, socioeconomic status, and water-saving technology and equipment also play roles in residents' domestic water-use behavior and demand, whereas the natural environment does not [12–17]. However, past research has mainly focused on the domestic water-use behavior of residents, and little work has been undertaken on the mechanism determining water-use behavior under the influence of different public services and social activities. In recent years, some scholars have proposed that, in contrast to domestic water-use behavior, social activities can affect people's water-use behavior. The conclusions were confirmed in research on college students' water-use behavior, which showed that psychological characteristics were significant factors, along with scenario-moderating factors, while there was no significant correlation with sociodemographic characteristics [18,19].

The hotel industry is a service industry [20] that plays an important role in business and leisure activities. Since 2005, with the rapid growth in China's GDP, the numbers of rooms and beds, among other indicators of hotels above a typical size, have been increasing [21]. As they are part of a typical high-water-consumption service industry, hotels have complex water-use structures. It is estimated that hotel water consumption in 2020 was approximately 1.613 billion m^3, accounting for 1.87% of the country's domestic water consumption. Hotel water is mainly used for guest rooms, catering, fitness, entertainment, cleaning, and greeneries. Studies have shown that there are significant differences in the water-use structures of hotels in different countries. In China, guest rooms consume the most water [22–24]. By analyzing the factors influencing hotel water use, scholars have proposed that there are significant differences in the proportions of water used in hotels with different service qualities and in star-rated hotel rooms [25]. As a result, the water-saving index and the water-use index (WUI) have been proposed to evaluate the efficiency of water use in hotels based on water-use structure analysis, and evaluation and optimization methods for the hotel water-use footprint have been explored from the perspective of the whole life cycle [26–28].

Previous research on hotel water use has mainly focused on four aspects: water-use structure, influencing factors, evaluation of water savings, and resident-based water-use behavior. There is a research gap regarding the behavioral mechanisms that determine hotel water use. Moreover, the current research has mainly looked at significance analyses and has missed multidimensional influencing factors, which has led to the failure to build a typical water-use behavior model based on important factors due to incomplete analysis of indicators. Therefore, this study proposes a multidimensional hypothesis for the first time to research the influencing factors for hotel consumers' water-use behavior and analyzes the mechanism determining water-use behavior, which is of great significance for exploring the potential for hotel water conservation, water management, and water efficiency.

2. Research and Analysis Methods

2.1. Questionnaire

The structure of hotel consumers' water-use behavior was the basis for the design of the questionnaire. Consumer water-use behavior in hotels and other public service spaces is accompanied by the service itself [29]. Based on the type of service, consumers' water-use behavior can be divided into direct and indirect categories (see Figure 1). Direct water-use behaviors include washing, hand washing, bathing, toilet flushing, and drinking, which are mainly controlled by the consumers themselves. However, indirect water-use behaviors, such as water consumption for food and cookware washing, cleaning, and sanitation, occur during the service process; such behaviors have low correlation with consumers. This study mainly focuses on the controlled and highly related water-use behavior of consumers in hotels. Based on the water-use hierarchy constructed by domestic scholars, the water used in toilet flushing mainly considers the water-use efficiency of terminal water appliances [9], whereas drinking, as a basic consumer need, is less affected by individual water-use behavior. Unlike toilet flushing and drinking habits, washing, hand washing, and bathing are more subjectively affected by individual behavior and are therefore considered typical water-use behaviors in hotels.

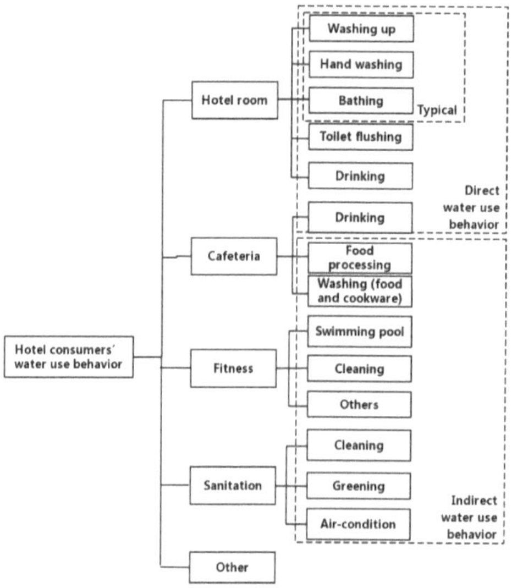

Figure 1. Structure of hotel consumers' water-use behavior.

To cover comprehensive and representative data, and to recover and analyze the questionnaires more easily, this study used web-based questionnaires. The survey questionnaire covered more than 25 provinces in China. As hotel consumers are highly mobile, thereby possessing significantly different features from household water users, regional factors were not considered in this study. As people in hotels come from different walks of life and come in great numbers, the questionnaire for this study included gender, age, income, and other individual characteristics of hotel consumers as well as their water-conservation awareness, consumption behavior, and personal water-use behavior, as detailed in Table 1. The level of water-conservation awareness was analyzed and scored through questions such as "daily water conservation behaviors and perceptions".

Table 1. Questionnaire content of hotel consumers' water-use behavior.

Type	Survey Content
Individual characteristics	Gender, age, education, income, etc.
Water-saving awareness	Degree of water-saving awareness
Consumption behavior	Type of hotel, travel purpose, length of stay in hotel, etc.
Water-use behavior	The single washing time of bathing, washing up, and hand-washing time, toilet-flushing frequency, drinking habits, etc.

A total of 292 valid questionnaires were collected, and the distribution of the valid sample data is shown in Figure 2. The ratio of men to women was 68% to 32%. Most participants were between 36 and 50 years old, accounting for 43% of the total; this was followed by participants between the ages of 21 and 35 and 51 and 60, which accounted for 36% and 17% of the total, respectively. Finally, consumers under 20 years old or above 60 years accounted for less than 3%. Regarding the distribution of education, more than half of the population had an undergraduate degree followed by people with a bachelor's degree (23%) and people with master's and doctorate degrees (more than 14% and 10%, respectively). In terms of income, more than one-fifth of the participants had a monthly income between 3000 and CNY 17,000, and a few participants had incomes below CNY 3000 and higher than CNY 17,000, accounting for 6% and 9%, respectively.

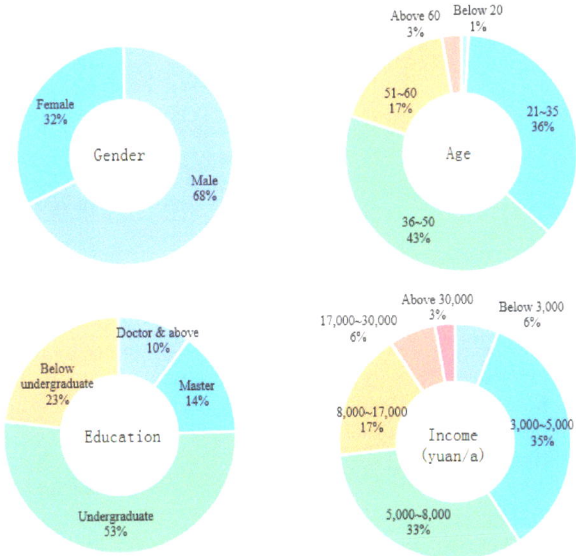

Figure 2. The distribution for sample data regarding gender, age, education, and income.

2.2. *Research Methods*

2.2.1. Research Hypotheses

To scientifically identify the factors influencing the water-use behavior of different people in the hotel industry and to deeply analyze their response mechanisms, this study identified influencing factors from multiple dimensions, such as single factors (H_{1a}–H_{1h}), multi-factors (H_{2a}), and comprehensive factors (H_{3a} and H_{3b}), and quantitatively analyzed the importance of the multi-factor influencing factors. Consequently, several hypotheses were proposed by referring to a previous study on water-use behavior, as shown in Table 2.

2.2.2. Research Methods

Spearman Rank Correlation Coefficient

The Spearman correlation coefficient is equivalent to the nonparametric form of the Pearson correlation coefficient, which does not require the distribution of variables, and is suitable for ordinal variables or continuous variables that do not follow a normal distribution [30–32]. According to the characteristics of the dependent variable in this study, SPSS software was used to calculate the Spearman rank correlation coefficient to identify the correlation between single hotel consumers' water-use behavior and the research factors, which entailed identifying the correlation coefficient between different factors and a single water-use behavior.

Table 2. Research hypotheses.

S/N	Hypotheses
H_{1a}	Gender significantly affects the water-use behavior of hotel consumers
H_{1b}	Age significantly affects the water-use behavior of hotel consumers
H_{1c}	There are significant effects of the educational level on the water-use behavior of hotel consumers
H_{1d}	Income significantly affects the water-use behavior of hotel consumers
H_{1e}	The degree of awareness of water conservation significantly affects the water-consumption behavior of hotel consumers
H_{1f}	The type of hotel stay significantly affects the water-use behavior of hotel consumers
H_{1g}	Travel purpose significantly affects the water-use behavior of hotel consumers
H_{1h}	The length of daily hotel stay significantly affects the water-use behavior of hotel consumers
H_{2a}	The degree of influence of each influencing factor on different water-use behaviors of hotel consumers is different
H_{3a}	Typical individual characteristics have a significant impact on the typical water-use behavior of hotel consumers
H_{3b}	Typical consumption behavior has a significant impact on the typical water-use behavior of hotel consumers

MLP Neural Network Model

A multi-layer perceptron (MLP) model, also known as a multi-layer feedforward neural network [33], consists of an input layer, hidden layer (one or more layers), and output layer, as shown in Figure 3. It is a neural network trained with an error backpropagation algorithm (BP algorithm) [34], which aims to simulate the structure and function of the nervous system for data processing, constantly adjusting the weights of the chains between the simulated neurons so that the whole network can adapt well to the relationships of the training data [35]. Owing to their strong nonlinear processing ability, high fault tolerance, and self-learning ability, neural systems have been widely applied for the identification and analysis of influencing factors [36–38]. In this study, the MLP model was used to analyze the importance of different influencing factors on water-use behavior. Compared to the analysis of the Spearman correlation coefficient, the MLP model focuses more on considering various factors as a whole to illustrate intrinsic effects.

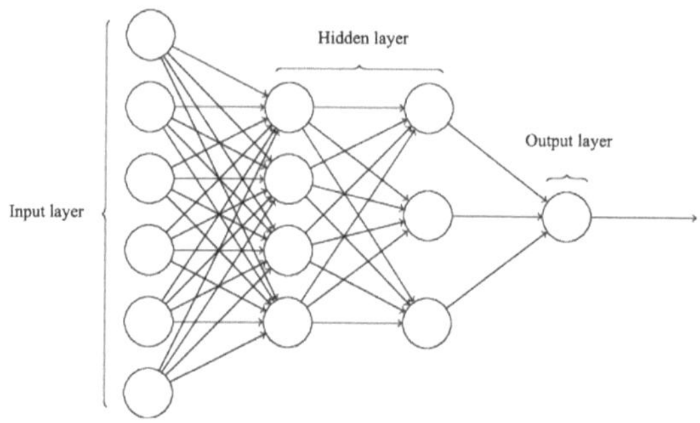

Figure 3. Multi-layer perceptron (MLP) model structure.

Structural Equation Model

The structural equation model is a statistical method based on the covariance matrix of variables to analyze the relationships between variables and is an important tool for multivariate data analysis [39–41]. It is often used to analyze the direct or indirect impact of one variable on another [42]. Structural equation analysis processes multiple dependent variables simultaneously and explains the relationship between variables in combination with methods such as route factor analysis [43]. Therefore, the structural equation model was used in this study to analyze the influence mechanisms by which typical individual characteristics and consumption behaviors influence the typical water-use behavior of hotel consumers and to verify the consistency between the hypotheses (H_{3a}, H_{3b}) and the sample data.

3. Identification and Analysis of Influencing Factors of Hotel Consumers' Water-Use Behavior

3.1. Identification of Single-Factor Influencing Factors

The Spearman correlation coefficient was used to analyze the correlation among individual characteristics, water-saving awareness, consumption behavior, and water-use behavior of hotel consumers; the results are shown in Table 3. There were significant differences based on gender in terms of washing up, bathing, and toilet flushing, and female consumers had a higher water-use behavior duration or frequency than male consumers. There were age-dependent differences in terms of toilet flushing; the frequency among older consumers was higher than that among younger consumers. Education had a significant impact on bathing and drinking and had a certain correlation with hand washing and toilet flushing. Income was positively correlated with water drinking and toilet flushing, and the correlation with drinking was significant. There was no significant correlation between the degree of water-conservation awareness and the hotel type. Travel purpose had a significant impact on drinking and for consumers on business travel: drinking > leisure tourism > visiting relatives and friends. The daily hotel stay had a general impact on the water-use behaviors of hotel consumers, and there were significant differences in other water-use behaviors.

Table 3. Identification of influencing single factors of hotel consumers' water-use behavior.

Type		Washing-Up Time	Hand-Washing Time	Bathing Time	Toilet-Flushing Frequency	Drinking Habit
Individual characteristics	Gender	0.143 **	0.081	0.304 ***	0.272 ***	0.041
	Age	0.075	0.075	0.046	0.107 *	0.074
	Education	0.053	0.105 *	0.153 ***	0.111 *	0.183 ***
	Income	0.03	0.068	0.067	0.125 **	0.216 ***
Water-saving awareness	Degree of water-saving awareness	−0.026	0.005	−0.004	0.061	−0.005
Consumption behavior	Type of hotel	−0.084	−0.004	0.096	0.082	0.074
	Travel purpose	0.009	0.001	0.004	0.01	−0.202 ***
	Daily stay in the hotel	0.152 ***	0.095	0.238 ***	0.209 ***	0.11 *

Note: ***, **, and * represent significance at the 1%, 5%, and 10% levels, respectively.

3.2. Identification of Influencing Multi-Factors and Quantitative Analysis of Importance

Of the 292 sample data points, 202 training and 90 test samples were randomly selected to construct the MLP neural network model of hotel consumers' water-use behavior. In this model, the input layer variables were gender, age, education, income, water-saving awareness, hotel type, travel purpose, and daily stay at the hotel. For all the water-use behaviors, the number of hidden layers was one, and the number of hidden layer units was five, three, five, five, and four, respectively. The Receiver Operating Characteristic (ROC) analysis uses the actual hotel water-use behavior of the sample individuals as the criterion for determining the optimal solution point. Using the probability of the sample case classification results given by the model as the classification judgment basis, the ROC curve is obtained. The fitness of the model was judged according to the sensitivity (1-specificity) of the classification test, as shown in Figure 4, and the area under each curve is shown in the table. The results show that the area under the curve was greater than 0.5, indicating that the fitting model was effective.

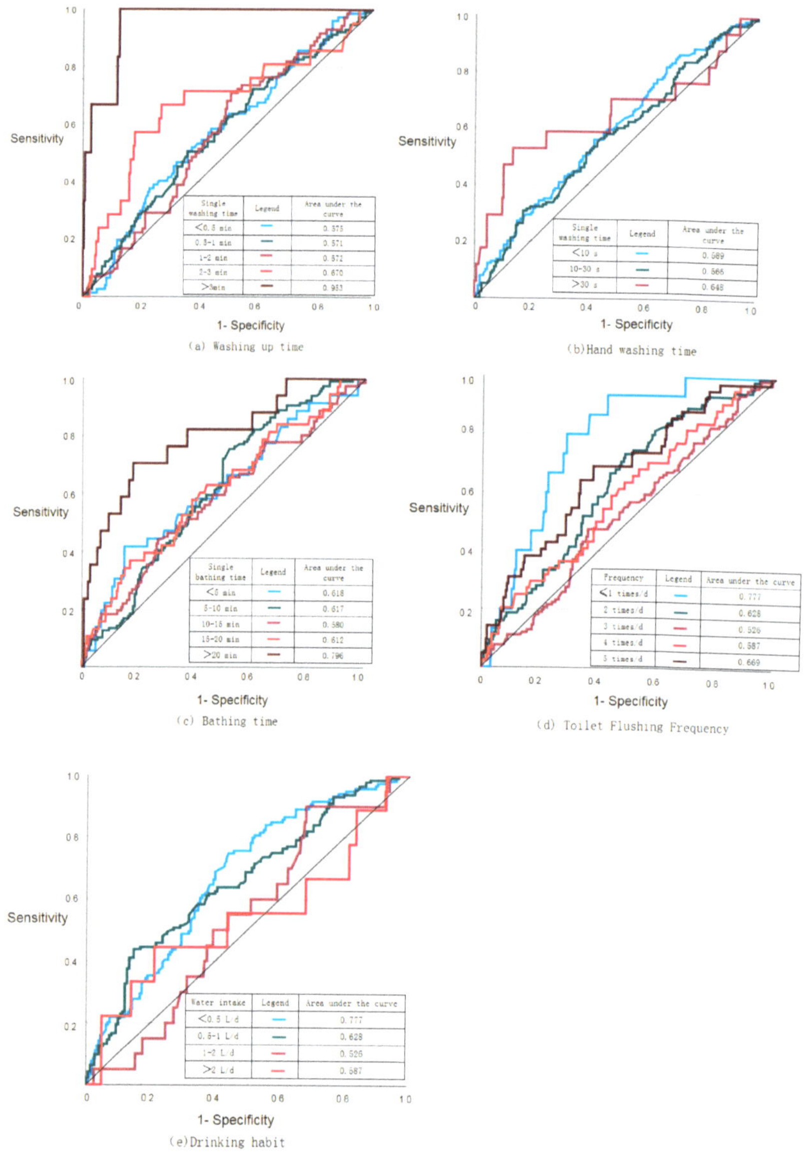

Figure 4. Receiver Operating Characteristic (ROC) curve of the MLP neural network model for the water-use behavior of hotel consumers: (**a**) Washing-up time; (**b**) Hand-washing time; (**c**) Bathing time; (**d**) Toilet-Flushing Frequency; and (**e**) Drinking habit.

The constructed neural network model of hotel consumers' water-use behavior was applied to quantitatively analyze the importance of various influencing factors on different water-use behaviors, as shown in Table 4. According to the analysis results, the degree of the influencing factors differs for the water-use behaviors of hotel consumers. Daily stay in hotels was the most important factor for washing up, hand washing, and bathing. The importance of the other influencing factors varied significantly. The second most influential factors on these behaviors were education, income, and gender. The order of importance of toilet flushing frequency was income > gender > education > daily stay in the hotel >

age > travel purpose > degree of water-saving awareness > hotel type. Drinking habits were more significantly influenced by income than other factors, with gender having the least impact.

Table 4. Importance of influencing factors of water-consumption behavior in the hotel.

Type		Washing-Up Time		Hand-Washing Time		Bathing Time		Toilet-Flushing Frequency		Drinking Habit	
		Importance	Normalization Importance	Importance	Normalization Importance	Importance	Normalization Importance	Importance	Normalization Importance	Importance	Normalization Importance
Individual characteristics	Gender	0.138	82.9%	0.076	38.6%	0.172	86.2%	0.165	80.1%	0.047	14.4%
	Age	0.118	70.9%	0.123	62.3%	0.081	40.8%	0.128	62.4%	0.142	43.1%
	Education	0.152	91.4%	0.082	41.4%	0.143	72.0%	0.141	68.5%	0.072	22.0%
	Income	0.118	71.1%	0.150	76.1%	0.096	48.3%	0.206	100.0%	0.329	100.0%
Water-saving awareness	Degree of water-saving awareness	0.086	51.9%	0.098	49.4%	0.087	43.6%	0.058	28.2%	0.080	24.2%
Consumption behavior	Hotel type	0.099	59.6%	0.145	73.5%	0.108	54.1%	0.046	22.5%	0.103	31.3%
	Travel purpose	0.123	74.3%	0.128	64.7%	0.114	57.5%	0.118	57.4%	0.153	46.4%
	Daily stay	0.166	100.0%	0.198	100%	0.199	100.0%	0.137	66.5%	0.075	22.8%

Overall, the direct water-use behaviors were affected by the duration of the hotel stay, followed by the type of hotel, income, age, gender, education, as well as the degree of water-saving awareness and travel purpose.

As shown in Figure 5, the factors influencing direct water-use behavior are ranked in order of importance as daily hotel stay, type of hotel, income, age, gender, education, degree of water-saving awareness, and travel purpose. Based on the analysis, the most important factor was the daily hotel stay, and the value of its importance was 0.181.

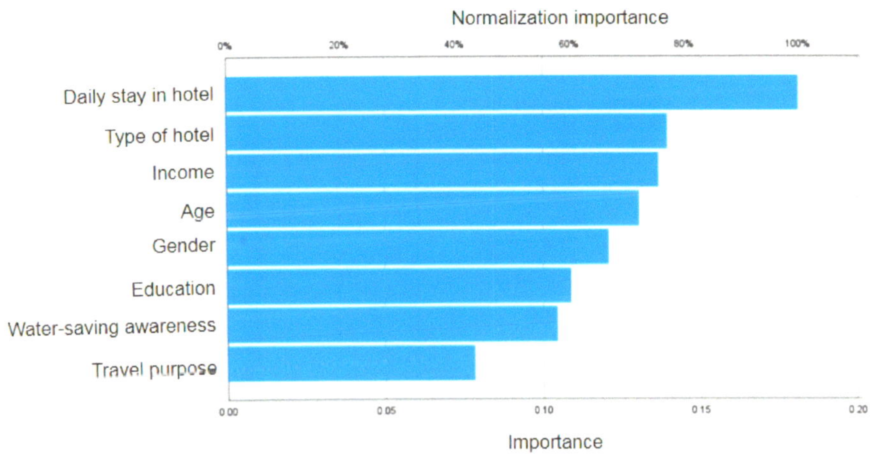

Figure 5. Importance of influencing factors on the comprehensive direct water-use behavior of hotel consumers.

3.3. Identification of Comprehensive Influencing Factors

Based on the above analysis, the typical individual characteristics of gender, age, income, type of hotel, and daily hotel stay were selected to analyze their impact on hotel consumers' water-use behavior using the structural equation model.

A parameter diagram of hotel consumers' typical individual characteristics and typical consumption behaviors on typical water-use behaviors was obtained, as shown in Figure 6. The structural equation model intuitively describes the relationship between latent variables through a path relationship diagram. As the latent variables represent phenomena that could not be directly measured, several relevant explicit variables were selected for interpretation. The latent and explicit variables were expressed as ellipses and squares, respectively. For example, gender, age, and income were selected to express typical individual characteristics.

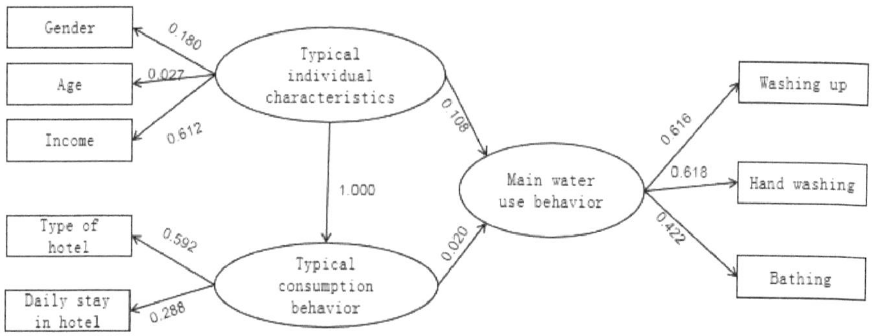

Figure 6. Parameter diagram of typical water-use behavior model of hotel consumers.

The chi-square ratio to the degree of freedom ratio was less than 5 [44], which is within a reasonable range of the evaluation criteria. The chi-square degree of freedom ratio of the fitting index of the model was $x^2/df = 3.962$, which meets the requirements of good model fitting. The regression coefficients between the variables in the constructed structural equation model are listed in Table 5. The significance test ($p < 0.05$) was used to analyze whether there is an influence relationship between the model variables; that is, in the construction of the model, the significance p value of the typical individual characteristics on the main water-use behavior was 0.000 < 0.05, which means that the typical individual characteristics have a significant impact on the main water-use behavior, and the path standardized coefficient was 0.108. The p value of the other paths was greater than 0.05, indicating that these paths were invalid, whereas their typical consumption behavior had no significant impact on the main water-use behavior.

Table 5. Regression coefficient of typical water-use behavior model of hotel consumers.

Factor (Latent Variable)	→	Item (Explicit Variable)	Non-Standardized Coefficient	Standardized Coefficient	Standard Error	p
Typical Individual Characteristics	→	Main Water-Use Behavior	0.545	0.108	0.027	0.000 ***
Typical Consumption Behavior	→	Main Water-Use Behavior	0.023	0.020	0.121	0.851
Typical Individual Characteristics	→	Typical Consumption Behavior	4.496	1.000	2.679	0.093 *

Note: *** and * represent significance at the 1% and 10% levels, respectively.

3.4. Hypothesis Test Results

The research hypotheses were tested and analyzed using Spearman's rank correlation coefficient, an MLP neural network model, and a structural equation model. According to the analysis results in Table 5, assuming that H_{2a} and H_{3a} are valid, each influencing factor has a different degree of influence on the water-use behaviors of hotel consumers, and typical individual characteristics have a significant influence on typical water-use behaviors of hotel consumers. H_{1a}, H_{1b}, H_{1c}, H_{1d}, H_{1g}, and H_{1h} were partially valid; gender, age, education, income, travel purpose, and daily stay in hotels had a significant impact on some water-use behaviors; H_{1e}, H_{1f}, and H_{3b} were not valid; and the degree of water-saving awareness and the type of hotel had no significant impact on the typical consumption behavior of hotel consumers.

4. Conclusions

This study proposes the hypothesis that individual characteristics, water-saving awareness, and consumption behavior have different degrees of influence on hotel consumers' water-use behavior. Through the questionnaire, these hypotheses were tested and analyzed using Spearman's rank correlation coefficient, the MLP neural network model, and structural equation model to identify the multi-dimensional influencing factors of hotel consumers' water-use behavior, and the importance of each influencing factor was analyzed. According to the results, the correlation and impact of water-use behavior on individual characteristics, water-saving awareness, and consumption behavior differed. Gender, age, education, income, travel purpose, daily stay in hotels, and other individual characteristics and consumption behaviors significantly affected water-use behaviors. The degree of water-saving awareness and the type of hotel had no significant correlation with the water-use behavior of hotel consumers. The influence degree on different water-use behaviors was quite different, and the direct water-use behavior was most affected by the length of stay in the hotel. From the perspective of comprehensive influencing factors, individual characteristics were the main influencing factors of typical water-use behavior, indicating that the water-use behavior of hotel consumers was less affected by water-saving awareness and consumption behavior.

This study constructed a water-use behavior model for hotel consumers to explore the water-use behavior mechanism, which serves as an important analysis method for water demand-side management and provides a scientific basis for water-resource management and policy formulation in this field. In addition, the model lays the foundation for a water-use predictive model for hotel water consumption based on water-use behavior. The research only focused on hotels and could not illustrate the general mechanism for the entire service industry but could provide a reference for subsequent studies on other industries in terms of the identification of influencing factors and mechanism research on water-use behavior.

Author Contributions: Conceptualization, X.B. and R.C.; methodology, X.B.; software, R.C.; validation, J.L. and M.H.; formal analysis, R.C.; investigation, R.C.; resources, X.B.; data curation, R.C.; writing—original draft preparation, R.C.; writing—review and editing, J.L.; visualization, M.H.; supervision, X.B.; project administration, X.B.; funding acquisition, X.B. All authors have read and agreed to the published version of the manuscript.

Funding: This research was funded by the Dean's Fund Project of China National Institute of Standardization (No. 542023Y-10361).

Data Availability Statement: Data are available on request from the authors.

Conflicts of Interest: The authors declare no conflict of interest.

References

1. Song, L. *Outliers Identification of Water Intake Monioring Data in Industrial and Domestic Service Industry Based on LOF and CEEMD*; Hebei University of Engineering: Handan, China, 2021.
2. Chen, M.; Yi, Y.N.; Qin, J. On management of and measures for planned water use management in China. *Water Resour. Dev. Res.* **2022**, *22*, 57–61.
3. Zhong, F.L.; Guo, A.J.; Jiang, D.W.; Yang, X.; Yao, W.G.; Lu, J. Research progress regarding residents' water consumption behavior as relates to water demancmanagement: A literature review. *Adv. Water Sci.* **2018**, *29*, 9.
4. Wegelinschuringa, M. *Water Demand Management and the Urban Poor*; Delft Netherlands International Water & Sanitation Centre: Delft, The Netherlands, 2002.
5. Zhong, F.; Li, L.; Guo, A.; Song, X.; Cheng, Q.; Zhang, Y.; Ding, X. Quantifying the influence path of water conservation awareness on water-saving irrigation behavior based on the theory of planned behavior and structural equation modeling: A case study from northwest China. *Sustainability* **2019**, *11*, 4967. [CrossRef]
6. Addo, I.B.; Thoms, M.C.; Parsons, M. Barriers and drivers of household water-conservation behavior: A profiling approach. *Water* **2018**, *10*, 1794. [CrossRef]
7. Shi, L.; Zhu, Y.; Li, H.; Zhao, Y.; Wang, J.; Wang, L.; Qu, J.; Zhen, L. Simulation of residential water consumption and analysis of water-saving potential in Beijing. *South—North Water Transf. Water Sci. Technol.* **2022**, *20*, 851–861.

8. Nyong, A.O.; Kanaroglou, P.S. A survey of household domestic water-use patterns in rural semi-arid Nigeria. *J. Arid. Environ.* **2001**, *49*, 387–400. [CrossRef]
9. Nauges, C.; Thomas, A. Privately operated water utilities, municipal price negotiation, and estimation of residential water demand: The case of France. *Land Econ.* **2000**, *76*, 68–85. [CrossRef]
10. Dandy, G.; Nguyen, T.; Davies, C. Estimating residential water demand in the presence of free allowances. *Land Econ.* **1997**, *73*, 125–139. [CrossRef]
11. Chen, X.; Xu, J.; Ji, Y. Research on determinant factors in residential water demand. *J. Econ. Water Resour.* **2005**, *25–26*, 73.
12. Zhao, W. An empirical analysis of the factors influencing residential household water consumption-an examination of survey data based on the water consumption behavior of Beijing residents. *J. Arid. Land Resour. Environ.* **2015**, *29*, 137–142. [CrossRef]
13. Lin, R.; Qi, Y.; Fan, H.; Yang, S.; Zhang, C. An analysis of the present situation and influencing factors of household water consumption in Beijing. *China Rural. Water Hydropower* **2021**, 6.
14. Sha, K.; Dorssen, A.V.; Brouwer, S. Enhancing domestic water conservation behaviour: A review of empirical studies on influencing tactics. *J. Environ. Manag.* **2019**, *247*, 867–876.
15. David, S. Water conservation: Theory and evidence in urban areas of the developed world. *Annu. Rev. Environ. Resour.* **2013**, *38*, 227–248.
16. Lee, M.; Berrin, T.; Maribel, B. Urban sustainability incentives for residential water conservation: Adoption of multiple high efficiency appliance. *Water Resour. Manag.* **2013**, *27*, 2531–2540. [CrossRef]
17. Chang, G. Factors influencing water conservation behavior among urban residents in China's arid areas. *Water Policy* **2013**, *15*, 691–704. [CrossRef]
18. Tortella, B.D.; Tirado, D. Hotel water consumption at a seasonal mass tourist destination. The case of the island of Mallorca. *J. Environ. Manag.* **2021**, *92*, 2568–2579.
19. Zhu, J.; Zhao, X.; Zhu, T.; Li, L. Which factors determine students water-saving behaviors? Evidence from China colleges. *Urban Water J.* **2021**, *18*, 860–872. [CrossRef]
20. Du, J. Model analysis of hotel customer satisfaction based on two-factor theory. *Mark. Mod.* **2008**, 40–41.
21. National Bureau of Statistics. *China Statistical Yearbook*; China Statistics Press: Beijing, China, 2021.
22. Antakyali, D.; Krampe, J.; Steinmetz, H. Practical application of wastewater reuse in tourist resorts. *Water Sci. Technol.* **2008**, *57*, 2051–2057. [CrossRef]
23. Gössling, S.; Peeters, P.; Hall, C.M.; Ceron, J.P.; Dubois, G.; Lehmann, L.V.; Scott, D. Tourism and water use: Supply, demand, and security. An international review. *Tour. Manag.* **2012**, *33*, 1–15.
24. Huang, X.; Zhao, T.; Yu, S.; Cao, W.; Wu, D.; Yu, K. Research on China's tertiary industry development and water saving. *China Water Wastewater* **2022**, *38*, 49–56.
25. Bohdanowicz, P.; Martinac, I. Determinants and benchmarking of resource consumption in hotels—Case study of Hilton International and Scandic in Europe. *Energy Build.* **2007**, *39*, 82–95. [CrossRef]
26. Zuo, J.; Liu, C.; Zhang, H. Assessment of water use efficiency of hotels in Beijing Municipality. *China Water Resour.* **2009**, 52–55.
27. Deng, S. Energy and water uses and their performance explanatory indicators in hotels in Hong Kong. *Energy Build.* **2003**, *35*, 775–784. [CrossRef]
28. Wang, Z. *Assessment and Reduction of Water Footprint in Hotel Service Industry: A Case Study of Shanghai*; Tianjin University: Tianjin, China, 2020.
29. Wang, Y.; Chen, Y.; Weng, J.; Jiang, Y. Analysis of the characteristics of urban public domestic water use in Beijing. *Water Wastewater Eng.* **2008**, 138–143.
30. Jiang, Y. *Study on Factor Analysis and Prediction Model of Water Consumption in Typical Public Buildings*; Chongqing University: Chongqing, China, 2021.
31. Li, R.; Yu, N.; Wang, X.; Liu, Y.; Cai, Z.; Wang, E. Model-Based synthetic geoelectric sampling for magnetotelluric inversion with deep neural networks. *IEEE Trans. Geosci. Remote Sens.* **2022**, *60*, 1–14. [CrossRef]
32. Zhu, X.; Xu, Z.; Liu, Z.; Liu, M.; Yin, Z.; Yin, L.; Zheng, W. Impact of dam construction on precipitation: A regional perspective. *Mar. Freshw. Res.* **2022**. [CrossRef]
33. Wu, L.; He, D.; Ai, B.; Wang, J.; Guan, K.; Zhong, Z. Path loss prediction based on multi-layer perceptron artificial neural network. *Chinese J. Radio Sci.* **2021**, *36*, 396–404.
34. Do, M.; Li, J.; Cui, L.; Wei, Y.; Su, H.; Li, C. Application of correlation analysis – Neural network model in water consumption prediction in Ningxia. *Pearl River* **2022**, *43*, 71–77.
35. Gardner, M.W.; Dorling, S.R. Artificial neural networks (the multilayer perceptron)—A review of applications in the atmospheric sciences. *Atmos. Environ.* **1998**, *32*, 2627–2636. [CrossRef]
36. Agatonovic-Kustrin, S.; Beresford, R. Basic concepts of artificial neural network (ANN) modeling and its application in pharmaceutical research. *J. Pharm. Biomed. Anal.* **2000**, *22*, 717–727. [CrossRef] [PubMed]
37. Wang, N.; Li, D. Research on influencing factors of forestry enterprise management performance based on MLP Model. *China For. Econ.* **2019**, 37–40.
38. Li, Y.; Dang, Y.; Tang, R. Analysis of hospotalization expenses and influencing factors of malignant tumor patients based BP neural network model. *China J. Pharm. Econ.* **2022**, *17*, 10–13.

39. Xie, X.; Xie, B.; Cheng, J.; Chu, Q.; Dooling, T. A simple Monte Carlo method for estimating the chance of a cyclone impact. *Nat. Hazards* **2021**, *107*, 2573–2582. [CrossRef]
40. Yin, L.; Wang, L.; Tian, J.; Yin, Z.; Liu, M.; Zheng, W. Atmospheric density inversion based on Swarm-C satellite accelerometer. *Appl. Sci.* **2023**, *13*, 3610. [CrossRef]
41. Liu, Z.; Xu, J.; Liu, M.; Yin, Z.; Liu, X.; Yin, L.; Zheng, W. Remote sensing and geostatistics in urban water-resource monitoring: A review. *Mar. Freshw. Res.* **2023**. [CrossRef]
42. Reisinger, Y.; Turner, L. Structural equation modeling with Lisrel: Application in tourism. *Tour. Manag.* **1999**, *20*, 71–88. [CrossRef]
43. Wu, B. *Pilot study of Structural Equation Modeling*; Tianjin University: Tianjin, China, 2006.
44. Raykov, T.; Marcoulides, G.A. *First Course in Structural Equation Modeling*, 2nd ed.; Routledge: New York, NY, USA, 2006.

water

MDPI

Article

Combined Forecasting Model of Precipitation Based on the CEEMD-ELM-FFOA Coupling Model

Xianqi Zhang [1,2,3] and Xiaoyan Wu [1,*]

1 Water Conservancy College, North China University of Water Resources and Electric Power, Zhengzhou 450046, China
2 Collaborative Innovation Center of Water Resources Efficient Utilization and Protection Engineering, Zhengzhou 450046, China
3 Technology Research Center of Water Conservancy and Marine Traffic Engineering, Zhengzhou 450046, China
* Correspondence: wuxy725@163.com

Abstract: Precipitation prediction is an important technical mean for flood and drought disaster early warning, rational utilization, and the development of water resources. Complementary ensemble empirical mode decomposition (CEEMD) can effectively reduce mode aliasing and white noise interference; extreme learning machines (ELM) can predict non-stationary data quickly and easily; and the fruit fly optimization algorithm (FFOA) has better local optimization ability. According to the multi-scale and non-stationary characteristics of precipitation time series, a new prediction approach based on the combination of complementary ensemble empirical mode decomposition (CEEMD), extreme learning machine (ELM), and the fruit fly optimization algorithm (FFOA) is proposed. The monthly precipitation data measured in Zhengzhou City from 1951 to 2020 was taken as an example to conduct a prediction experiment and compared with three prediction models: ELM, EMD-HHT, and CEEMD-ELM. The research results show that the sum of annual precipitation predicted by the CEEMD-ELM-FFOA model is 577.33 mm, which is higher than the measured value of 572.53 mm with an error of 4.80 mm. The average absolute error is 0.81 and the average relative error is 1.39%. The prediction value of the CEEMD-ELM-FFOA model can closely follow the changing trend of precipitation, which shows a better prediction effect than the other three models and can be used for regional precipitation prediction.

Keywords: CEEMD; phase space reconstruction; FFOA; precipitation; forecasting

check for updates

Citation: Zhang, X.; Wu, X. Combined Forecasting Model of Precipitation Based on the CEEMD-ELM-FFOA Coupling Model. *Water* **2023**, *15*, 1485. https://doi.org/10.3390/w15081485

Academic Editor: Paul Kucera

Received: 28 February 2023
Revised: 31 March 2023
Accepted: 6 April 2023
Published: 11 April 2023

1. Introduction

Precipitation is an important way to supply water resources to a basin or region. The accurate precipitation forecasts are valuable and rather important for the integration of natural hazards forecasting [1,2]. Precipitation is affected by many factors [3,4], such as topography, atmospheric circulation, the underlying surface, and human activities. Precipitation time series often have the characteristics of being multi-scale, nonlinear, and unstable.

With the development of machine learning [5,6], scholars at home and abroad have done a lot of related research on the accurate prediction of precipitation by using machine learning algorithms and achieved fruitful results. Partal et al. used a wavelet fuzzy neural network to predict the daily precipitation of three stations in Turkey [7], and the results show that the prediction accuracy of the neural network model is better than that of the classical multiple regression model. Aksoy et al. studied the prediction of monthly precipitation in arid and semi-arid areas through feedforward back propagation (FFBP), radial basis function, and generalized regression artificial neural network (ANN), and the results show that ANN is effective in predicting precipitation in dry months [8]. Alizamir et al. used an extreme learning machine (ELM), a single hidden layer feedforward neural network, an artificial neural network, genetic programming, and quantile mapping to predict large-scale global precipitation; ELM was superior to all other methods in predicting monthly precipitation [9]. The above research is mainly aimed at the traditional neural network, which is not capable of processing non-stationary data and high-frequency abrupt data, and the prediction error is generally between 5 and 20%, which is the bottleneck to further improving the accuracy. Precipitation data are affected by many factors, and most of them show nonlinear and non-stationarity characteristics in the time scale. Therefore, using a coupling model to reduce the non-stationarity of the original series has become a new way to increase the prediction accuracy of precipitation.

At the end of the last century, Huang proposed a new method of processing non-stationary signals, empirical mode decomposition [10], which has been widely used in various fields of signal processing [11,12]. The CEEMD model [13] is an adaptive EMD derived from empirical mode, which can be decomposed into stationary signals with different characteristic scales depending on the characteristics of the signal itself. In the process of signal reconstruction, two Gaussian white noises with the same amplitude and opposite phase are added at the same time, which solves the prediction error of the high-frequency component of the EMD model and also solves the reconstruction error of the EEMD model and restrains the influence of mode aliasing and residual white noise. Wang et al. constructed the CEEMD-SE-HS-KELM prediction model and applied it to the short-term wind power prediction of a wind farm in China [14]. The RMSE and MAE were 2.16 and 0.39, respectively, which were superior to the EMD-SE-HS-KELM, HS-KELM, KELM, and extreme learning machine (ELM) models. Wang et al. constructed a CEEMD-ARIMA prediction model and conducted experiments with precipitation data from 1960 to 2010 in Ningxia Hui Autonomous Region. The results showed that the accuracy of the CEEMD-ARIMA model was higher than that of the ARIMA model at all time scales. All the above studies show that the time series data preprocessed by the CEEMD model can make a certain contribution to improving the prediction accuracy of a traditional neural network model.

At present, there are two deficiencies in the research on the combination of the CEEMD model and neural networks. First, modeling studies on typical non-stationary series of hydrological data such as precipitation are not comprehensive, and the practicability of constructing coupling models between more types of neural network models and CEEMD models needs to be further studied. Secondly, on the basis of existing studies, a new data analysis model is introduced to further enrich the coupling prediction model. Whether it can improve the accuracy of precipitation prediction is worth further exploration. In order to solve the above problems, the swarm optimization algorithm, the fruit fly optimization algorithm (FFOA), is introduced in this paper, which has the characteristics of simple operation and strong local search ability. Combining CEEMD, ELM, and FFOA, a coupling model to further improve the prediction accuracy is sought. In the first mock exam, the CEEMD is used to decompose the precipitation time series into several intrinsic modal components (IMF components). Then, the hidden layer feedforward neural network is constructed for each IMF component, and the extreme learning machine is used for simulation and prediction. Finally, the Drosophila algorithm is used to optimize the accumulation coefficient between IMF components so that the predicted value is as close as possible to the true value, further improving the accuracy of precipitation prediction. In order to verify the validity of the prediction model, the monthly precipitation at Zhengzhou Station is forecasted, and good results are obtained, which provides a new way for precipitation prediction in the future.

2. Research Method

The combined prediction model decomposes complex precipitation prediction problems into relatively simple component prediction problems by CEEMD decomposition, which reduces the difficulty of analysis. At the same time, the model fully considers the contribution of time series information of different scales to the target results and the advantages of ELM in time series prediction, which is conducive to improving precipitation prediction accuracy. In addition, the FFOA method is introduced in the fusion process of the prediction results of each sub-model, and the fusion coefficients of each sub-model are optimized, which further improves the prediction accuracy of the model.

The specific steps of modeling are as follows:

(1) Data preprocessing. Multi-scale decomposition of the original precipitation time series. The CEEMD was used to multi-scale decompose the time series of the seasonal value of the repaired precipitation to obtain Q intrinsic model components $(IMF1 - IMFQ)$ with different frequencies and a residual term R;

(2) Intrinsic model components and residual normalization. Phase space reconstruction of each decomposition subsequence. (a) The chaos of each IMF and residual term R are the premise of constructing the prediction model by phase space reconstruction. Therefore, before the reconstruction of phase space, it is necessary to determine whether the Lyapunov exponent of each IMF and residual term R is greater than 0. If it is greater than 0, it means that the time series has chaotic characteristics. This article will use the Wolf method to calculate the maximum Lyapunov exponent of each IMF and residual term R. (b) The delay time τ_q of each IMF and residual term R is determined by mutual information method. (c) Determine the embedding dimension m_q of each IMF and residual term R by Cao method. (d) Phase space reconstruction of the one-dimensional time series data set of each IMF and residual term R is performed to obtain the dataset $D_q = \{X_i(t), Y_i(t), t = 1, 2 \cdots, M\}$ in the phase space domain. Normalize the datasets reconstructed in each phase space (Figure 1);

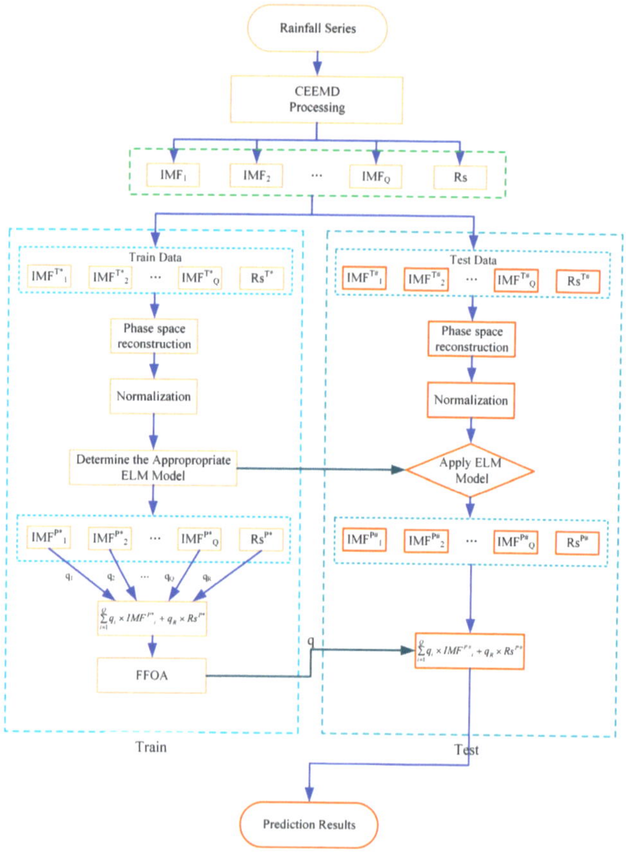

Figure 1. The technical route of the CEEMD-ELM-FFOA Coupling Prediction Model (* Marks training data and # marks test data).

(3) Construction of the ELM prediction model. Establish prediction models for each IMF and residual term R. Because the time delay τ_q and the embedded dimension m_q obtained by each decomposition in the phase space reconstruction process are different, it is necessary to establish the prediction model based on the ELM method, respectively, and to reverse normalize the obtained prediction values;

(4) Fusion parameter calculation. The results of each sub-prediction are integrated, and the correlation coefficient of each sub-prediction model is optimized. Using FFOA to optimize the variable coefficients of each sub-prediction model. In optimization, the objective is to minimize the sum of squares of errors, and the variable coefficient optimization problem can be expressed as $\min \sum_{t=1}^{T} \left(y_t - \sum_{q=1}^{Q+1} \omega_q \bullet \overline{y}_{qt} \right)^2$.

2.1. Complementary Ensemble Empirical Mode Decomposition

For the analysis and processing of non-stationary signals, Huang et al. proposed the empirical mode decomposition method (EMD) and continuous mean screening method in 1998 [15,16].

CEEMD is a new adaptive decomposition algorithm based on EMD [17] theory and improved on EEMD [18], which was proposed by Yeh et al. [19] in 2010. It can not only effectively overcome the mode aliasing phenomenon in EMD but also eliminate the residual white auxiliary noise added in EEMD to a great extent and improve the computational efficiency of decomposition [20]. The specific steps are as follows:

(1) For a set of raw time series signals $P(t)$, add a pair of Gaussian white noises with the same amplitude and phase $\omega^n(t)$, denoting the noise amplitude as β_0, Acquire a new signal $x_i^+(t)$ and $x_i^-(t)$.

$$\begin{cases} x_i^+(t) = P(t) + \beta_0 \omega^n(t) \\ x_i^-(t) = P(t) - \beta_0 \omega^n(t) \end{cases} \tag{1}$$

(2) Using EMD to modal decomposition of new signal group information, A set of intrinsic modal functions (IMF) and residual $r(t)$ is obtained. N is the number of intrinsic modal functions.

$$\begin{cases} x_i^+(t) = \sum_{j=1}^{N} IMF_j^{+i} + r_i^+(t) \\ x_i^-(t) = \sum_{j=1}^{N} IMF_j^{-i} + r_i^-(t) \end{cases} \tag{2}$$

(3) Varying noise amplitude β_0, Repeat the steps (1) and (2). The mean value of each IMF is calculated according to EMMD. M is the number of positive and negative white noise added.

$$IMF_j = \frac{1}{2M} \sum_{i=1}^{M} \left(IMF_j^{+i} + IMF_j^{-i} \right) \tag{3}$$

(4) Calculate the residual difference term of CEEMD decomposition. $R(t)$ is the residual component of the original sequence.

$$R(t) = P(t) - \sum_{j=1}^{N} IMF_j \tag{4}$$

From the above process, the CEEMD decomposition is a process of reconstructing the original signal through multiple Eigen mode extraction. It retains the advantage of EMD in processing non-stationary sequences and makes large noise in the high-frequency components of EMD, thus reducing the reconstruction error caused by the introduction of white noise in EMD. Therefore, it is more suitable for predictive analysis using machine learning.

2.2. Extreme Learning Machine

ELM is a machine learning algorithm based on a feedforward neural network [21,22]. ELM can initialize the input weight matrix and bias matrix randomly. Compared with the traditional neural network, ELM can randomly initialize the input weight matrix and bias matrix and has the advantages of strong generalization ability, less manual operation, and fast training speed on the premise of ensuring learning accuracy [23–26]. The network structure [27] of ELM is shown in Figure 2.

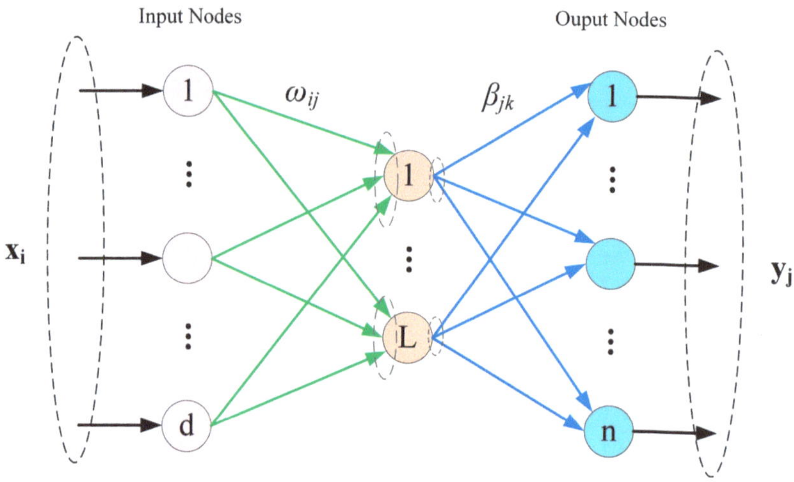

Figure 2. The ELM network structure.

The algorithm has three layers: input layer with d input neurons, hidden layer with L hidden neurons, and output layer with n output neurons.

Give a set of sample data (x, y), where $x_i = [x_{i1}, \cdots, x_{in}]^T \in R^n, y_i = [t_{i1}, \cdots, t_{im}]^T \in R^m$ has L hidden neurons in the above figure, its network structure can be expressed as [28–30]:

$$\sum_{i=1}^{L} \beta_i \bullet g(\omega_{ij} x_j + b_i) = O_j \tag{5}$$

where $\beta_i = [\beta_1, \beta_2, \cdots, \beta_n]$ is the output weight matrix, $g(z)$ is the activation function, ω_{ij} is the input weight, b_i is the threshold value of the hidden neuron, and O_j is the output result of the extreme learning machine.

A mathematical fitting regression algorithm is to predict the value by infinitely reducing the error. There are β_i, ω_{ij} and b_i, that is:

$$\sum_{i=1}^{n} \|O_j - y_j\| = 0 \tag{6}$$

$$\sum_{i=1}^{L} \beta_i \bullet g(\omega_{ij} x_j + b_i) = y_j \tag{7}$$

Equation (7) can be abbreviated as:

$$H\beta = Y \tag{8}$$

where

$$H = \begin{bmatrix} g(x_1;\omega_1;b_1) & \cdots & g(x_1;\omega_L;b_L) \\ \vdots & \cdots & \vdots \\ g(x_n;\omega_1;b_1) & \cdots & g(x_n;\omega_L;b_L) \end{bmatrix}_{n\times L}, \beta = \begin{bmatrix} \beta_1^T \\ \vdots \\ \beta_L^T \end{bmatrix}_{L\times m}, T = \begin{bmatrix} y_1^T \\ \vdots \\ y_n^T \end{bmatrix}_{n\times m}$$

It can be proved that when the excitation function $g(z)$ is infinitely differentiable, it is not necessary to adjust all the network parameters.

ω_{ij} and b_i are randomly selected at the beginning of training, and it is fixed during training. The output weight matrix β_i can be obtained by solving the least squares solution of the following equations of linear equations:

$$\beta = H^+ Y \tag{9}$$

where H^+ is the Moore–Penrose generalized inverse of the hidden layer matrix H.

2.3. Fruit Fly Optimization Algorithm

Inspired by the foraging behavior of fruit flies, Pan et al. proposed the fruit fly optimization algorithm (FFOA) [31,32]. The basic idea is to use flies superior visual and olfactory senses to locate food; the optimal solution to the problem is searched by iteration. The basic optimization process can be divided into the following steps [33–35].

Step (1): Initialize the parameters, set the population size N, the maximum number of iterations max, and the position of the fruit fly population X_axis, Y_axis, and give the random direction and distance of each fruit fly individual; then the fruit fly individual begins to search for food using the sense of smell [36–38]:

$$\begin{aligned} X_i &= X\_axis + Rand() \\ Y_i &= Y\_axis + Rand() \end{aligned} \tag{10}$$

where Rand () is the Drosophila flight range, that is, the iterative step size.

Step (2): Preliminary calculation; calculate the distance between each individual fruit fly and the origin of the coordinates $Dist_i$; then calculate the judgment value of the taste concentration of each fruit fly individual S_i:

$$\begin{aligned} Dist_i &= \sqrt{X_i^2 + Y_i^2} \\ S_i &= 1/Dist_i \end{aligned} \tag{11}$$

Step (3): Localization of olfaction; substituting the taste concentration judgment value S_i in step (2) into fitness function to find the taste concentration of each individual position of the fruit fly $Smell_i$, and find out the fruit fly with the best taste concentration in the fruit fly population (find the maximum value) [39–41]:

$$\begin{aligned} Smell_i &= Function(S_i) \\ [bestSmell \quad bestIndex] &= \max(Smell_i) \end{aligned} \tag{12}$$

Step (4): Visual orientation; record the taste concentration value and position coordinates of the fruit fly with the best taste concentration. At the same time, the fruit fly population will fly to this position by exerting their visual advantage:

$$\begin{aligned} bestSmell &= bestIndex \\ X\_axis &= X(bestIndex) \\ Y\_axis &= Y(bestIndex) \end{aligned} \tag{13}$$

Step (5): Iterative optimization; repeat steps (2) to (3), and determine whether the taste concentration value is bigger than the taste concentration of the previous iteration. If not, repeat the above steps (2) to (3) within the maximum number of iterations; if so, go to step (4).

2.4. Evaluation Method

RE represents the relative percentage error, MAE represents the mean absolute error, RMSE represents the root mean square error, and MAPE represents the mean relative percentage error [42].

$$\mathrm{RE} = \frac{|y_t - \overline{y}_t|}{y_t} \times 100\% \tag{14}$$

$$\mathrm{MAE} = \frac{1}{N}\sum_{t=1}^{N}|y_t - \overline{y}_t| \tag{15}$$

$$\mathrm{RMSE} = \sqrt{\frac{1}{N}\sum_{t=1}^{N}(y_t - \overline{y}_t)^2} \tag{16}$$

$$\mathrm{MAPE} = \frac{1}{N}\sum_{t=1}^{N}\left|\frac{y_t - \overline{y}_t}{y_t}\right| \times 100\% \tag{17}$$

where y_t represents the original value, $\overline{y}_t$ represents forecasting value.

RE represents the relative error between a single set of simulated data and the real data. Compared with MAE, RMSE, and MAPE, it can reflect the accuracy of a single predicted value. MAE reaction simulates the average absolute error of multiple data points at one time, which is convenient for comparison between multiple simulations and multiple model simulations. RMSE is squared before calculating the error stack, which is conducive to magnifying the error display. It is convenient to show whether there is excessive error in a set of forecast data. MAPE shows the average relative error of a set of data, is an important parameter to compare the accuracy of prediction, and is suitable for a horizontal comparison of the accuracy of different models. It is worth noting that the smaller the value of these calculation parameters, the smaller the prediction accuracy of the model.

3. Case Study

3.1. Research Area Survey

Zhengzhou was chosen as the research area to verify the validity and accuracy of the prediction model. Zhengzhou is the capital city of Henan Province, a megalopolis and the central city of the central plain city, an important central city in central China, and an important comprehensive transportation hub of the country, as approved by the state council. Zhengzhou belongs to the northern temperate continental monsoon climate. The territory of 124 rivers is divided into the Yellow River and Huaihe River systems. The average annual precipitation in Zhengzhou is 636.7 mm; the amount of surface water resources is 494 million cubic meters; and the amount of groundwater resources is 953 million cubic meters. The total amount of water resources is 1.124 billion cubic meters, the amount of water resources per capita is 179 cubic meters, and the amount of water resources per mu is 256 cubic meters. It is an area of severe water shortage. Therefore, the prediction of precipitation in Zhengzhou is of great significance to the regional economic distribution and the effective development and utilization of water resources.

The monthly precipitation sequence of Zhengzhou city in this study is based on the monthly precipitation data of Zhengzhou station from 1951 to 2020 provided by the National Data Center for Meteorological Sciences and Water Resources Bulletin of Zhengzhou. The sequence length was 840 months, the mean value was 48.03 mm, the standard deviation was 67.92 mm, and the maximum precipitation value was 692.2 mm in August 1963. The inter-annual variation of surface precipitation in the study area is shown in Figures 3–5. It can be seen from the figure that precipitation in Zhengzhou is generally stable and fluctuates around the mean value of several years in each year. Although precipitation has a decreasing trend, it is not significant, which can be regarded as a non-stationary time series with a weak trend, and this also reflected the reasonability of the selected CEEMD method.

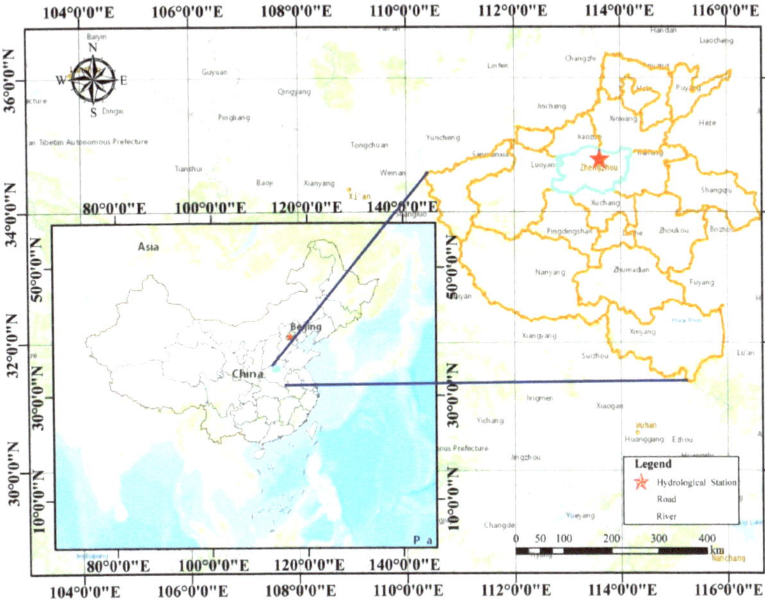

Figure 3. Location map of the study area.

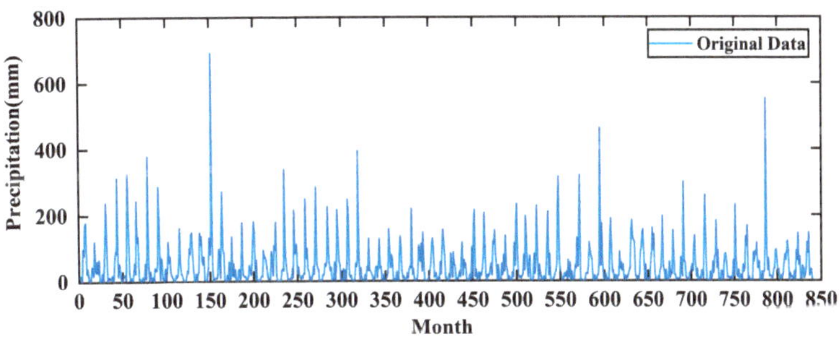

Figure 4. Monthly precipitation of Zhengzhou Station during 1951–2020.

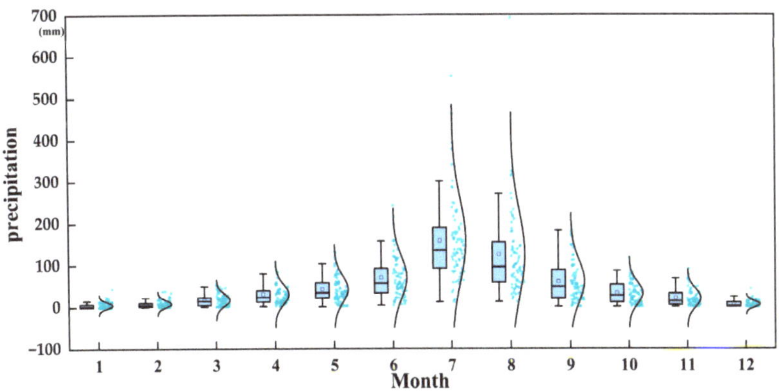

Figure 5. Boxplot of monthly precipitation at Zhengzhou Station during 1951–2020.

The operating system used in this experiment is Win10, and the deep learning framework is MATLAB2021 b. In terms of hardware, the CPU is eight-core Intel Xeon E5-2630 v4, the memory is 48 G, the GPU is a Nvidia Tesla P100, and the video memory is 16 G.

3.2. Multi-Scale Decomposition of Precipitation Time Series Data Based on CEEMD

Using the CEEMD algorithm to decompose the original data of the monthly precipitation time series of Zhengzhou from 1951 to 2019, it is found that when the noise amplitude is 0.2 and the noise logarithm is 50, the decomposition effect is ideal. After CEEMD decomposes the time series, eight IMF components and one trend component are obtained, as shown in Figure 5.

As shown in Figure 6, the precipitation time series was divided into eight IMF components and one corresponding trend term, where the IMF1 component underwent the greatest fluctuation with high frequency and the shortest wavelength; the amplitude of IMF2, IMF8, and the trend term were gradually reduced, as were their frequencies, but their wavelengths were gradually increased. After EMD processing, the fluctuation and non-stationarity of the precipitation time series of Zhengzhou were reduced to a great degree, and the original series was decomposed into periodic IMF components in order to relieve the prediction difficulty.

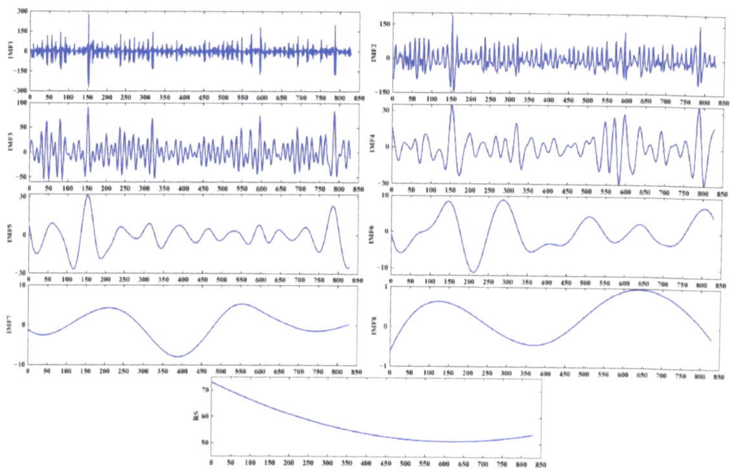

Figure 6. Precipitation CEEMD decomposition in Zhengzhou from 1951 to 2019.

3.3. Model Prediction

Whether the eight IMF components and one trend item obtained by CEEMD decomposition are chaotic time series can be identified by the Lyapunov index method. This paper adopts the mutual information method and the Cao method to obtain the time delay τ_q of each decomposition item and the embedded dimension m_q wolf method to calculate the maximum Lyapunov index of each decomposition amount. The calculation results are shown in Table 1. The table of the Lyapunov index value greater than 0 illustrates that the decomposition sequence has chaotic characteristics.

Table 1. Phase space reconstruction information table.

	IMF1	IMF2	IMF3	IMF4	IMF5	IMF6	IMF7	IMF8
τ_q	2	1	3	6	11	12	15	20
m_q	13	12	7	4	2	4	5	2
Lyapunov	0.105	0.0583	0.048	0.0524	0.0936	0.0208	0.0232	0.0335

Figure 6 shows that after CEEMD decomposition, the volatility and non-stationarity of the time series of annual precipitation in Zhengzhou are greatly reduced, and the training effect of IMF1-IMF8, the real value and predicted value of trend items are getting better and better; the relative error and average relative error of IMF1-IMF8, trend items show a decreasing trend; the training effect of the decomposed high-frequency component IMF1 is slightly poor, while the training effect of the low-frequency component IMF8 and trend item is very good. After the time series of annual precipitation in Zhengzhou is decomposed, the non-stationary nature of the time series is reduced so that ELM can better predict its components and trend term (Figure 7).

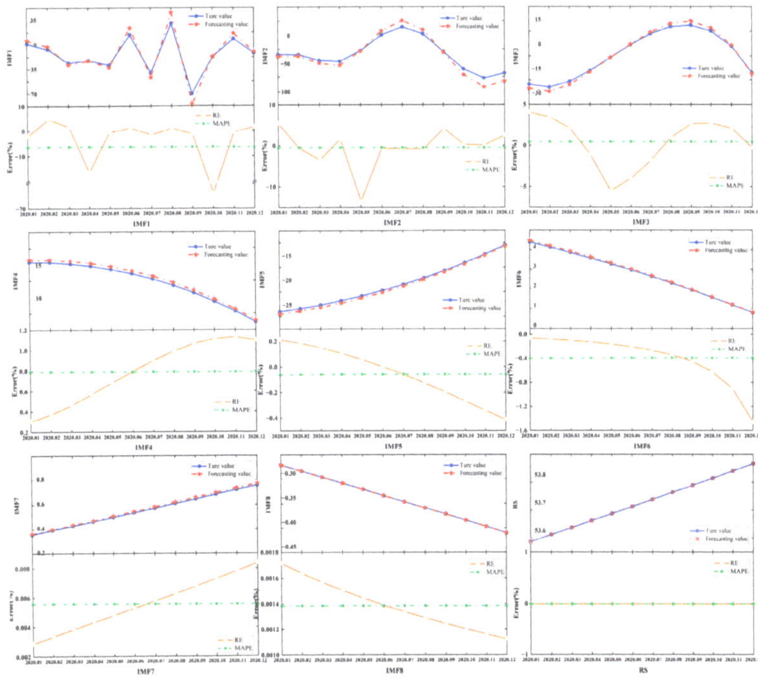

Figure 7. Prediction results of IMF1-IMF8 and trend term and the errors.

3.4. Determining the Correlation Coefficient of Combination of Decomposed Sequences

When the prediction results of each decomposition series are fused, the value of the variable coefficient of each sub-prediction model is related to the influence of the prediction output value of each decomposition series on the final prediction results and determines the final prediction accuracy and performance of the combined model. Therefore, this paper uses Matlab2021b to write the simulation program, the precipitation series of different scales after CEEMD decomposition as the training and test sets, and the variable coefficients of each sub-prediction model are then adaptively trained and optimized by the FFOA algorithm. FFOA initialization: Drosophila population size: Pop = 500; maximum iterations: Maxgen = 10,000. After several experiments, FFOA has achieved better optimization performance, obtained the optimal combination variable coefficients of eight different IMF and residual R sub-prediction models, as shown in Table 2.

Table 2. Optimization coefficient table.

	IMF1	IMF2	IMF3	IMF4	IMF5	IMF6	IMF7	IMF8	R
Optimization coefficient	1.005	0.987	0.971	1.131	1.126	0.897	0.999	1.073	1.001

3.5. Model Validation

The predicted results of IMF1~IMF8 and the trend term were reconstructed into the predicted value of monthly precipitation and compared with the original value of monthly precipitation. The calculated prediction error is shown in Table 3.

Table 3. Relative error indexes table.

Month	Precipitation		Absolute Error /mm	RE /%
	True Value	Forecasting Value		
2020.01	44.40	42.45	1.95	4.40
2020.02	34.60	34.27	0.33	0.96
2020.03	8.40	8.42	0.02	0.19
2020.04	17.00	17.32	0.32	1.87
2020.05	37.50	37.38	0.12	0.32
2020.06	116.90	115.99	0.91	0.78
2020.07	83.70	82.12	1.58	1.89
2020.08	146.03	145.83	0.47	0.32
2020.09	15.60	15.47	0.13	0.81
2020.10	33.90	33.73	0.17	0.49
2020.11	33.50	33.95	0.45	1.35
2020.12	5.80	5.60	0.20	3.38
Mean relative error = 1.39%				

Table 3 illustrates that the maximum value, minimum value, and average value of the relative error of the CEEMD-ELM-FFOA coupling prediction model were 4.40%, 0.19%, and 1.39%, respectively, so the relative prediction error of the model was small with a high eligible rate.

Figure 8 displays the prediction curves of precipitation at Zhengzhou Station during 2020.01–2020.12. It can be seen from Figure 7 that the predicted values are basically consistent with the true values. Therefore, the goodness of fit of the CEEMD-ELM-FFOA coupling model is high and it can be used for regional precipitation prediction.

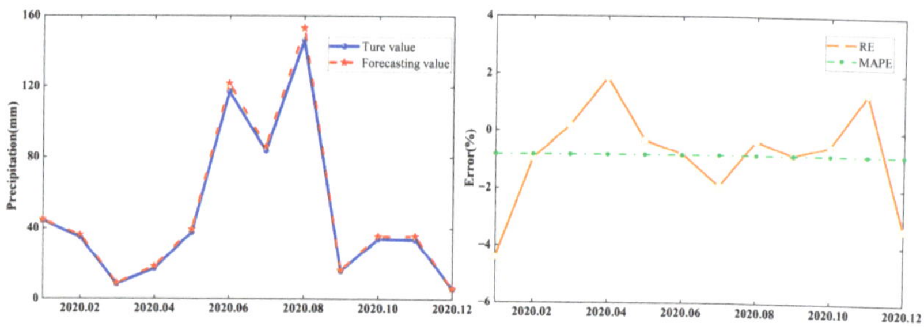

Figure 8. Precipitation prediction curve of Zhengzhou Station in 2020.

4. Discussion

In the same period, there has been little research on precipitation using the "decomposition-prediction-reconstruction" coupling method, so reference is made to literature using similar mathematical structure models for comparison. Bo H, et al. proposed a short-term load forecasting method for parks based on complementary integrated empirical mode decomposition (CEEMD), sample entropy, the SBO optimization algorithm, and the least squares support vector regression (LSSVR) model. Taking a park in Liaoning Province as an example, the results show that MAPE is 2.03 and RMSE is 3.14. The calculated errors of MAPE and RMSE in this paper are 1.39% and 0.81, respectively, which are close to each other, confirming the feasibility of establishing such a coupling model.

In order to verify the superiority of the CEEMD-ELM-FFOA coupling model in precipitation prediction, the ELM prediction model, the EMD-HHT prediction model, the CEEMD-ELM coupling model, and the CEEMD-ELM-FFOA coupling model were respectively used for prediction, followed by a comparison of their prediction effects. The comparison results of the CEEMD-ELM-FFOA model with other models in prediction error are listed in Figures 9–11 and Table 4.

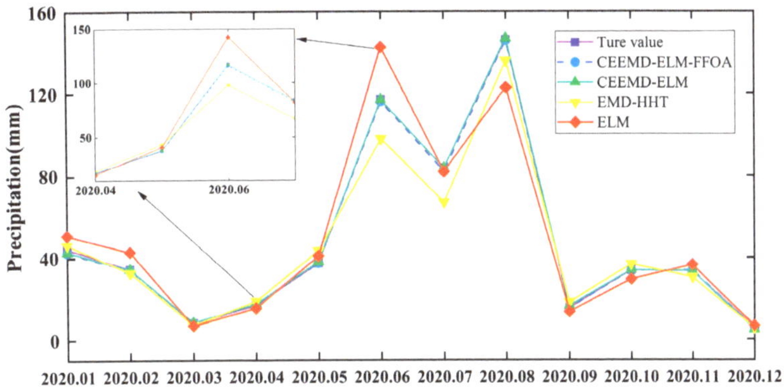

Figure 9. Prediction results of different models.

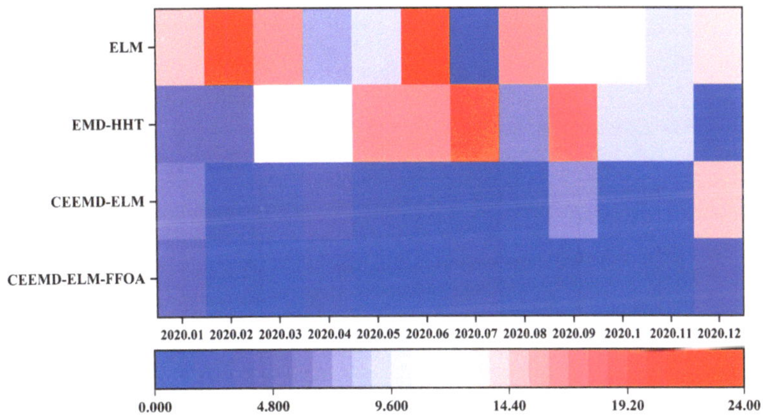

Figure 10. Hot spots of prediction errors of different models.

Table 4. Comparison of evaluation indexes of different models.

Predictive Model	MAE (mm)	RMSE (mm)	MAPE (%)
CEEMD-ELM-FFOA	0.55	0.81	1.39
CEEMD-ELM	0.63	0.92	3.23
EMD-HHT	5.64	8.22	10.92
ELM	6.83	10.70	13.33

As shown in Table 4 and Figures 9–11, the CEEMD-ELM-FFOA coupling model achieved the highest goodness of fit for predicting precipitation, and its MAE, RMSE, and MAPE were lower than those of the other three prediction models; the goodness of fit of the CEEMD-ELM prediction model was higher than that of the EMD-HHT prediction model. In the meantime, it could be seen that the prediction effect of the "decomposition-prediction-reconstruction" model was remarkably superior to that of a single neural network ELM prediction model. After the decomposition of the original signal, the non-stationarity of the sequence is reduced, the influence of extreme weather conditions on the prediction is weakened, and the prediction accuracy is improved. The prediction accuracy of extreme value is greatly elevated after FFOA optimization and reorganization of data.

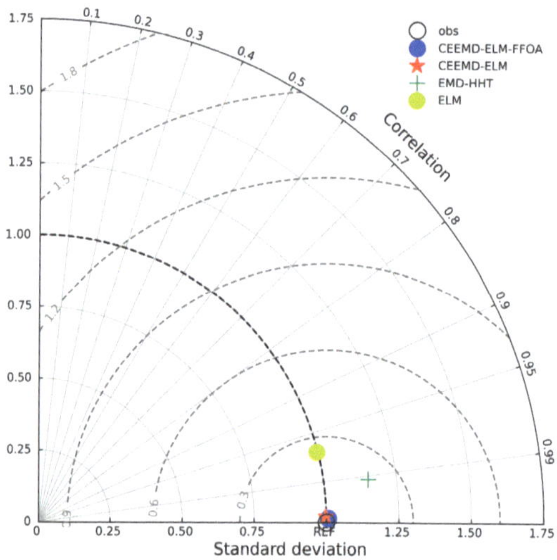

Figure 11. Taylor diagrams for statistical comparison of different models.

The advantages of the CEEMD-ELM-FFOA coupling model mainly focus on its high prediction accuracy, which is consistent with its establishment aim to improve the prediction accuracy on the basis of existing studies. In addition, ELM and FFOA models have the characteristics of simple structure, few parameters, and easy operation, which are also brought into the coupling model. At present, the limitations of this model are mainly reflected in the fact that, compared with the traditional neural network prediction model, although the prediction accuracy is improved, the complexity of the model is increased. For some scenarios that require simple prediction as well as qualitative prediction analysis, the operation is relatively complicated [43]. The applicability of this model in the prediction of other non-stationary hydrological data and the further research focus are to enrich the application range of the model. At the same time, physical parameters affecting precipitation (such as temperature, evaporation, wind speed, etc.) are not considered in this study, and precipitation prediction with these parameters will also become the direction and focus in the future.

5. Conclusions

(1) In order to improve the accuracy of precipitation prediction, this paper uses the empirical model decomposition method, extreme learning machine, and fruit fly optimization algorithm to build a precipitation prediction model based on CEEMD-ELM-FFOA and predicts the monthly precipitation in Zhengzhou. The prediction results show that the model effectively improves the accuracy of precipitation prediction and will predict the change of regional precipitation better;

(2) Multiple sub-time series will be generated after the precipitation time series are decomposed by CEEMD. These sub-time series need to establish their own prediction models to carry out prediction work, and the overall prediction results of the model will be affected by the prediction results of these sub-models. In this paper, FFOA is applied to the fusion of prediction results from sub-models, and the problem of variable coefficient optimization of each sub-model is reasonably solved. The experimental results show that FFOA has a good adaptive optimization ability for variable coefficient optimization and is an effective algorithm for model combination variable coefficient optimization;

(3) The results show that the maximum, minimum, and average relative errors of the CEEMD-ELM-FFOA coupling prediction model are 4.40%, 0.19%, and 1.39%, respectively. The model has a small relative prediction error and a high qualification rate;

(4) Although the overall prediction accuracy of the CEEMD-ELM-FFOA coupling model is high, the phase-space reconstruction method only expands the data dimension from the perspective of statistics. As the input data of the ELM model, it does not consider the physical parameters affecting precipitation (such as temperature, evaporation, wind speed, etc.), which will be the research direction and focus in the next step.

Author Contributions: X.Z.: software; validation; supervision; writing—reviewing and editing. X.W.: conceptualization; data curation; methodology; software; visualization; writing—original draft. All authors have read and agreed to the published version of the manuscript.

Funding: This research received no external funding.

Data Availability Statement: The datasets used and/or analyzed during the current study are available from the corresponding author upon reasonable request.

Conflicts of Interest: The authors declare no conflict of interest.

References

1. Giannaros, C.; Dafis, S.; Stefanidis, S.; Giannaros, T.M.; Koletsis, I.; Oikonomou, C. Hydrometeorological analysis of a flash flood event in an ungauged Mediterranean watershed under an operational forecasting and monitoring context. *Meteorol. Appl.* **2022**, *29*, 2079. [CrossRef]
2. Kotroni, V.; Cartalis, C.; Michaelides, S.; Stoyanova, J.; Tymvios, F.; Bezes, A.; Christoudias, T.; Dafis, S.; Giannakopoulos, C.; Giannaros, T.; et al. DISARM early warning system for wildfires in the eastern Mediterranean. *Sustainability* **2022**, *12*, 6670. [CrossRef]
3. Trenberth, K.E.; Dai, A.; Rasmuss, R.M.; Parsons, D. The changing character of precipitation. *Bull. Am. Meteorol. Soc.* **2003**, *84*, 1205–1218. [CrossRef]
4. Alexander, L.V.; Zhang, X.B.; Peterson, T.C.; Caesar, J.; Gleason, B.; Tank, A.M.G.K.; Haylock, M.; Collins, D.; Trewin, B.; Rahimzadeh, F.; et al. Global observed changes in daily climate extremes of temperature and precipitation. *J. Geophys. Res. Atmos.* **2006**, *111*, 1042–1063. [CrossRef]
5. Hao, H.; Zhu, H. Application of improved grey waveform prediction method in precipitation prediction. *Water Sav. Irrig.* **2021**, *313*, 41–44+50.
6. Liu, X.; Zhao, N.; Guo, J.Y.; Guo, B. Monthly precipitation prediction of Qinghai Xizang Plateau based on LSTM neural network. *J. Earth Inf. Sci.* **2020**, *22*, 1617–1629.
7. Partal, T.; Kisi, O. Wavelet and neuro-fuzzy conjunction model for precipitation forecasting. *J. Hydrol.* **2007**, *342*, 199–212. [CrossRef]
8. Aksoy, H.; Dahamsheh, A. Markov chain-incorporated and synthetic data-supported conditional artificial neural network models for forecasting monthly precipitation in arid regions. *J. Hydrol.* **2018**, *562*, 758–779. [CrossRef]
9. Alizamir, M.; Moghadam, M.A.; Monfared, A.H.; Shamsipour, A. Statistical downscaling of global climate model outputs to monthly precipitation via extreme learning machine: A case study. *Environ. Prog. Sustain. Energy* **2018**, *37*, 1853–1862. [CrossRef]
10. Huang, N.E.; Shen, Z.; Long, S.R.; Wu, M.C.; Shih, H.H.; Zheng, Q.; Yen, N.-C.; Tung, C.C.; Liu, H.H. The empirical mode decomposition and the Hilbert spectrum for nonlinear and nonstationary time series analysis. *Proc. A* **1998**, *454*, 903–995.
11. Wang, Y.; Dong, R. Low frequency oscillation analysis of multi signal Prony power system with Improved EMD. *Control Eng.* **2019**, 1335–1340.
12. Xing, W.; Wang, L.P.; Luo, P.P.; Zhao, L.; Weng, Y.; Gao, B. Time frequency matrix DEM noise reduction method based on Wavelet and EMD. *J. Xi'an Inst. Aeronaut.* **2019**, *37*, 43–47.
13. Zhang, J.L.; Liu, Z.Y.; Wang, M.X. Research on natural gas price prediction model based on CEEMD-ELM-ARIMA. *Nat. Gas Oil* **2021**, *39*, 129–136.
14. Wang, K.; Niu, D.; Sun, L.; Zhen, H.; Liu, J.; De, G.; Xu, X. Wind power short-term forecasting hybrid model based on CEEMD-SE Method. *Processes* **2019**, *7*, 843. [CrossRef]
15. Wu, Z.; Huang, N.E.; Long, S.R.; Peng, C.K. On the trend, detrending, and variability of nonlinear and nonstationary time series. *Proc. Natl. Acad. Sci. USA* **2007**, *104*, 14889–14894. [CrossRef] [PubMed]
16. Flandrin, P.; Rilling, G.; Goncalves, P. Empirical mode decomposition as a filter bank. *IEEE Signal Process. Lett.* **2004**, *11*, 112–114. [CrossRef]
17. Damerval, C.; Meignen, S.; Perrier, V. A fast algorithm for bidimensional EMD. *IEEE Signal Process. Lett.* **2005**, *12*, 701–704. [CrossRef]
18. Wu, Z.; Huang, N.E. Ensemble empirical mode decomposition: A noise-assisted data analysis method. *Adv. Adapt. Data Anal.* **2011**, *1*, 1–41. [CrossRef]

19. Yeh, J.R.; Shieh, J.S.; Huang, N.E. Complementary ensemble empirical mode decomposition: A novel noise enhanced data analysis method. *Adv. Adapt. Data Anal.* **2010**, *2*, 135–156. [CrossRef]
20. Wang, D.; Wei, S.; Luo, H.; Yue, C.; Grunder, O. A novel hybrid model for air quality index forecasting based on two-phase decomposition technique and modified extreme learning machine. *Sci. Total Environ.* **2017**, *580*, 719–733. [CrossRef]
21. Huang, G.B.; Zhu, Q.Y.; Siew, C.K. Extreme learning machine: A new learning scheme of feedforward neural networks. In Proceedings of the 2004 IEEE International Joint Conference on Neural Networks (IEEE Cat. No.04CH37541), Budapest, Hungary, 25–29 July 2004; IEEE: Piscataway, NJ, USA, 2005.
22. Huang, G.B.; Zhu, Q.Y.; Siew, C.K. Extreme learning machine: Theory and applications. *Neurocomputing* **2006**, *70*, 489–501. [CrossRef]
23. Luo, Z.S.; Pan, K.C. Wax deposition rate prediction of waxy crude oil pipelines based on LASSO-ISAPSO-ELM algorithm. *Saf. Environ. Eng.* **2022**, *29*, 69–77.
24. Li, X.; Dong, Z.; Wang, L.; Niu, X.; Yamaguchi, H.; Li, D.; Yu, P. A magnetic field coupling fractional step lattice Boltzmann model for the complex interfacial behavior in magnetic multiphase flows. *Appl. Math. Model.* **2023**, *117*, 219–250. [CrossRef]
25. Gao, C.; Hao, M.; Chen, J.; Gu, C. Simulation and design of joint distribution of rainfall and tide level in Wuchengxiyu Region, China. *Urban Clim.* **2021**, *40*, 101005. [CrossRef]
26. Liu, Y.; Zhang, K.; Li, Z.; Liu, Z.; Wang, J.; Huang, P. A hybrid runoff generation modelling framework based on spatial combination of three runoff generation schemes for semi-humid and semi-arid watersheds. *J. Hydrol.* **2020**, *590*, 125440. [CrossRef]
27. Chen, L.; Wang, S.; Zhang, Y.-H.; Wei, L.; Xu, X.; Huang, T.; Cai, Y.-D. Prediction of nitrated tyrosine residues in protein sequences by extreme learning machine and feature selection methods. *Comb. Chem. High Throughput Screen.* **2018**, *21*, 393–402. [CrossRef]
28. Pan, H.X.; Cheng, G.J.; Cai, L. Comparison of the extreme learning machine with the support vector machine for reservoir permeability prediction. *Comput. Eng. Sci.* **2010**, *32*, 131–134.
29. Mohammed, E.; Hossam, F.; Nadim, O. Improving extreme learning machine by competitive swarm optimization and its application for medical diagnosis problems. *Expert Syst. Appl.* **2018**, *104*, 134–152.
30. Li, L.D.; Cui, D.W. SSA-ELM hydrological time series prediction model based on wavelet packet decomposition and phase space reconstruction. *People's Pearl River*, 2022; *in process*.
31. Pan, W.T. A new fruit fly optimization algorithm: Taking the financial distress model as an example. *Knowl. Based Syst.* **2012**, *26*, 69–74. [CrossRef]
32. Yue, Z.; Zhou, W.; Li, T. Impact of the Indian Ocean Dipole on Evolution of the Subsequent ENSO: Relative Roles of Dynamic and Thermodynamic Processes. *J. Clim.* **2021**, *34*, 3591–3607. [CrossRef]
33. Huo, H.H. Research on Fruit Fly Optimization Algorithm and Its Applications. Master's Thesis, Taiyuan University of Technology, Taiyuan, China, 2015.
34. Hu, R.; Wen, S.; Zeng, Z.; Huang, T. A short-term power load forecasting model based on the generalized regression neural network with decreasing step fruit fly optimization algorithm. *Neurocomputing* **2017**, *221*, 24–31. [CrossRef]
35. Lv, S.X.; Zeng, Y.R.; Wang, L. An effective fruit fly optimization algorithm with hybrid information exchange and its applications. *Int. J. Mach. Learn. Cybern.* **2018**, *9*, 1623–1648. [CrossRef]
36. Xu, D.; Li, J.; Liu, J.; Qu, X.; Ma, H. Advances in continuous flow aerobic granular sludge: A review. *Process Saf. Environ. Prot.* **2022**, *163*, 27–35. [CrossRef]
37. Ge, D.; Yuan, H.; Xiao, J.; Zhu, N. Insight into the enhanced sludge dewaterability by tannic acid conditioning and pH regulation. *Sci. Total Environ.* **2019**, *679*, 298–306. [CrossRef]
38. Yuan, L.; Yang, D.; Wu, X.; He, W.; Kong, Y.; Ramsey, T.S.; Degefu, D.M. Development of multidimensional water poverty in the Yangtze River Economic Belt, China. *J. Environ. Manag.* **2023**, *325*, 116608. [CrossRef] [PubMed]
39. Li, J.; Wang, Z.; Wu, X.; Xu, C.; Guo, S.; Chen, X. Toward Monitoring Short-Term Droughts Using a Novel Daily Scale, Standardized Antecedent Precipitation Evapotranspiration Index. *J. Hydrometeorol.* **2020**, *21*, 891–908. [CrossRef]
40. Wu, X.; Guo, S.; Qian, S.; Wang, Z.; Lai, C.; Li, J.; Liu, P. Long-range precipitation forecast based on multipole and preceding fluctuations of sea surface temperature. *Int. J. Climatol.* **2022**, *42*, 8024–8039. [CrossRef]
41. Yin, L.; Wang, L.; Tian, J.; Yin, Z.; Liu, M.; Zheng, W. Atmospheric Density Inversion Based on Swarm-C Satellite Accelerometer. *Appl. Sci.* **2023**, *13*, 3610. [CrossRef]
42. Stefanos, S.; Stavros, D.; Dimitrios, S. Evaluation of Regional Climate Models (RCMs) Performance in Simulating Seasonal Precipitation over Mountainous Central Pindus (Greece). *Water* **2020**, *12*, 2750. [CrossRef]
43. Xu, K.; Ding, Y.; Liu, H.; Zhang, Q.; Zhang, D. Applicability of a CEEMD-ARIMA Combined Model for Drought Forecasting: A Case Study in the Ningxia Hui Autonomous Region. *Atmosphere* **2020**, *13*, 1109. [CrossRef]

MDPI AG
Grosspeteranlage 5
4052 Basel
Switzerland
Tel.: +41 61 683 77 34

Water Editorial Office
E-mail: water@mdpi.com
www.mdpi.com/journal/water

www.ingramcontent.com/pod-product-compliance
Lightning Source LLC
Chambersburg PA
CBHW041352210926
51850CB00013B/3059
9783725836543